HIGH-SPEED DREAMS

NEW SERIES IN **NASA** HISTORY

Roger D. Launius
SERIES EDITOR

RELATED BOOKS IN THE SERIES

Single Stage to Orbit: Politics, Space Technology, and the Quest for Reusable Rocketry
Andrew J. Butrica

NASA and the Space Industry
Joan Lisa Bromberg

Space Policy in the Twenty-First Century
edited by W. Henry Lambright

The Space Station Decision: Incremental Politics and Technological Choice
Howard E. McCurdy

Faster, Better, Cheaper: Low-Cost Innovation in the U.S. Space Program
Howard E. McCurdy

HIGH-SPEED DREAMS

NASA AND THE TECHNOPOLITICS OF
SUPERSONIC TRANSPORTATION, 1945–1999

ERIK M. CONWAY

THE JOHNS HOPKINS
UNIVERSITY PRESS
BALTIMORE

Johns Hopkins Paperback edition, 2008
9 8 7 6 5 4 3 2 1

The Johns Hopkins University Press
2715 North Charles Street
Baltimore, Maryland 21218-4363
www.press.jhu.edu

The Library of Congress has catalogued the hardcover edition of this book as follows:

Conway, Erik M., 1965–
High-speed dreams : NASA and the technopolitics of super-
sonic transportation, 1945–1999 / Erik M. Conway.
    p.   cm. — (New series in NASA history)
Includes bibliographical references and index.
ISBN 0-8018-8067-X (hardcover : alk. paper)
1. Supersonic transport planes—Political aspects—United
States—History—20th century. 2. United States National
Aeronautics and Space Administration—Research—History
—20th century. I. Title. II. Series.
TL685.7.C695 2005
387.7'3349—dc22   2004017271

ISBN 13: 978-0-8018-9081-9
ISBN 10: 0-8018-9081-0

A catalog record for this book is available from the British Library.

Title page illustration: Concept drawing of Boeing Reference H,
1992. Photo 92-03919, courtesy NASA and The Boeing
Company.

TO MY DEAR FRIENDS from the University of Minnesota's Program in the History of Science and Technology: Kai-Henrik Barth, Eric Boyles, Michael Buckley, Juliet Burba, Robert Ferguson, Steve Fifield, Kevin Francis, Eric Hinsdale, Mark Largent, Karin Matchett, Michael Reidy, Ioanna Semendeferi, Mary Thomas, Effy Vayena, and Chris Young.

Thanks for great times.

CONTENTS

Preface   ix

Acknowledgments   xi

List of Abbreviations Used in the Text   xv

Introduction   1

1   Constructing the Supersonic Age   14

2   Technological Rivalry and the Cold War 48

3   Engineering the National Champion   82

4   Of Noise, Jumbos, and SSTs   118

5   Of Ozone, the Concorde, and SSTs   157

6   The Airbus, the Orient Express, and
    the Renaissance of Speed   189

7   Toward a Green SST   224

8   Sic Transit HSCT   259

    Conclusion   301

    Notes   307

    Essay on Sources   353

    Index   359

PREFACE

This book grew out of a contract from the National Aeronautics and Space Administration's Langley Research Center, which has been my home since August 1999. I moved from a postdoctoral fellowship at the National Air and Space Museum in Washington, D.C., to Langley after the original awardee, Deborah G. Douglas, left Langley to become curator at the MIT Museum. In an unexpected telephone call, NASA Chief Historian Roger D. Launius invited me to submit a proposal for the rebidding. Without his willingness to take a risk on an unknown, recently graduated scholar, I would never have written this book.

The contract itself came about after NASA's High Speed Research program was cancelled. HSR, a $1.8 billion project to prepare new technologies for a second-generation supersonic transport, had run from 1990 until late in 1998. The program's leadership team then decided to fund a history of NASA's supersonic transport research over the past several decades, including HSR. Because their project office was closing down late in 1999, the Langley Research Center's Office of External Affairs administered the contract. Their decision to fund a history produced a rewarding but also frustrating experience for me and for my principal point of contact in the former HSR program, Virginia "Gin" Marks. HSR had been conducted under a new "Limited Exclusive Rights" doctrine to protect the intellectual property rights of the various contractors, and one ramification of that doctrine for us proved to be seemingly endless reviews of the two HSR chapters to ensure I did not violate that agreement. A second ramification was that, while I had full access to the program's documents, I was not allowed to cite any of them in my notes. That restriction, which I did not discover until after I had finished the first drafts of the final two chapters of the book, forced me to rewrite chapter 8 based almost entirely on oral histories to ensure an acceptable level of documentation. It had not been NASA policy prior to the late 1980s to restrict access to its publications except when they relate to military technologies; hence there is a sharp difference between the level of documentation in this book's first six chapters, which explicitly draw on a portion of more than a thousand publicly available technical reports, and the final two. Political conditions change, and one corollary of the increasing corporatization of the American government has been its gradual abandonment of *public* purpose to better serve corporate demands.

Despite the restrictions and reviews, however, the opinions and arguments expressed herein are mine. I was not pressured in any way by my reviewers at NASA, at General Electric Aircraft Engines, or at the Boeing Company to change nontechnical material. Not everyone who read the many drafts of this work agrees with me—or with each other—and errors of omission, commission, and interpretation are mine alone.

ACKNOWLEDGMENTS

In writing this book, I have been aided by many people who gave willingly of their time and effort. The readers of the manuscript made the largest sacrifices and were the greatest aid to me: NASA Chief Historian Roger Launius and his assistant, Stephen Garber; Langley Chief Scientist Dennis Bushnell; retired Boeing program managers Holden R. "Bob" Withington and Malcolm MacKinnon and still active Boeing engineer Robert Welge; GE Aircraft Engine managers Samuel Gilkey and Leigh Koops; NASA retirees Howard Wesoky, Louis J. Williams, Laurence K. Loftin, Jr., Cornelius Driver, William Aiken, Jack Suddreth, Dominic Maglieri, and Richard T. Whitcomb; and my former dissertation advisor Arthur L. Norberg. Their comments improved the text enormously and saved me from many errors. So too did the comments of three anonymous reviewers recruited by the NASA History Office and several other persons who read smaller portions of the work, including Robert "Joe" Shaw, Robert Golub, Rodney Ricketts and William Gilbert of NASA, and Michael McCloskey from the Sierra Club. Thank you all.

My long-distance intellectual companion in this quest was Deborah G. Douglas, curator of the MIT Museum. Many thanks for getting a better job and leaving this one to me, for many hours of conversation about NASA and the aircraft industry, and for the occasional spot on the floor for Boston-area research visits. Stuart McCook of the University of Guelph was a sounding board for thoughts about environmental history, aviation, and the space agency's role in both. Andrew Butrica's insights on the conservative space movement were very helpful in formulating the last few chapters of this book. James R. Hansen of Auburn University, who started a similar project many years ago, left a variety of documents and provided some excellent early advice. Jessica Wang at the University of California at Los Angeles hosted a lengthy California stay, and our conversations about the politics of modern science over the years have been very valuable.

I interviewed a great many people in researching this book—too many to acknowledge here. A complete list is in the bibliography. I want to thank a few non-NASA people whose testimony was critical but who could easily have used institutional barriers to avoid me. Michael Henderson, Malcolm MacKinnon, Bob Welge, Bruce Bunin, Bob Withington, and John Swihart of the Boeing Company; Sam Gilkey, Leigh Koops, and Andrew Johnson of GE Aircraft Engines; Harold Johnston of the University of California, Berkeley, and Michael Prather of the University of California, Irvine; and Leo Beranek of Bolt, Beranek and Neumann all gave unstintingly of their time, and I'm grateful. NASA histories rarely manage to integrate the contractor end of NASA's programs due to the sheer difficulty of the task, and it's the contractors who do most of the work and spend most of the money. This is a far better story due to their aid.

Alone among U.S. aerospace manufacturers, the Boeing Company maintains an archive that is accessible to researchers. I gleaned much material about the national SST program there, and thanks belong to Boeing's archivists, Michael Lombardi and Thomas Lubesmeyer, for digging out lots of material for me. Also warm thanks go to Moran Tompkins for hosting my extended Seattle research visit. Finally, Pat McGinnis at Boeing's Long Beach facility gave excellent aid during my California expedition.

Research libraries are crucial to all historians, and the Floyd Thompson Technical Library at the NASA Langley Research Center has been my home for the duration of this project. Library head Carolyn Helmetsie and facilities manager Garland Gouger provided an excellent working environment. The interlibrary loan staff, Kenneth Carroll and Cecelia Grzeskowiak, rounded up veritable stacks of history books and articles that engineering libraries don't carry. The reference staff, Jannie Davis, John Ferrainolo, Jason Jacobs, Greta Lowe, and Susan Miller, cheerfully tracked down scores of obscure technical documents. And the circulation department, Katrina Cooperwood, Christine Gwaltney, Theresa Hornbuckle, Deborah Palmer, Sarah Schwaner, and Andrea Wilson, spent uncounted hours obtaining and photocopying documents for me. Other libraries have also been of aid. At the Massachusetts Institute of Technology Archives, Elizabeth Andrews and Lois Beattie gave me access to the Citizen's League against the Sonic Boom collection. And finally, Lauren

Lassleben at the University of California, Berkeley's Bancroft Library has done signal service in organizing and serving the Sierra Club collection.

I have also been lucky to receive assistance from other federal agencies. At the Air Force Historical Research Agency at Maxwell Air Force Base, Archie DiFante declassified several useful documents for me. Diana Cornelisse and Bruce Hess of the Air Force Material Command history office lent me several manuscript histories and formerly classified documents related to the XB-70 "Valkyrie" bomber program. Matthew Dibiase, at the National Archives and Records Administration's Philadelphia Branch, was very helpful over several week-long research visits. And former Federal Aviation Administration historian Ned Preston graciously allowed me access to the agency's small remaining collection of SST project office documents.

I am indebted to the staff of the NASA History Office for aid during many research visits. Archivist Jane Odom provided access to NASA administrator Daniel Goldin's papers. Colin Fries, Mark Kahn, and John Hargenrader served many files from the office's reference collection, and Nadine Andreassen coordinated my visits. At the Glenn Research Center, Kevin Coleman, Lori Manthey, Mary Joyce Moran, and Bonita Smith provided value research assistance during two visits. Finally, Terrence Hertz of NASA headquarters provided several key documents from the High Speed Research program's last days.

This history is the result of a subcontract worked out between A. Gary Price and Virginia B. Marks at the Langley Research Center, and I thank both for their efforts. Gin Marks also served as my point of contact for the former HSR program and coordinated the review process, with the aid of Cheryl Harrell. I hope the result is worth all the trouble. I also want to thank Carol-Ann Courtney, Karen Credeur, Michael Finneran, Carole Hancock, Keith Henry, Marion Kidwell, Donna Lawson, James Meehan, Margarette Pitts, and Marny Skora of Langley Research Center's Office of External Affairs for providing a friendly, welcoming working environment. Last, Carla Coombs of Science and Technology Corporation deserves special mention for her help in working out the details of this subcontract and its several modifications.

Finally, I want to thank Michael Neufeld and the attendees of his monthly Historical Seminar on Contemporary Science and Technology, held at the National Air and Space Museum. The evening talks and dinner conversation have been a great place for keeping in touch with the field and forming and discussing new ideas. I wish it weren't so far away.

ABBREVIATIONS USED IN THE TEXT

AAF   Army Air Forces
AAOE   Airborne Antarctic Ozone Experiment
ACEE   Aircraft Energy Efficiency program
AEC   Atomic Energy Commission
AESA   Atmospheric Effects of Stratospheric Aviation (program)
AOA   angle of attack
ASEB   Aeronautics and Space Engineering Board
ASHOE   Antarctic Southern Hemisphere Ozone Experiment
AST   Advanced Supersonic Transport
ATA   Air Transport Association
BAC   Boeing Aircraft Company
BAE   British Aerospace
BOAC   British Overseas Airways Corporation
CAB   Civil Aeronautics Board
CAEP   Committee on Aviation Environmental Protection
CAP   Combat Air Patrol
CIAP   Climate Impact Assessment Program
CMC   ceramic-matrix composite
DARPA   Defense Advanced Research Agency
DOT   Department of Transportation
E^3 Energy Efficient Engine project
EI   environmental impact
EPM   Enabling Propulsion Materials program
EPNdB   effective perceived noise level in decibels
FAA   Federal Aviation Agency (later Federal Aviation Administration)

FLADE	Fan-on-blade
GEAE	GE Aircraft Engines
GEBO	Generalized Bomber program
HOTAL	Horizontal Take-Off and Landing
HSCT	high speed civil transport
HST	high-speed transport
HSR	High Speed Research program
IATA	International Air Transport Association
ICAO	International Civil Aviation Organization
IMC	intermetallic-matrix composite
IPT	Integrated Planning Team
ITD	Integrated Technology Development Team
L/D	lift-to-drag ratio
LDI	lean direct injection
LFC	laminar flow control
LPP	lean premixed, prevaporized
MAESA	Measurements for Assessing the Effects of Stratospheric Aircraft program
MWDP	Mutual Weapons Development Program
NAA	North American Aviation
NACA	National Advisory Committee for Aeronautics
NADC	Naval Air Development Center
NASA	National Aeronautics and Space Administration
NASP	National AeroSpace Plane
NNEP	Navy-NASA Engine Program
NOAA	National Oceanic and Atmospheric Administration
OSTP	Office of Science and Technology Policy
OTA	Office of Technology Assessment
PSAC	President's Science Advisory Committee
PTC	Preliminary Technology Concept
PLdB	perceived level in decibels
psf	pounds per square foot
RAE	Royal Aircraft Establishment
RQL	rich-burn, quick quench, lean burn
SAC	Strategic Air Command
SCAR	Supersonic Cruise Aircraft Research program
SCAT	Supersonic Commercial Air Transport program

SCEP	"Man's Impact on the Global Environment: Report of the Study of Critical Environmental Problems"
SDI	Strategic Defense Initiative
SEEDS	Stratospheric Emissions Effects Database
SFC	specific fuel consumption
SLFC	supersonic laminar flow control
SR	study requirement
SRI	Stanford Research Institute
STAC	Supersonic Transport Advisory Committee (Britain)
STAG	Supersonic Transport Advisory Group
STEM	shaped tube electrolytic machining
TBC	thermal barrier coat
TBO	time between overhaul
TCA	Technology Concept Aircraft
TFX	Tactical Fighter, Experimental
TI	technology integration
TMT	Technology Management Team
TOMS	Total Ozone Mapping Spectrometer
UARO	Upper Atmosphere Research Office
UNEP	United Nations Environment Program
USAF	United States Air Force
VCE	Variable Cycle Engine program
VLA	European Very Large Aircraft
VSCE	Variable Stream Control Engine

HIGH-SPEED DREAMS

INTRODUCTION

This book is a history of American attempts to design and build a "supersonic transport," or SST, an airliner capable of flying faster than the speed of sound. Technologically, SSTs have been possible since the late 1950s, and the United States government has tried three times to foster one. In 1963, the Kennedy administration approved a program to design and build two prototype SSTs, with the expectation that aircraft companies would then manufacture large numbers of them. After cancellation of this "national SST" program in 1971, the National Aeronautics and Space Administration funded a much smaller research program into advanced SST technologies, the Supersonic Cruise Aircraft Research program (SCAR), hoping that the resulting technological advancements would make a production SST possible during the 1980s. This too was cancelled, in 1981. In 1989, however, NASA was able to win approval of a third SST research program, this time named "High Speed Research," or HSR. The HSR program's goal was to produce the technologies necessary for an environmentally acceptable, economically viable "High Speed Civil Transport," a new euphemism for an SST. The High Speed Research program was cancelled in 1998, still many years short of a production HSCT. This is therefore a history of three programs that did not result in aircraft. While SSTs do exist, in the form of the Anglo-French Concorde and Russian TU-144, both of these 1960s-era aircraft were commercial failures and now survive as museum displays. And they are not American aircraft in any case.

The long SST saga reveals how national politics and business interests interact in the realm of high technology. All three American SST programs were rooted in national and international politics, products of state-sponsored drives to achieve and sustain dominance in the commercial aircraft industry and in its vital subcontracting industries. All three collapsed when their political alliances disintegrated. In the two largest programs, a lack of corporate interest in actually manufacturing an aircraft played a major role in unraveling the programs' political backing. Corporate executives found the technical risk too high, and the economic return too improbable, to invest the billions of dollars necessary to put an SST in production. Hence one focus of this book is the politics of the aircraft business.

The other major focus is on environmental politics; more specifically, the impact of environmentalism on NASA, aircraft technologies, and SST economics. "New environmentalists" led a political battle against the first U.S. SST program that succeeded in killing the program in 1971. But they accomplished far more than the demise of a noisy, polluting, and uneconomical airplane. They produced a corporate demand that the government finance the technologies necessary for environmental mitigation, and NASA became the vehicle for this effort. In short, the collision between dreams of speed and hopes of a healthier environment forced engineers to transform aircraft technologies.

PROGRESS AND SUPERSONICS

Writer John Newhouse has referred to the commercial aircraft trade as a "sporty game."[1] When he wrote in 1982, there were four manufacturers of large commercial airliners in the world, not including Soviet builders; since then, that total has been reduced to two: the Boeing Company and Airbus Industrie. The business Newhouse described is very high risk. A new airplane requires many billions of dollars to design, build, test, and certify. Only large companies, or companies backed by government financing, could enter the commercial jet airliner business, and most of the companies that have tried have failed. The point at which companies "break even" and begin to see a return on a new aircraft is seven to ten years, if the company survives that long. The airliner market is small compared to other industrial products, such as automobiles, and a design's rejection by only a handful of major airlines can destroy the design's market and bankrupt the builder. The industry is also subject to an unusually intense business cycle whose period is generally shorter than the investment horizon for new aircraft. The industry has never developed business plans or manufacturing processes that permit relatively painless accommodation to the cycle's variations, and both downturns and upswings have tended to impose very high costs on the manufacturers. Finally, contrary to conventional financial expectations, the high risk is not mirrored by high returns. Of the many companies that have tried to make money on commercial jet airliners, only the Boeing Company has ever consistently succeeded. But its profit margins have been small, in most years hovering between 2 and 3 percent.

Adding to the high risk of the business is that nation-states have chosen to use commercial aircraft manufacturing as a means of economic redevelopment. After World War II, Britain and France each subsidized the reconstruction of their domestic aircraft industries with their eyes on earning export revenues (i.e., dollars with which to pay their import bills). Eventually discovering insufficient market to support more than one "European" manufacturer, they banded together with West Germany to create Airbus Industrie.[2] Other nations also have used state financing to create industries in smaller-sized aircraft. The aircraft business is therefore not only high risk in normal commercial terms; it is also very highly politicized. The fact that every other nation in the business subsidizes airliner manufacture in one way or another has given proponents of SST technology their "in" to the American political system. Commercial products do not normally receive help from the U.S. government, and the existence of foreign "unfair competition" allows SST supporters to claim public monies they would not otherwise get. The U.S. government, in turn, seeks to protect the export revenue the aircraft trade earns by funding commercially relevant aircraft research to ensure U.S. aircraft continue to be technologically superior. In 1999, for example, Boeing delivered about $20 billion worth of airliners to non-U.S. customers, 55 percent of its aircraft production that year.[3] Sometimes, the government has agreed to fund SST research as part of its defense of the industry. SST projects are thus creatures of international competition.

If the U.S. government's interest was in protecting and promoting the commercial aircraft trade, why did it choose to support SST projects instead of other possibilities? Alternatives existed, chief among them very large aircraft, "jumbos" and "superjumbos." In addition, there are many individual technologies possible that could make subsonic commercial aviation more efficient but have not drawn the government's favor. Given the well-known technical challenges dominating supersonic flight, which include not only high noise levels but high aerodynamic drag (and higher fuel consumption) and high skin temperatures (200°F to 650°F), what causes promoters to push, and politicians to accept, SST programs instead of other, less challenging and expensive, technologies?

The SST promoters' answer is typically "productivity," but that claim deserves examination. There are three ways to increase aircraft productivity, of which only two are under aircraft manufacturers' control. The first is to raise the aircraft's cruising speed, the supersonic approach. The second is to increase aircraft capacity, the "superjumbo" approach. And the third is to reduce aircraft ground time, which is primarily an issue of airline scheduling, not aircraft technology. Hence when SST promoters say "productivity," what they actually mean is "speed." Speed is the value that makes SSTs attractive, and it is speed that promoters use to sell the SST concept to politicians. The value of speed thus deserves some discussion.

Speed, according to historian Christian Gelzer, is a virtue in American society.[4] By virtue, he means a concept that is believed to be good in and of itself. Things that are virtuous require no independent justification. As far back as the eighteenth century, Americans could be found who celebrated the acceleration of business, commerce, and life, but during the nineteenth century, Americans as a people came to embrace speed as a virtue. Faster was uncritically accepted as better, and entrepreneurs and inventors schemed to produce faster steamboats, faster railroads, faster automobiles, and in the twentieth century faster airplanes. Because early aircraft did not have the range to cross North America unrefueled, and they could not fly at night, enterprising businessmen even constructed an "air-rail" system in 1929 to speed travelers across the nation in forty-eight hours—a financial disaster, as there simply were not enough travelers to justify the investment. But the idea that speed was good remained deeply entrenched in American society and culture throughout the twentieth century.

Speed was also inextricably linked to commerce. In 1929, President Herbert Hoover's Committee on Recent Economic Changes argued that the long economic boom the nation had experienced during the previous decade had been the product of acceleration in the American economy.[5] The Hoover Commission's conclusion was given wide circulation through advertisements, such as the magazine ad by one of the nation's most famous companies found in figure I.1. GE's ad proclaims that saving time lengthens life, but speed had a significant economic value, albeit one impossible to assign a precise measure. For example, overnight airmail between New York in Chicago, inaugurated in 1926, saved banks millions of dollars in "float" costs on bank drafts—interest that they would otherwise have lost had the bank drafts gone by some slower means.[6] Faster transportation also meant businessmen could make more deals

"To save time is to lengthen life—"

"ACCELERATION, rather than structural change, is the key to an understanding of our recent economic developments."
—From the report of President Hoover's Committee on Recent Economic Changes

THE PLOD of the ox-cart. The jog trot of the horse and buggy. The rush of the high-powered motor car. The zoom of the airplane. Acceleration. *Faster* speed all the time.

Speed and more speed in production, transportation, communication, and as a result, more wealth, more happiness, and, yes, more leisure for us all.

Scientific research has been the pacemaker of this faster, yet more leisurely, existence. At a steadily

increasing rate it is giving us hundreds of inventions and improvements which speed up work, save time and money, revolutionize life and labor in the modern age.

Conceive how much time modern electric lighting has saved the American people—not to mention the billion dollars a year in lighting bills saved by the repeatedly improved efficiency of the MAZDA lamp. Think of the extraordinary democratization of entertainment and education made possible by the radio tube!

Both these benefits to the public owe much to the steady flow of discovery and invention from General Electric laboratories. So do the x-ray and cathode-ray tubes, the calorizing of steel, atomic-hydrogen welding, the generation of power for home and industry at steadily lower costs.

The G-E monogram is a symbol of research. Every product bearing this monogram represents to-day and will represent to-morrow the highest standard of electrical correctness and dependability.

JOIN US IN THE GENERAL ELECTRIC HOUR, BROADCAST EVERY SATURDAY EVENING ON A NATION-WIDE N.B.C. NETWORK
95-719H

GENERAL ⊛ ELECTRIC

FIG. I.1.   "To save time is to lengthen life," 1930. GE did not yet make airplane engines but nonetheless chose the airplane to represent the future of speed. The text argues that scientific research has led to speedup of production, transportation, and communication, resulting in more wealth, happiness, and leisure. [From *Technology Review* 32 (May 1930)] in the same amount of time, or seek new markets in regions once closed by sheer slowness of transport. When this GE ad was created, the company was not significantly involved with the aircraft business, but it nonetheless chose the airplane to represent the future of speed.

Nowhere is the American quest for speed more strongly reflected than in aviation history. In 1915, the United States founded an agency whose purpose was to advance the science of aeronautics. The National Advisory Committee for Aeronautics (always referred to as "the N-A-C-A") pursued many different research areas.[7] But its extensive research in aerodynamics before World War II was focused on improving aircraft speed, and after the war, using the new turbojet and rocket engines, it continued to promote the advancement of aircraft speed. During the late 1940s and through the end of its life in 1958, the NACA, in cooperation with the U.S. Air Force, carried out a quest for speed and altitude immortalized by novelist Tom Wolfe in *The Right Stuff,* by engineer Laurence K. Loftin, Jr., in *Quest for Performance,* and by historian Richard Hallion in *On the Frontier,* among many, many others.[8] While the NACA did many other things in its lifetime, its efforts to promote the advancement of aircraft speed gained the agency most of its fame and much of its support. And nothing harmed the agency's reputation more than its "failure" to invent either the turbojet or rocket engines—two technologies whose value to aviation was greater speed potential.[9] While there are other possible measures of progress in aviation—Loftin, the engineer, considered efficiency at least as worthwhile a goal—speed is what has been celebrated.

"Progress" in aviation, then, has traditionally been measured in the achievement of higher speeds and higher altitudes. The SST was the "next logical step" for commercial aviation under this paradigm. In the 1950s, success at producing large supersonic military aircraft inspired even the most pragmatic of engineers to dream of making supersonic travel possible, setting off an international race to be first. National governments in the United Kingdom, France, the Soviet Union, and the United States all financed SST projects in the 1960s, believing that failure to do so would result in dire consequences for their nation's balance of trade, technological competitiveness, and most importantly, national prestige. Hence the SST was not just an engineer's dream. It represented British, French, and Soviet hopes of technological parity—or superiority—to the United States, and conversely the U.S. aerospace industry's desire to remain on the cutting edge of high technology.[10]

In his excellent book *Concorde and the Americans,* Kenneth Owen has explicated the importance of transatlantic politics to the construction of the Anglo-French Concorde during the 1960s.[11] Following up an earlier work, Owen examines the Concorde project in terms of the transatlantic interactions it spawned: competition to be first, alliances between anti-SST groups, high-level diplomatic missions, and industrial espionage. Central to his study is the idea that international competition for supremacy in aerospace technologies and the prestige expected to flow from success drove the SST projects in both Western Europe and the United States, rather than rational calculus based on markets. Trapped in their alliance by a treaty lacking an escape clause, France and Britain were forced to complete Concorde despite the evaporation of its market. The two governments then gave the fourteen completed Concordes to their two national airlines, and subsidized their operation.

Howard Moon built a book on the Soviet Union's Tupolev TU-144 around a similar idea. Using recently opened Soviet records, Moon constructed a history of the TU-144 project around the idea that the SST flourished in the Soviet Union because "it incarnated cultural myths—largely irrational and unexamined—of prestige, power, and supreme speed."[12] The West therefore did not have a monopoly on dreams of speed. And like the Soviet space program under Nikita Khrushchev, the TU-144 was supposed to demonstrate Soviet technological competence in the cold war competition with the West. But while the Soviets lost the moon race, they won the SST competition. Tupolev beat Concorde into the air, but flew for barely a decade, primarily carrying mail.

The U.S. national SST program has also attracted a history, Mel Horwitch's *Clipped Wings.*[13] While Moon and Owen focused on the international competition to be first, Horwitch based *Clipped Wings* on the internal conflicts in the United States that the SST competition spawned. Horwitch examined in detail conflicts in the executive branch over whether an SST would be marketable, conflicts between executive branch supporters and hostile legislators, and conflicts between supporters and newly powerful environmental groups opposed not just to the American SST but to the very idea of an SST. He finally credits the environmental groups with defeating the American SST, by making the plane so unpopular that the Senate finally voted it down in 1971.

Taken together, Owen, Moon, and Horwitch explicate the international and domestic politics of the SST on both sides of the Atlantic in fine fashion.

But the American SST lost in the Senate by only two votes, and those, as Horwitch admits, were cast by senators who did not care one way or the other about the SST—they sought retribution against Washington State's senators for anti-tobacco votes they had cast.[14] And that detail reinforces an implication that Horwitch and Owen both noted in passing but *Aviation Week and Space Technology*'s editor made a great deal of at the time: Boeing, the SST contractor, did not lobby very hard to save its project. Historian James Hansen has expanded upon this, and upon conversations he had with NASA personnel, to question whether Boeing wanted the program to continue at all after 1967.[15] The issue Hansen raises is fundamentally an economic one: the aerospace industry recession of 1969–71, combined with Boeing's 1966 launching of the 747 program, had left the company nearly bankrupt. From the company's financial perspective, cancellation of the program was better than continuation. Missing from the previous examinations of the American SST program has been a detailed examination of the aircraft industry during these years from an economic standpoint, a point raised by Walter McDougall in his review of Horwitch's book.[16] Further, Hansen pointed out that NASA engineers believe that the final design (Boeing had so much trouble that it went through three designs) was unworkable—and Boeing's management knew it. Boeing management may well have wished the program terminated, too, suggesting economics played as important a role in the program's demise as environmentalism. Hence the economic and technological perspectives bear investigation in a new history of the SST project.

These perspectives are especially important given that the SST did not die with the 1971 program cancellation. Instead, in the United States the SST concept moved to NASA, where engineers have kept it alive (at least unofficially) ever since. Owen remarked upon a new NASA project to build an SST, renamed "High Speed Civil Transport," in the conclusion to *Concorde and the Americans*. This project, begun in the late 1980s, ran through 1998, when it too collapsed. Boeing was once again the major contractor, and this time the major competitor was not Concorde but the French state-owned company Aerospatiale, a major member of the European Airbus consortium. Aerospatiale had threatened to build a "Concorde II," prompting new U.S., Soviet, and Japanese supersonic research programs. This time, there were no environmentalists involved in the program's demise. Instead, in some striking parallels with the earlier American SST, the program ran into technical and financial problems that prompted Boeing to pull out of it. This book will thus contribute an investigation of the second major American SST program to the historical literature.

If SSTs are inextricably linked to the politics of the aircraft trade, they are also irretrievably linked to environmental politics. When SST research began in the 1950s, modern environmentalism did not exist.[17] Nor did the environmental regulatory structure that is now a well-established, though hardly uncontroversial, part of the modern nation-state. In aviation, the key environmental regulations are those that govern aircraft noise in and around airports. Like the laws governing water and air pollution, aircraft noise pollution regulations are products of environmental politics of the late 1960s.[18] Noise regulations have had profound effects on aircraft design. The first generation of subsonic jet aircraft were both loud and relatively inefficient. But, while the airlines complained mightily about the costs of noise reduction, when the government got around to regulating noise, the engineering reality quickly proved to be that noise reduction and fuel economy went hand in glove for the subsonic jets. Making the jet engine a more efficient propulsion system for these aircraft also made it substantially quieter. This was not true of the SST, where the relationship between vehicle efficiency and noise is reversed. The new noise rules introduced in the late 1960s had enormously negative impacts on the SST the United States was designing, rendering its already marginal economics worse than unacceptable. Noise concerns had damaging impacts on all of the SST designs considered in this book, and they are the chief reason that no one has so far succeeded in designing an economical SST.

The centrality of environmental politics to the SST story is also reflected in the fact that environmental activists chose to make the United States's first SST program the target of what they called a "symbolic campaign" against American technological practice.[19] This was the first such campaign, and probably the most successful. "Symbolic campaigns" were conceived as high-profile attacks on particularly egregious technologies, aimed less at defeating them than at changing public attitudes about the kinds of technologies Americans should accept. The SST fostered speed at the cost of fuel efficiency, noisier airport environments, greater pollution, and higher ticket prices. It was precisely the sort of technology that critics of American technological practice despised. They sought to instill an ethic of environmental responsibility in the public that would help drive technologists away from such environmentally destructive approaches. Implicitly, at least, they recognized that technologies were themselves political, and they hoped that campaigns against particular instances would redirect research and development toward more environmentally benign technologies.

The anti-SST movement's 1971 victory over the pro-SST forces is fondly remembered among their ranks—and it is also remembered, not quite as fondly, within the aerospace industry. Memories of defeat at the hands the "ecologists" directly affected the planning for and structure of the nation's next two SST research programs, the Supersonic Cruise Aircraft Research program, or SCAR, and the High Speed Research program, started in 1989. Boeing HSR manager Michael Henderson framed the problem this way: "There was no way we were going to try to build this thing over the bodies of the environmentalists. They had to be on board or we weren't going to do it."[20] The technological content of the 1960s American SST program and that of its successors thus differed dramatically. The earlier project had simply sought to produce an economically viable vehicle, and dealt seriously with noise only at the very end of the program in reaction to growing political pressure. Its successors focused directly on environmentally relevant technologies from their beginnings. They sought to design quieter engines, noise-suppressing nozzles, and reduced-emissions combustors in the hope they could eventually build an environmentally acceptable SST. The effort to devise "greener" aircraft technologies was quite successful during the 1990s, demonstrating that environmental politics could be transformed into hardware.

SUPERSONICS AND NATIONALISM

In *The Winged Gospel,* historian Joseph Corn has illuminated what he describes as "America's romance with aviation" in the first half of the twentieth century.[21] Americans had greeted the Wright brothers' success at Kitty Hawk in 1903 with skepticism at first, but as the two became more public with their achievements and flew their airplane before ever larger crowds, they were increasingly greeted with enthusiasm that verged on the religious. The Wrights, and then all pilots, became popular heroes, masters of a machine that appeared miraculous in its ability to bring humans closer to the heavens—and to destroy distance. The airplane became the root of a redemptive, messianic faith. Americans perceived in the airplane's ability to consume distance a way to bring people closer together, ensuring mutual understanding and thus preserving peace.

Corn's story is one of technological enthusiasm surrounding the airplane in America, a theme expanded upon by Peter Fritzsche in his history of flight in interwar Germany. In Germany the airplane, and specifically gliding, became linked to nationalism. The aerial age, he argues, introduced a powerful fear into the nation's culture.[22] The easy way the airplane could transcend national borders rendered all nations supremely vulnerable to whichever nation best mastered the new technologies of flight. Hence the airplane became a vessel of both hope and fear, and one that drew substantial national investment by European powers between the two world wars. It was also deployed in nationalist propaganda, particularly after Versailles Treaty restrictions were lifted in 1926 and Germany openly pursued the aerial age.

Germany was hardly the only nation that recognized the airplane's ability to easily cross national boundaries. In the United States Army, a small group of believers in the airplane's power to be an ultimate weapon were able to create the modern strategic bomber shortly before World War II. During that war, in what historian Michael Sherry has described as an act of technological fanaticism, they were able to achieve a sheer, destructive power that gave the nation the ability to produce Armageddon—the ability to destroy not merely its enemy but everything. Merged with the atomic bomb, the airplane represented destruction on a scale no previous century could contemplate. Its newly global reach rendered the United States as vulnerable to destruction as Germany had been in the interwar period, and Paul Boyer has brilliantly revealed the culture of fear that followed the Manhattan Project's success in 1945.[23] After the Soviet Union's test of an atomic bomb in August 1949, the United States began an enormous peacetime buildup of its nuclear weapons stockpiles and of its strategic air arm, driven far less by need than by fear of its remote, secretive, but apparently technologically advanced enemy.

The culture of nuclear fear was the context in which the race for the supersonic future began. The U.S. Air Force and the National Advisory Committee for Aeronautics collaborated during the 1940s and 1950s to produce supersonic fighters for national defense and supersonic strategic bombers for national offense. These programs consumed billions of dollars and the engineering talents of thousands of designers. By 1958, they were beginning to work on an intercontinental-range bomber that could fly at three times the speed of sound. The Air Force sought ever higher speeds partly because during World War II speed *had* mattered. Intercepting incoming bombers with fighters was a matter of relative speed. Fighters had a better chance of making an interception, all other things being equal, if they were faster than their targets. But the practical uses of speed were not the only reason the organization pursued supersonic technologies. Air Force leaders remained committed to the belief that speed equaled progress—faster airplanes were automatically superior ones.

Yet speed ceased to be useful in military aircraft at the end of the 1950s. The development of guided antiaircraft missiles and of intercontinental ballistic missiles rendered raw speed pointless. No aircraft could be faster than missiles were, and their development threatened not only the quest for higher aircraft speeds but aircraft—particularly strategic bombers—as a general class of weapon. Intercontinental ballistic missiles could reach their targets in slightly under thirty minutes and there was no feasible defense against them. Because of this, by the late 1950s the prospect of a new supersonic strategic bomber was already gone. It was cancelled by President Dwight Eisenhower, and then again by his successor, John F. Kennedy. Neither president saw the use for the very expensive supersonic bomber when the remaining tasks available to bombers could be carried out by much more efficient, inexpensive subsonic aircraft.

Cancellation of the supersonic bomber infuriated its proponents, who immediately tried to preserve the technology and its design teams by arguing that the bomber was the route to a supersonic *commercial* future. Supporters linked supersonic military technology to civilian technological progress and to nationalism, insisting across decades that the United States had to develop supersonic, and later hypersonic, technology first or be "left behind." America without supersonic airliners, they insisted, would become a technological backwater. It would cease to be an industrial power, becoming a poor, defeated, agrarian beggar nation.

In her history of the French nuclear power industry, Gabrielle Hecht has argued that nuclear power became an essential part of French nationalism after World War II. The industry's development represented reconstruction of the nation as well as its future return to economic and technological independence. French nuclear engineers, she shows, exerted themselves to construct an industry that was distinctly French while simultaneously trying to construct a France that could not be non-nuclear.[24] They sought to link

French *radiance*, French greatness and uniqueness in the world, with nuclear energy to create a French national identity that rested, in part, on the nation's nuclear prowess.

Hecht's argument rests on the idea that national identity can be based upon technological prowess, and the United States has long considered itself the world's foremost technological nation. Despite enormous efforts by "nuclear enthusiasts," nuclear power did not become a dominant technology. Neither did supersonic transports. But aerospace technology, as it did in Germany between the wars, has, and aided by the culture of nuclear fear that waxed and waned throughout the cold war, its promoters have relied heavily on the discursive strategy of linking aerospace technology to national defense and thus to national greatness. Aerospace technology is part of the "American technological sublime."[25] Space-related accidents, the Apollo 204 fire, explosion of the Space Shuttle Challenger, and disintegration of the Space Shuttle Columbia, provoke periods of national mourning because they are highly public failures of one thing most Americans accept as part of the nation's identity (and destiny): its absolute superiority in aerospace technology. These accidents shake the nation's self-confidence because they undermine its faith in its technological greatness and raise questions about the perfectibility of technology that most citizens do not wish to face.

The promoters of the supersonic future have quite deliberately justified their own work in terms of maintaining the nation's leadership in aerospace technology. Yet they have not succeeded in linking supersonics so tightly to national greatness that an America devoid of supersonic airliners cannot still be America. They have failed because "Americanness" is complicated by an equally strong national devotion to market economics. It is not enough that America be the world's foremost technological nation. To remain America, it must also allow free markets to choose which advanced technologies become embedded in the nation's infrastructure and which remain only the dreams of their promoters. And the market has not been kind to those who dream of speed.

1 CONSTRUCTING THE
SUPERSONIC AGE

World War II witnessed the introduction of two key technologies that promised to revolutionize air transportation in the postwar period, the turbojet engine and the swept wing. Both of these permitted substantial increases in aircraft speed. Even before war's end, the U.S. Army Air Forces and the National Advisory Committee for Aeronautics (NACA) had taken steps to develop and integrate these technologies into faster aircraft. By 1947, the two agencies were preparing several different aircraft that they hoped would fly at supersonic speeds.

Their attempt to break the sound barrier was driven in equal measure by a belief that faster aircraft would have great military utility and by the NACA's desire to investigate the exotic conditions of supersonic flight. Above about Mach .8, shockwave formation on aircraft surfaces caused control problems and a rapid increase in drag. Above the speed of sound, estimated drag would be twice that experienced by a subsonic jet airliner at its cruising speed. As speed increased further, aircraft would experience heating, although aerodynamicists had no good estimates of this effect. Using an experimental rocket plane, the two organizations succeeded at "going supersonic" in October 1947; over the next fifteen years, working with several different contractors, they pursued ever larger, faster supersonic aircraft. By 1958, the Air Force's Wright Air Development Center had produced a supersonic bomber roughly the same size as the Boeing B-47 and had a far larger supersonic bomber prototype being built by North American Aviation. The rapid progress in supersonics led to tremendous enthusiasm for the technology. Aircraft engineers who had entered the 1950s highly skeptical of the commercial potential of supersonic aircraft met the next decade believing that commercial supersonic flight was just around the corner.

Accomplishing supersonic flight required massive investments of both money and skills. While the first breaching of the sound barrier proved surprisingly easy, the rocket plane the United States had used could not stay supersonic—or even airborne—for more than a few minutes. There were a variety of reasons for this: its rocket engine burned fuel prodigiously, the craft's aerodynamics proved not to be particularly efficient, and the aircraft had not been designed to withstand the heat supersonic flight produced over long periods of time. Replacing the voracious rocket engine with a more fuel-efficient air-breathing jet meant coming to an understanding of how supersonic air would behave in an engine. Jet engines capable of operating efficiently at the high temperatures involved in supersonic flight had to be developed. Militarily useful supersonic aircraft would also have to be more aerodynamically efficient. These problems all appeared solvable in 1950, but they would require time and money.

The Air Force and the NACA developed a wide variety of experimental aircraft and prototypes in their effort to make supersonic flight routine. The two agencies did not formally collaborate outside the X-plane research they performed at Muroc Dry Lake, in the California desert.[1] Instead, research activities at the NACA's major laboratories in Virginia, Ohio, and California produced new concepts that the Air Force had its contractors immediately apply to new aircraft programs. The new ideas were often communicated informally by contacts between engineers at the two agencies well before they were fully explored and published. This led to very rapid innovation and also high technical risk. Applying a new idea that was not yet fully understood could result in a superior aircraft or a bad one. In the hothouse aeronautical R&D environment that existed during the 1950s, success and failure came in roughly equal measure. But engineers learn at least as much from failure as from success, and by the late 1950s, the aeronautical engineering community had learned a great deal about supersonics.[2]

Yet while the engineering community had learned a great deal during their military quest for ever higher speed, it ended the decade with a substantial gap in its collective knowledge. Engineers still had not developed the ability to accurately estimate the performance of supersonic aircraft from paper studies and wind tunnel experiments, and one historian has argued that the Air Force's technological ambitions had run far ahead of aeronautical science.[3] This helped lead to enormous contract overruns as companies struggled to meet performance guarantees they had given. The quest for speed also caused military aircraft to run up against a "heat barrier," forcing companies to engineer new materials and new production processes. This only enhanced the "cost growth" of military aircraft programs during the decade. Hence, by decade's end, the contractors faced a major credibility gap with civilian officials in the Defense Department.

None of this, however, deterred aviation enthusiasts, who began pushing for the application of supersonic technologies to civilian airliners even before they had been proven on military aircraft. The first flight of Convair's B-58 "Hustler" Mach 2–capable bomber in November 1956 initiated a stampede of SST studies in the United States and in Europe. These were reported on in highly favorable terms in the trade and mainstream presses even as the instigator, the B-58, proved to have very serious performance and safety problems in addition to cost overruns that are still legendary. Ignoring, and in some cases actively denying, that there were major uncertainties facing them, aeronautical engineers used the existence of the B-58 and another Air Force Mach 3 bomber program to promote a vision of a supersonic future that could arrive in the mid-1960s.

In their promotion of a supersonic future the enthusiasts were inestimably helped by the Soviet Union. The 3 October 1957 launch of Sputnik, the world's first artificial satellite, initiated what one historian has called a "media riot."[4] This media riot quickly became a continuous attack on the Eisenhower administration's defense policy, something presidential aspirant and Senate Majority Leader Lyndon Baines Johnson was able to turn to his party's favor. The political turmoil that followed Sputnik also allowed the aircraft manufacturers to present their demands for federal financing of a commercial supersonic transport in the rubric of national prestige. Carefully linking the Sputnik embarrassment with Eisenhower's determined efforts to end supersonic bomber programs (and equally determined effort to head off an SST), the enthusiasts were by 1960 arguing that failure to build the world's first supersonic transport would have apocalyptic consequences for the nation's future.

GOING SUPERSONIC

In his history of the NACA's Langley Aeronautical Laboratory, historian James Hansen examined the agency's quest for a "research airplane" during the late phases of World War II.[5] The lab's goal was an aircraft capable of exploring the possibilities of transonic flight. This was the one area in which aerodynamic theory of the day did not provide significant clues to aircraft behavior. Above Mach .8 and below Mach 1.2, airflow over wings was unstable and thus could not be resolved mathematically. In a recent article, engineer-turned-historian Walter Vincenti has reported how the transonic issue was eventually resolved analytically, but he makes clear that this was accomplished well after aircraft began routinely passing through the transonic region.[6] Without theory as a guide, the NACA's John Stack, an aerodynamicist who had made significant contributions to supersonic theory during the 1930s, believed that an aircraft designed to examine the transonic region was the only way to obtain information necessary to make the design of efficient supersonic aircraft possible.

Stack began campaigning for such a craft in 1944, starting with Ezra Kotcher at the U.S. Army Air Force's Materiel Command (which was responsible for aircraft purchasing), arguing that the Army Air Forces should explore the possibilities of such an aircraft themselves. Kotcher agreed with Stack, and at Wright Field in Dayton, Ohio, where the agency housed its research organizations, AAF researchers began a "Mach .999" study. The study's primary purpose was to evaluate propulsion options for the aircraft. Two options existed for transonic propulsion: a turbojet engine or a rocket engine. Kotcher's report argued that experience with one of the AAF's experimental jets suggested that turbojets could not supply the power necessary to overcome the transonic drag and hence a rocket had to power the craft. Stack did not believe rockets safe enough for manned aircraft, and his opinion was widely shared by Langley's staff. A rocket plane also could not stay airborne long enough to collect the kinds and volume of data that Stack wanted. This opened a rift between the NACA and Wright Field, which was relatively easily resolved in December 1944. Since the Air Force was paying for the aircraft, its leaders chose to follow Wright Field's recommendation and buy a rocket plane. The AAF chose the Bell Aircraft Company to build the "XS-1," as it was designated, at Bell's plant in Buffalo, New York.[7]

Stack was not the sort to give up easily, and having been rebuffed by the Air Force, he turned to friends at the Navy Bureau of Aeronautics for support. There he had better luck, and during early 1945 the bureau assigned engineers at the Douglas Aircraft Company in Long Beach, California, to study his idea for a transonic jet-powered research aircraft. This airplane became the D-558. Thus by war's end, the NACA had inspired not one but two research aircraft that would enable the organization to obtain information on transonic flight.[8]

The Langley laboratory supported the development of both aircraft, helping Bell and Douglas determine such things as appropriate wing thickness, which would affect the aircrafts' top speeds and their stability at high speeds. For the XS-1, Langley's researchers advised Bell to build two sets of wings of different thicknesses, intending to use them to explore how the different wing thicknesses correlated with wind tunnel results between Mach .9 and 1.0.[9] More significantly, Stack and Robert Gilruth, head of Langley's flight-testing group, convinced Bell that the entire horizontal tail surface had to move. Normal practice was a fixed surface with the trailing edge hinged to permit pitching the nose up and down (the hinged part was the "elevator"). But the two men believed that the elevator might become blanketed by a shockwave and lose effectiveness, while it was highly unlikely that the entire surface of the horizontal stabilizer would. Finally, and interestingly, Langley did not advocate building the XS-1 with a swept wing.[10] While R. T. Jones's version of the swept-wing theory had been verified early in the XS-1's design phase and it had been described for some of the Army Air Forces' project officials, the NACA had not pushed the AAF to make the design a swept-wing aircraft. When called on the carpet for this omission, the agency's defense was that experimental evidence was insufficient to make such a recommendation at the design reviews in 1945.[11] While criticism of the Langley group was perhaps justified, in retrospect the high- and low-speed pitch-up problems that the first swept-wing aircraft encountered suggest that a swept-wing XS-1 would have been a treacherous aircraft indeed.

The D-558 project proceeded rather differently, with Douglas initially proposing a multifaceted program. There were to be six "phase one" aircraft powered by the General Electric TG-180 axial-flow turbojet to explore speeds up to Mach .9. Two of these aircraft would be converted later on to the more powerful Westinghouse 24C turbojet with supplementary rocket units. These aircraft would be used for investigating speeds up to Mach 1. In July, Stack had discussed with the Navy's representative Captain Walter Diehl designing a later version of the D-558 with swept wings to verify the Busemann/Jones theory in flight. The two agreed that some "phase 2" aircraft should have new wings swept thirty-five degrees added later in the program. Both XS-1 and D-558 were to be flight tested at the Air Force's wartime test facility, at Muroc Dry Lake.

The Air Force had agreed to build several XS-1s, and to lend the second one to the NACA for its own use. The primary reason for this arrangement was that the two organizations had different goals for the test program. The Air Force wanted the XS-1 to achieve supersonic speeds quickly, so that it could be used as a prototype for a supersonic combat plane (and, of course, earn it some excellent publicity once the flights were made public). The NACA wanted a longer-duration, more systematic test program to collect data for comparison with wind tunnel data and to inform later designs.[12] The AAF accepted the first two XS-1s from Bell in May 1947, and using the number one aircraft, made eight powered flights between May and October to investigate the craft's characteristics. On 14 October, test pilot Charles "Chuck" Yeager reached a speed of Mach 1.06, the first pilot to reach supersonic speeds and live to tell about it.[13]

Whereas the XS-1 earned the renown of breaking the sound barrier first, Stack's preferred aircraft, the Douglas D-558, was obsolete before it ever flew. As Hansen has pointed out, the D-558-1 (with straight wings) was intended to provide data on transonic airflow that wind tunnels could not provide. The tunnels were considered useless for this application in 1944 because wall effects between Mach .8 and Mach 1 invalidated the data. But the NACA had generated several solutions before the D-558-1 was completed. Small models launched on rockets from the agency's Wallops Island facility could easily provide transonic data. Small models mounted on the wing of a special P-51 Mustang could, too, because the Mustang's wing generated flows up to Mach 1.4 on its upper surface. Tests using instrumented "bombs" dropped from a B-29 were similarly useful. Finally, in 1946 Langley physicist Ray Wright had designed a modification to high-speed wind tunnels that permitted them to operate free of the choking problem.[14] Hence the D-558 was finished too late to serve its intended purpose.

The problem the NACA flight research program faced in keeping up with the state of the art was simply lack of money. The Army Air Forces had started a nearly bewildering variety of aircraft programs shortly before the end of World War II using its vastly greater resources. As these programs matured during the late 1940s and early 1950s, a veritable flood of new designs sporting turbojets, straight and swept wings, delta wings, and speeds from high subsonic to Mach 2 poured off the drafting boards of the half-dozen aircraft manufacturers that had in-house design capabilities. At one point or another, the NACA's flight research station experimented with using all these aircraft, building upon what it learned from its more famous X planes. These proto-type and experimental aircraft have been widely documented in the vast enthusiast literature and do not need revisiting here.[15] To aviation enthusiasts, however, they appeared to represent a pattern of progress, attaining higher speeds in a systematic manner, almost as if on a schedule.

That enthusiasm for speed was first reflected in *Life* magazine. Even before the XS-1 flew, *Life* had gotten wind of the NACA's impending assault on the sound barrier and asked Langley's aerodynamicists what a supersonic airliner of the future might look like. John Stack got a group together to lay out a commercial supersonic aircraft using 1947 state-of-the-art technology. The design they produced for the magazine, pictured in figure 1.1, featured R. T. Jones's swept wings on a bullet-shaped, X-1–like body. A large turbojet engine in the tail provided power for takeoff and landing, but turbojets were not powerful enough to pierce the sound barrier so the group added rocket engines in detachable pods that would drop away once the aircraft was supersonic.[16] Above Mach 1.2, ramjets in the wing roots took over and powered the plane. But the design had a range of only 1,500 miles, permitting it to fly the famed New York to Havana casino run but not much else. And it could carry only ten passengers, hardly enough to provide economical operation.

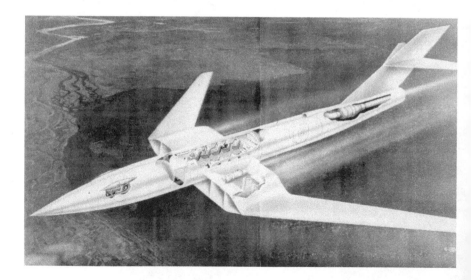

FIG. 1.1. A very early SST sketch. The concept relied on three turbojets for takeoff, two disposable rockets for transonic acceleration, and four ramjet engines for cruise. Fuel consumed most of the fuselage space. The vehicle would carry only a handful of passengers to a distance of 1,500 miles. [Courtesy John V. Becker]

The design was the beginning, however, of an effort that aeronautical engineers both at the NACA's labs and in the aircraft manufacturers' design groups returned to repeatedly during the next decade. While supersonic transportation was at best a distant dream in 1947, it represented an obvious future outcome of the quest for higher speeds and altitudes, and as engineers developed new knowledge about supersonic flight, they revisited the SST issue to see how much closer to the dream they had come. The next assessment of the state of SST art was published five years later, following a concerted attempt by the NACA, the Air Force, and the Navy to produce jet-powered aircraft capable of supersonic speeds.

OF DELTA WINGS AND DRAG PREDICTIONS

In its quest for ever higher aircraft speeds, the Army Air Forces had not waited for Stack's research airplanes to get airborne before pursing supersonic military aircraft. Even before World War II's sudden end it had begun exploring the supersonic possibilities offered by the new turbojet engines and liquid-fueled rockets. But because aerodynamic theory had a "gap" in its analytical power between Mach .8 and Mach 1.2, the agency and its contractors had no means of assessing precisely how severe the transonic drag rise would be. They therefore could not determine how much thrust an aircraft would need to pass through the sound barrier. This was one of the things the XS-1 was supposed to help determine experimentally. But because the Army Air Forces did not wait for it, the agency was, in the words of one historian, "flying blind."[17] As its allegedly supersonic airplanes matured, they collided with the dread barrier instead of flying through it.

In September 1945, the Army Air Forces had launched a program to produce a supersonic interceptor intended to defend the United States from enemy bombers. The aircraft was to achieve 700 miles per hour in level flight, using (originally) a composite rocket and turbojet power plant. This ambitious goal led Convair, the AAF's chosen contractor, to investigate the delta wing. Essentially a highly swept wing "filled in" to look like a complete triangle (or the Greek letter "delta," from which the name derived), the delta wing had been independently invented in both the United States and Germany. In the 1930s, the NACA's Langley laboratory had conducted an exhaustive investigation of various low-aspect ratio wings, including several of triangular shape. These, the researchers found, had poor subsonic flying qualities. The triangular wings generated very little lift at low speeds, meaning that aircraft using them would have very high takeoff and landing speeds. Their poor low-speed performance also meant that landing and takeoff would have to be performed at very high angles of attack, limiting pilot visibility and forcing the use of long, heavy landing gear. The sole benefit the triangular wing provided was good high-speed performance. And since no propulsion system existed to power a supersonic plane, the NACA had not pursued the delta wing idea further. In Germany during the 1930s, Alexander Lippisch, who later designed the rocket-powered Me-163 "Komet" fighter, had also explored delta wings. He had actually designed a delta-winged ramjet-powered aircraft during the war that was never built, and Lippisch's work fell into Soviet hands after VE day, while Lippisch wound up at Wright Field.[18]

Building on the NACA's prewar work, Convair initially came up with a bizarre design. The aircraft was completely wrapped about its huge engine, with the cockpit, such as it was, not on top of the fuselage as in traditional aircraft but in a conical spike protruding out of the engine's air intake (which formed the "nose" of the plane.) The Air Force was not pleased with the design, which even if flyable gave the pilot almost no visibility. Convair redesigned it as a more "conventional" pure turbojet aircraft with the cockpit in the traditional position, and this aircraft became the XF-92A. The delta wing was swept to sixty degrees, which Convair studies in late 1946 showed was the most efficient angle for the aircraft's intended top speed. The Air Force approved the design in November 1946, but only as a research aircraft.[19] North American Aviation's F-86 "Saber" was already capable of transonic speeds, and by this time it was already clear that the XF-92A would not go supersonic. The turbojet engines that were available were not powerful enough to overcome the drag rise the aircraft would encounter. But the XF-92A could still serve a useful purpose in determining the delta wing's real-world performance.

The plane's unusual wing attracted attention quickly. After fifty hours of testing by Convair, the aircraft was turned over to the Air Force, which flew it a few hours before handing it off to the NACA.[20] After installing a new engine that incorporated an afterburner designed by the Lewis Flight Propulsion Lab (which dramatically increased thrust by spraying fuel into the hot engine exhaust and igniting it), the NACA Flight Research Station instrumented the plane and used it for delta-wing stability research. Even with the additional power delivered by the afterburner, however, the XF-92A was still not capable of exceeding Mach 1 in level flight. The transonic drag rise was significantly greater than the Air Force and Convair had expected. But it did demonstrate the basic soundness of the delta wing design, which Convair relied upon for its next supersonic proposal to the Air Force, the YF-102.

The YF-102 stemmed from the Soviet Union's surprise test of an atomic bomb in 1949, which prompted the Air Force to revive its pursuit of a supersonic interceptor capable of destroying bombers attacking U.S. cities. Convair's YF-102 proposal had two competitors, both of which proposed unique configurations that may have been too controversial in 1951, when the Air Force chose the YF-102. Lockheed's Model 99 had the straight, ultra-thin wing that eventually became the basis of the still controversial F-104 several years later, while Republic's design was placed in "long term" development, from which it did not emerge. Because of the extensive wind-tunnel and flight testing that the XF-92A had undergone, the basic design had the Air Force's confidence.

That confidence did not last long. Convair began doing extensive wind tunnel work on the YF-102 models it had built late in 1951, and by the middle of the following year had hit "a snag," in the words of one author.[21] Test results from the NACA's newly converted transonic tunnel at the Langley laboratory indicated that the aircraft would not, in fact, exceed Mach .9 in level flight, given the power available from Pratt & Whitney's J-57 engine. Convair's team was not entirely convinced that the NACA's interpretation of the data was correct, because new wind tunnels (especially one whose design was novel and untried) were rarely trustworthy. But the evidence was good enough to raise concern both at Convair and in the Air Force. This was not much faster than the F-86. Speed was this aircraft's raison d'être, and if the F-102 was not going to be faster than what the Air Force already had, it would be cancelled.

By 1951, therefore, nearly five years after the XS-1 had begun flying, aeronautical engineers still could not accurately predict the magnitude of the transonic drag rise. Aeronautical engineer turned historian Walter Vincenti has demonstrated that analytically resolving the nonlinear flow patterns that occurred in transonic flight took theoreticians at the NACA's Ames lab several years longer than it took to build and fly the XS-1.[22] Even these methods were so time consuming that they were not useful for aircraft design for another twenty years. Instead, drag magnitude estimates had to be based on experimental research conducted in wind tunnels and on the NACA's research aircraft. In the absence of electronic digital computers, deriving adequate "rules of thumb" from the data generated this way also took years. The Air Force's development programs were running ahead of the NACA's research results. In Convair's case, the inability to predict performance analytically threatened to cost the company a major production contract.

What followed was the NACA's first major contribution to supersonic aircraft design knowledge during the decade. An aerodynamicist at the Langley laboratory had conceived of a way to dramatically reduce the transonic drag that aircraft had to overcome to reach supersonic speeds during 1951. Working with photographs taken during experiments in the eight-foot High Speed Tunnel, Richard T. Whitcomb had the idea that if one tailored an aircraft's surface area so that air flowing around the plane "saw" a smooth variation in cross-sectional area from nose to tail, the drag might be much less. Not everyone at Langley was convinced that Whitcomb was right, but the doubters were forced to support testing the hypothesis by Adolph Busemann, the originator of swept wings. Busemann thought Whitcomb's idea was brilliant, and his support caused John Stack (who found the idea far less convincing) to order Whitcomb to prove his theory. By April 1952, Whitcomb had tested enough wind tunnel models to convince his critics. The NACA's director, Hugh Dryden, initially did not want to publish the results, in order to keep them secret, and instead the agency invited the manufacturers to send engineers to Langley to learn about it from Whitcomb directly. Whitcomb's "transonic area rule," as the idea came to be called, saved the F-102 program.[23]

The original YF-102 was already being built, and Convair decided to complete it and the other nine planned test aircraft as originally designed in case the NACA was wrong. The YF-102 was test flown late in 1953, and its poor performance became a matter of record. Although the plane slightly exceeded Mach .9, it could not break the sound barrier in level flight. The company had hedged its bets in the interim, however, by working with Whitcomb to redo the aircraft's shape in accordance with his insight. The new aircraft emerged from the plant as the YF-102A late in 1954, and featured what has become the classic shape for supersonic aircraft: a narrow "wasp waist" that makes the aircraft look like the traditional glass "Coke" bottle with wings. This aircraft pierced the dreaded sound barrier on its second test flight—while in a thirty-degree climb. The aircraft consistently achieved Mach 1.2 in level flight, and the Air Force ordered nearly a thousand of them, though it had Convair completely redesign the aircraft to more efficiently utilize Whitcomb's rule. The result of the thorough redesign was the F-106, which with a more powerful engine reached Mach 1.9 in December 1956.

Whitcomb's transonic area rule also led to the redesign of another Convair supersonic aircraft, the B-58 "Hustler" bomber. The B-58 had grown out of a set of bomber studies known collectively as the "Generalized Bomber" program (GEBO). The program's goal was less to produce a specific aircraft than to understand how the rapidly changing technologies of flight affected potential future aircraft. The study had resulted in literally thousands of designs, many of which were fanciful to say the least.[24] The work nonetheless formed the basis of both supersonic strategic bombers built during the decade. The first of these was to achieve supersonic speed in mid-decade and begin replacing the B-47 at decade's end.

To meet that ambitious schedule, in 1950 the Wright Air Development Center had selected Convair and Boeing to compete for a design contract. The USAF specified a bomber that was to cruise at a high subsonic speed most of the distance to its target and then "dash" at supersonic speed for at least two hundred miles to and from the target area.[25] The two proposals the companies delivered in 1951 were very different. Boeing relied on a conservative approach, using a higher wing sweep than on the B-47 and using the increased power available from the J-57 engine to reach Mach 1.3. Convair, on the other hand, proposed a delta wing bomber that might reach Mach 2. During 1951 and 1952, the two designs were tested using rocket models launched from the NACA's Wallops Island facility. The rocket models demonstrated that Convair's non-area-ruled design had twice the allowable drag, and that result caused John Stack to have a model sized for the sixteen-foot transonic tunnel at Langley made and tested, verifying the results. Convair's design had to be "area ruled" just as the YF-102 had to be.[26] Once it had been, however, its delta wing achieved a superiority over Boeing's design that the Air Force's project managers at Wright Field found compelling, and in late 1952 they proposed that Convair be selected. In December, the Air Force approved Convair's selection and during early 1953 negotiated the necessary contracts.

The new bomber was to weigh nearly 200,000 pounds, about the same as the B-47, achieve Mach 2.1 at 50,000 feet, and to have a range of 4,000 miles. It was to fly at subsonic speed to within a thousand miles of the target, then accelerate to supersonic speed to dash to the target, attack, and withdraw.

Return home was to be at subsonic cruise speed. Even this, however, rapidly proved more than Convair could manage. The prediction problem had not been resolved by Whitcomb's area rule. There were other "unknowns" in supersonic flight that researchers were just beginning to encounter. And due to a lack of an agreed-upon method of prediction, controversy over whose methods were the most accurate extended into the next decade.[27] In the B-58's case, the Air Force and the NACA became increasingly skeptical of Convair's ability to deliver the performance it had promised.[28]

Area ruling of the B-58 had reduced the transonic drag of the aircraft enough to get it past Mach 1, but by itself it did not provide enough drag reduction to permit sustained flight at supersonic speeds. In essence, the aircraft's drag was still so high that it could not carry enough fuel to meet the range guarantees Convair had given the Air Force. Research at the NACA's Langley laboratory revealed that the engine nozzles Convair was planning to use were the source of a substantial amount of drag. According to John Swihart, an assistant branch chief at the lab and a specialist in nozzle design, the original nozzles produced a large amount of "boat tail drag" that prevented the aircraft from achieving a useful range. Swihart's group assisted in designing new nozzles for the aircraft that gave it a tolerable range, but the improvement was not enough to convince the user agency, Curtis LeMay's Strategic Air Command, that the B-58 would be a useful aircraft.

Essentially, the range problem descended from the very low lift-to-drag (L/D) ratio the B-58 had at supersonic speed, a problem endemic to all supersonic aircraft. In supersonic cruise, the B-58 had an L/D about one-fourth that of the B-47's at its cruise speed. This was not a matter of poor design. All supersonic aircraft generate far more drag at their cruising speed than do subsonic aircraft. That means they burn considerably more fuel to fly a given distance. To put it another way, for the B-58 to match the B-47's range, it had to devote a much larger fraction of its total weight to fuel. The Air Force had only required a short supersonic dash to the target, so the B-58 could cruise subsonically most of the way to the target, reducing its fuel needs significantly.[29] But because the delta wing had comparatively poor subsonic performance, fuel still consumed the B-58's entire internal volume, forcing Convair to place the plane's weapon in a streamlined pod under the fuselage. The improvements NACA and Convair had made to the original configuration were not enough to turn it into an efficient airplane.

Yet the fact that the B-58 promised to develop the technologies necessary for supersonic speeds lasting more than the few minutes that the fighter-sized aircraft could manage caused aeronautical engineers to reinvestigate the potential for supersonic air travel. In September 1952, for example, two engineers at the Boeing Company published an analysis of supersonic transportation, while both the F-102 and the B-58 were being redesigned.[30] The two had previously argued that subsonic jets could be commercially profitable; in this article, they concluded that supersonic aircraft would not be in the foreseeable future.[31] Their analysis was based on a set of then-reasonable assumptions: supersonic aircraft had lift-to-drag ratios about one quarter those of subsonic aircraft, cruise altitude had to be kept to 40,000 feet to mitigate explosive decompression of the passenger cabin, the airplane had to operate off conventional 6,000-foot runways, and it had to be restricted to Mach 2 to prevent structural problems stemming from high temperatures generated by friction. They showed that given these assumptions and an 1,800-mile flight, a Mach 1.3 transport's costs would be four times those of a subsonic jet's.[32] Improvements in any of the performance areas they had cited would improve a supersonic transport's economics considerably, but the authors believed that no such improvements were on the immediate horizon. Instead, they argued that while a supersonic transport was not currently feasible, "the door may be reopened in a time period 30–40 years hence if some of the particular operational limitations imposed in this study are removed."[33] The supersonic transport was a "Buck Rogers transport of the 21st century," not something they expected to see in only a few years.[34]

TOWARD SUPERSONIC CRUISING

In a mere five years, however, aeronautical engineers came to believe that Buck Rogers's future would appear in the 1960s, not the remote twenty-first century. By late 1956, an SST operating at a mere Mach 1.3 would have been obsolete. Technological developments during the mid-1950s suggested that large, supersonic cruise aircraft had become possible, through new aerodynamic design principles and active cooling of engine components. With the NACA's help, the U.S. Air Force built its vision of the 1960s around a strategic bomber capable of cruising at Mach 3.

Air Force leaders had not let the inability to estimate an aircraft's performance stand in the way of higher speeds. The Strategic Air Command considered a bomber obsolete after ten years, because by that time, its leaders assumed, the Soviet Union would have deployed defenses against it. New aircraft programs took longer than that to move a design from the drawing boards to active deployment. To keep ahead of the enemy, SAC and the Air Force had to continually push the state of the art. Because SAC's bomber generals believed that only speed and altitude served as adequate defenses, the bomber replacing the B-58 would have to fly even higher and faster. Two years before the B-58 even flew, therefore, Air Force headquarters had already requested proposals for a bigger, faster, higher-flying strategic bomber to replace it.

SAC had been pursuing two ideas that might produce a supersonic aircraft with a true intercontinental range. One of these was a nuclear-powered bomber that would have an unlimited range. As early as 1945, the Fairchild Engine and Aircraft Corporation had begun examining the potential for nuclear-propelled aircraft. In March 1946, the Air Force created the "Nuclear Energy Propulsion for Aircraft" program with Fairchild as the program manager.[35] By 1949, Fairchild's efforts had been successful enough that the Air Force and the NACA had begun examining the specific problem of how to build a nuclear jet engine.[36] Fairchild did not have the jet engine testing facilities to carry out that part of the program, and the Air Force and the Atomic Energy Commission replaced Fairchild with General Electric in 1951.[37]

GE's proposed jet engine replaced the combustion chambers of a traditional jet with an air-cooled atomic reactor.[38] Air would still be compressed after being drawn into the engine, but then would pass through channels between the fuel elements of a small nuclear reactor, cooling the fuel by absorbing its heat. The air would then expand through a turbine just as in conventional jet engines. GE's program appeared to be succeeding in 1954, and because a bomber with a range limited only by the endurance of its human crew would be the perfect strategic bomber, this was the concept SAC favored.[39] Late that year, the Air Force awarded contracts to Convair and to Lockheed to design airplanes around these engines under a program named "Weapon System 125A."

The nuclear propulsion plant was still a high-risk program from the Air Force's point of view, however, because no one really understood all the problems that a nuclear-powered airplane might have. Radiation was the biggest unknown in 1954. No one knew at the time how much radiation constituted a dangerous level to humans, and hence no one knew how much shielding for the aircraft's crew might weigh. Since weight was a major issue for any aircraft, it could turn out to be fatal to the overall program. After several internal studies, the Air Force initiated a more conventionally powered long-range bomber program in April 1955 to serve as a backup plan.[40] This second strategic bomber program began its life as Weapon System 110A.

The WS-110A program office at Wright Field requested proposals for a bomber with a range of at least six thousand miles and a supersonic dash capability from six contractors in July 1955, but only Boeing and North American Aviation chose to participate in the competition.[41] Both companies received contracts to pursue their designs in November of that year. Relying heavily on the results of the earlier Generalized Bomber studies, both companies proposed giant aircraft to the Strategic Air Command in May 1956. To increase the range of their aircraft despite the poor aerodynamic efficiency of supersonic aircraft, both companies equipped their designs with what Boeing named "floating wing tips." These were large external fuel tanks attached to the tips of the airplane's wings that had wings of their own. Both aircraft would fly subsonically to about a thousand miles from the target, then drop the now-empty "floating wing tip" tanks to make their supersonic dashes to the target. SAC liked the floating wing tip idea at first, and found it worthy of an accelerated development program.[42] The aircraft that resulted from their use, however, were enormous. The companies estimated their aircraft would weigh more than 700,000 pounds. They were twice the weight of the not-exactly-tiny B-52 bomber.

On 13 August, the Air Force's Source Selection Board recommended that North American's design be selected, but this did not immediately happen. Instead, the Air Force kept both contractors going on reduced budgets while Air Force headquarters reconsidered the whole program. In October, headquarters ordered the WS-110A program office to reorient the program to produce an all-supersonic design. Headquarters was not convinced that the proposed bombers were sufficiently superior to what the Strategic Air Command already had to justify the development cost, which was then estimated at about $1 billion.[43] And by late 1956, new techniques appeared capable of making a supersonic cruise aircraft with a useful range possible.

The first major innovation was an extension of Whitcomb's transonic area rule into the supersonic speed range. In 1953, both Whitcomb and Robert T. Jones, now at the Ames laboratory, had published papers on a supersonic area rule. They had approached the problem from different directions. Whitcomb, who was primarily an experimentalist, had approached the problem through wind-tunnel research and had derived a highly efficient general configuration.[44] Jones, a theoretician, had instead revised linear aerodynamic theory to incorporate area ruling. The resulting rule was difficult to apply to real aircraft, because it required a great deal of calculation. In essence, aerodynamicists had to calculate the cross-sections of a configuration by finding the areas of planes passed through it, and then sculpt the configuration so that the changes in area were gradual. Sudden increases or decreases in area would cause excess drag. This method had weaknesses beyond its high computing requirements. It was Mach number specific. One had to tailor the vehicle to a specific speed to achieve the lowest drag. Further, it meant that aerodynamicists could not design an aircraft in pieces, putting most of their effort into the wing and adding it to a fuselage only late in the design process. As Jones put it at the end of his paper, aerodynamicists now had to "learn to design the wing and fuselage together," a major change for a profession that had grown up around airfoil design exclusively.[45] For the next decade, the NACA's aerodynamicists would spend an enormous amount of effort developing analytical tools to improve the nation's ability to design whole aircraft.

A second new idea that led to the WS-110 program's transformation came from the Ames laboratory in Sunnyvale, California, in 1956. The Ames lab had been maintaining a research program into speeds greater than Mach 5 that was directed at achieving higher L/D ratios, again through application of simultaneous wing-body design. A member of that program, aerodynamicist Alfred Eggers, had come up with a theoretical concept that promised to allow an aircraft to cruise supersonically at a reasonable aerodynamic efficiency. That principle was the result of an inspiration that he had while mowing his lawn.[46] He reasoned that if one placed the body of a hypersonic aircraft entirely under the wing, the shockwave produced by the body would act both on the body and on the wing. Since the shockwave was essentially just a high-pressure field, the additional pressure it placed on the bottom of the wing would produce additional lift.[47] He believed that most efficient use of this principle meant designing the wing and body so that the body's shockwave coincided with the bottom of the wing's leading edge, and the effect could be further increased by using wing tips that were angled downward.[48] The resulting additional lift could increase the aerodynamic efficiency of a Mach 3 aircraft by 20 to 30 percent.

Eggers's "supersonic wedge principle," as program managers at Wright Field decided to call it, inspired a great deal of enthusiasm. Aeronautical engineers judged an aircraft's overall efficiency by combining its speed, aerodynamic efficiency, and its propulsion efficiency in an equation called the Breguet factor. Raising speed to Mach 3 and aerodynamic efficiency (L/D) to around 7.0 made competitive supersonic cruise appear almost possible. What would complete the equation would be an improvement in propulsion efficiency. Engineers already believed that jet engines would be somewhat more efficient at Mach 3 than at Mach 2 (and certainly more than at Mach 1.3 as proposed earlier in the decade), but the improvement would not be enough. However, other technologies had matured during the 1950s that promised still greater engine efficiency.

The most important of these was the maturation of blade cooling technologies for the turbines of jet engines. Providing air cooling to the turbine blades of jet engines was not a new idea in the mid-1950s. It had first been done in Nazi Germany by the Junkers aircraft company, which had applied it to overcome wartime shortages of metals capable of withstanding jet engine internal temperatures.[49] Frank Whittle's engine had had the same problem in the early war years, but Britain had not lacked the metals necessary to make high-temperature steels (primarily chromium and nickel). The Nazi regime simply did not have access to these materials during the war, and thus Junkers's engineers had sought a different solution to their thermal problems. They had constructed hollow turbine blades and built the turbine rotors so that air could be forced through the centers of the blades. The air would remove heat from the blades, keeping the blades at a temperature they could tolerate, while the air expanding through the turbine could be much hotter. If done properly, these cooled blades would allow higher operating temperatures and thus better fuel economy (or increased thrust) while also extending the engine's operating life.

At the end of the war, the NACA's Lewis Flight Propulsion Lab had picked up that idea and its researchers spent the next decade examining it and several other methods of cooling turbine blades. They analyzed film, transpiration, and convection cooling, the use of liquid coolants in addition to air, and they developed analytical methods of predicting the performance of engines using various cooling methods.[50] In 1955, the agency produced a survey of its accomplishments in turbine cooling. The study's authors contended that using blade cooling on jet engines would permit turbine inlet temperatures to rise from 1,500°F to 2,000°F, which could increase thrust by more than a third without increasing fuel consumption.[51]

The lab's studies indicated that the most effective form of cooling would be transpiration cooling, which required that the blade be made porous, so that air provided by a central passage inside the blade could "leak" out through the blade walls, cooling them.[52] This was also the most difficult to achieve, however, because of the difficulty of manufacturing such blades and ensuring that they stayed clean (so the tiny pores did not become plugged). Convection cooling, which was essentially the method Junkers had pioneered during the war, suffered from the low surface area available for heat transfer from the blade to the air flowing inside the hollow blade passage, but the blades were much simpler to build. The lab's researchers had therefore investigated several methods of increasing the internal surface area of blades, including placing multiple tubes inside, inserting fins, and corrugating the inner surface of the blade. British researchers at the National Gas Turbine Institute had also focused on putting numerous small tubes into gas turbine blades, but using a very different method. While the NACA had built hollow blades and brazed bundles of tubes into them, the British group had designed a way to machine passages into the blade directly.[53] When U.S. companies began investigating blade cooling under pressure from the Air Force's Flight Propulsion Laboratory, they relied on the Lewis lab's theoretical investigations, but devised construction methods much more similar to the British approach.

The Air Force's propulsion laboratory at Wright Field had run its own investigations of blade cooling techniques during the 1950s by having engine manufacturers construct blade sets, installing them in test rigs, and conducting ground tests on them. Successful sets were installed in J43 engines and flight tested in various aircraft to verify their performance in the "real world." This final step of flight testing was necessarily part of the Air Force Flight Propulsion Lab's mission because the Lewis lab did not have the ability to "man rate" and fly complete engines.[54] Flightworthy engines differed substantially from the test rig engines that the Lewis lab, the Air Force, and the engine manufacturers built for ground testing, and therefore a concept that appeared to work well in a test rig might not perform well in a flyable engine. The B-58's J79 engine had introduced a novel transonic compressor, for example, that suffered from serious vibration problems in flight that had not appeared in ground testing. By 1956, the Air Force lab's efforts had made clear that blade cooling techniques could be applied safely to jet engines and would substantially improve a supersonic engine's efficiency over that of the J79 engine powering the B-58.

The conjunction of the NACA's "supersonic wedge principle," the supersonic area rule, and improvements in propulsion efficiency produced by blade
cooling suggested to Air Force leaders that there was no longer a need to pursue a subsonic bomber that was only capable of a short supersonic "dash." As
the NACA's director, Hugh Dryden, put it, "a strange and wonderful thing
[had] happened. . . . Almost simultaneously, research programs that had been
under way at the NACA labs in Virginia, California, and Ohio began to pay
off. The result . . . was that the companies and the Air Force suddenly realized
it would not be much harder to design a long range bomber that could fly
its whole mission supersonic."[55] This would, they thought, greatly accelerate
the Strategic Air Command's response to a Soviet attack while giving its aircraft better survivability in Soviet airspace. Air Force headquarters therefore
directed the Air Research and Development Center at Wright Field to redirect
the WS-110A program toward the construction of an intercontinental bomber
capable of cruising supersonically from the United States to the Soviet Union.

In March 1957 the Air Force notified both Boeing and North American Aviation that their initial designs were no longer acceptable, and they were to
explore a supersonic cruise aircraft incorporating the new technologies.[56] At
the same time, it asked the NACA to undertake its own examination of potentially efficient supersonic shapes. While the Air Force's reasoning in asking the
NACA to undertake the study is not recorded, the conflict with Convair over
the B-58's performance likely provided the motivation. The NACA would certainly produce test results based on the latest state-of-the-art technology that
the Air Force could then compare the contractors' designs with, giving it a
means of estimating how "optimistic" the contractors' proposals were. The
Air Force established a deadline of November 1957 for the two contractors.
The NACA program would not be completed until January 1958, but preliminary results from it would be available to the Air Force in time to choose
between the two contractors. The Ames and Langley labs embarked on their
study in June 1957.[57]

The NACA's supersonic bomber program began with the testing of a model
referred to as the "canard bomber." (See figure 1.2.) This model was a derivative of North American Aviation's "Navaho" missile configuration, which was
also intended to cruise at Mach 3. Donald Baals, Thomas Toll, and Owen Morris of the Langley laboratory had modified the configuration in accordance
with Egger's "supersonic wedge principle" by flattening the afterbody and
placing a "wedge" under it and the large delta wing. The wedge would be the

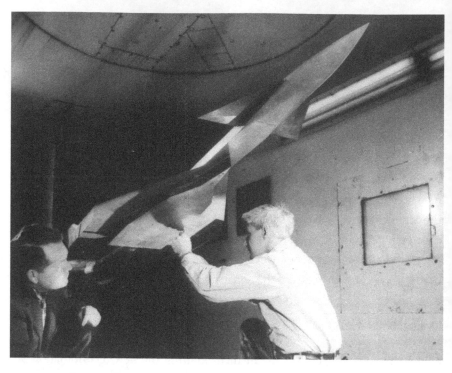

FIG. 1.2. The "Canard Bomber" model. Individuals are not identified. [Courtesy NASA, photo L57-5231]

air intake for the engines in a real airplane based upon the configuration, and in supersonic flight it would provide the "body shock" that Eggers's theory depended upon. The researchers had adjusted the wing sweep angle to conform to the shock angle, and they had incorporated the fold-down wing tips suggested by the "wedge principle," too.[58] They derived an estimated L/D for the model of 6.0 at Mach 2.87.

North American Aviation's design, as it evolved in mid-1957, was a ringer for the NACA's "canard bomber" model and Eggers's original drawing. North American (NAA) added six General Electric X279 engines to the large, flat afterbody and turned the "wedge" into a pair of two-dimensional variable-geometry air inlets to feed the engines, and it fleshed out the vexing problem of what to build the plane out of using the experience it had gained working on the Navaho missile's structure.[59] The resulting aircraft, NAA claimed, would have a lift-to-drag ratio at its Mach 3 cruising speed of over 6.9, a substantial improvement over Convair's B-58, and a considerable improvement over the NACA's own numbers on its similar configuration.[60]

North American's proposed bomber was to have a combat radius of 5,200 n.m. at a gross takeoff weight of 483,000 pounds, without refueling, and using a boron-based "High Energy Fuel" being developed in a separate program.[61] A fuel that released more heat than standard jet fuel per pound would increase an airplane's range in a direct proportion; early in the High Energy Fuels program the Air Force had aimed for a 50 percent increase, but by the late 1950s the fuels being examined by the Air Force propulsion lab and the Lewis lab promised an increase of about 30 percent. The engines that Wright Field had under development for 1960s aircraft were all being designed to use such a fuel, including the engines for the WS-110A.

The engine powering North American's proposed bomber was General Electric's X279 afterburning turbojet. GE had developed the engine privately as a major updating of the company's J79 engine, and its major innovation was the use of blade cooling in the first stage of the turbine. The blades were internally similar to those produced earlier in the decade by Britain's National Gas Turbine Establishment, with many small air passages machined directly into the blades, but the company had developed a proprietary electrochemical milling process for forming those passages. Called "shaped tube electrolytic machining," or STEM, the process could produce round or shaped tubes of extremely small size within the blades. STEM could produce the cooling passages much faster than mechanical milling could, and it could also space the tubes more closely together, providing better cooling.[62] These advantages had not been enough to win approval of the engine by the Air Force's propulsion laboratory, whose personnel believed an engine proposed by Allison was superior, but because both Boeing and North American preferred the GE engine Air Force headquarters had overridden them in May 1957. The Air Force designated the engine "J93" in September, and GE assigned individual model numbers to versions intended for standard jet fuel and for the proposed High Energy Fuel.

Boeing's revised configuration for the WS-110A program was Boeing model 804-1A. The aircraft was to have a gross takeoff weight of 530,000 pounds with a payload of 10,000 pounds and an unrefueled combat radius of 3,350 n.m. using conventional jet fuel or 4,775 n.m. using High Energy Fuel. With a single refueling, its combat radius would be 4,000 n.m. or, using High Energy Fuel, 5,540 n.m. It would have six General Electric X279 afterburning turbojet engines placed in individual nacelles underneath the trailing edge of a highly swept delta wing with "clipped" wing tips and a canard.[63]

While Boeing had incorporated the latest engine technology into its proposal, it had not used Eggers's "supersonic wedge principle" in its design. George Schairer's design crew did not believe Eggers was correct about the benefits of placing the shockwave under the wing's leading edge. Working primarily with Langley researchers, they had convinced themselves that placing the shockwave under the trailing edge of the wing was the proper approach.[64] Placing the shock under the forward-curved leading edge of the wing would produce drag in addition to lift, while locating the shock under the rearward-curved trailing edge would produce a small forward thrust instead. This would counteract a part of the drag produced by the six nacelles while still providing the lift benefit.

The very different approaches the two companies took reflected a controversy between the Ames and Langley laboratories over Eggers's theory. Langley researchers did not believe that the additional lift his approach promised would appear without also producing additional induced drag, also known as "drag-due-to-lift." This would reduce the lift improvement on a real airplane but would not show up on wind tunnel models, which were deliberately tested at conditions of zero lift. They believed their own interpretation of the concept was the correct approach to improving supersonic performance, and George Schairer's group at Boeing accepted the Langley position. The Ames lab, which did most of the testing on the North American configuration, defended the "home theory." This dispute between the labs was not resolved before the Air Force chose the winning design.

The companies submitted their proposals in November 1957. North American's proposal offered a higher aerodynamic efficiency than did Boeing's, and in fact it offered a higher efficiency than the two NACA labs had gotten on their "canard bomber" models despite the two designs' fundamental similarity. On that strength, at least officially, the Air Force awarded the WS-110 contract to North American on 23 December 1957, and designated the new design the XB-70. The Air Research and Development Command's 1959 history of the program makes clear, however, that the Air Force knew North American's L/D figures were highly "optimistic" when it made the award.[65] The WS-110A program managers had gotten the same briefing on the "canard bomber" from the NACA's researchers as the two companies had, after all, and after the controversy over Convair's B-58, they also knew that the NACA's evaluations tended to be more realistic (or more conservative, from the companies' points of view). They were also aware that the Ames and Langley labs were not in agreement on the issue.

But North American had a substantial edge over Boeing due to Boeing's own arrogance. Boeing's WS-110A program manager, Holden R. "Bob" Withington, recalled many years later that his company had refused to adopt systems engineering techniques in its management proposal, leading to substantial friction with the Air Force.[66] His belief had been that the Air Force should tell the company what it wanted the airplane to do and let Boeing design and build it; instead, the Air Force's new management technique gave it control over virtually every individual part of the aircraft. Boeing refused to change its management proposal, making it an "unresponsive" contractor. This was the real reason North American won the contract—it was happy to let the Air Force run the program. The Air Force's selection of North American in December was formalized in January 1958 with a development contract. The total procurement cost for fifteen test aircraft and an operational wing of thirty aircraft was to be $2.1 billion, and North American was to begin delivering them in 1965.[67] Procuring the 142 aircraft that SAC wanted was expected to cost $5.5 billion.[68]

To be sure, there were many unresolved issues. One mundane-seeming problem, for example, was what to seal the XB-70's fuel tanks with. No known material would tolerate the XB-70's hot environment. But this was the kind of problem industry engineers were used to solving during new aircraft development programs. For example, to make its revolutionary B-47 work Boeing had had to have jet fuel re-engineered because it tended to solidify after several hours in the intense cold that existed at 40,000 feet.[69] It had also gotten the steel industry to engineer a new grade of steel for the 707's landing gear.[70] The B-70 program would give the aeronautical engineers the opportunity to solve all such routine problems of "triplesonic cruising," making Mach 3 routine.

OF VALKYRIES, SPUTNIKS, AND THE CONSTRUCTION OF SUPERSONIC ENTHUSIASM

The innovations in supersonic technologies that had been generated by the NACA and that the Air Force was developing into working machines had, by 1957, changed the minds of many engineers about the timetable for entry into the supersonic age. They did not believe that anything like thirty or forty years was reasonable any longer. Ten years, at worst fifteen, would see the shrinking of the world through commercial supersonic flight. No one in the industry had missed the significance of the Air Force's October 1956 decision to redirect the WS-110 program toward sustained supersonic flight. If the Air Force and its contractors could make a bomber capable of that feat, the technologies that composed it could be transformed into an airliner capable of sustained supersonic flight. Such an aircraft would destroy the subsonic jet airliner market just as certainly as the subsonic jets had annihilated the market for piston-engined airliners.

Hence while the first flight of Convair's B-58 bomber received a great deal of press in late 1956, the Air Force's redirection of the WS-110 program was the more important event historically. Only a few weeks after the Air Force's order, Boeing inaugurated a company-funded project to study the development of a supersonic transport.[71] One of President Eisenhower's old friends, Elwood Quesada, his special assistant for aviation, also began championing a national SST program based upon WS-110. And John Stack, whose advocacy of a transonic research airplane had kicked off the quest for supersonic flight, put together an SST study group at the Langley laboratory to work on configurations more advanced than the B-70.[72] John Swihart and Laurence Loftin, another Langley engineer, also believe that Stack lobbied his counterparts in Great Britain to seriously examine supersonic transports, so as to spark some friendly competition.

Whatever their motivation, Stack's British colleagues had launched their own study program on 1 October 1956. Britain's Royal Aircraft Establishment, which served many of the same functions that the NACA did in the United States, had been running an independent research program into the problems of supersonic flight. The RAE had also achieved what it perceived as a breakthrough in supersonic aerodynamics during 1955, a so-called "slender delta" configuration that was more efficient at supersonic speeds than was the delta wing on the B-58. It also used a novel means of producing extra lift at low speeds that its pioneers called "controlled separation." Aerodynamicists typically designed airfoils to prevent the airflow from detaching from a wing's surface because the resulting turbulence increased drag substantially. But Eric Maskell and Dietrich Küchemann had reasoned that careful shaping of the leading edge of a low-aspect ratio delta wing could cause the formation of stable vortices over the wing that would increase lift.[73] This would permit an aircraft to fly more like a "normal" subsonic airliner during the critical takeoff and landing phases. This breakthrough suggested to British engineers the same thing Eggers's supersonic wedge principle had to their American counterparts. Further, because the RAE and the NACA participated in a program that encouraged them to share military aircraft-related research, the RAE also knew about Eggers's theory and the Langley lab's criticism of it.[74] The RAE therefore believed that an economically competitive supersonic transport might now be within the state of the art.

But Britain had no large supersonic bomber program with which to turn these research results into an operating aircraft. This would give the United States a substantial lead in the next aeronautical revolution unless the British government embarked rather quickly on a program to design a supersonic transport, hoping once again to be first to market. The British Ministry of Supply, which oversaw government financing of the British aircraft industry, thus formed a "Supersonic Transport Advisory Committee," STAC, to put together such a program. This group first met on 5 November 1956, eleven days before the Boeing Company initiated its in-house SST study.[75]

Similar worries about keeping up with the United States existed in France. Like Britain, France had developed supersonic fighter aircraft during the 1950s, but it had not pursued large, long-range bombers. The French government had used a state-funded commercial aircraft program, "Caravelle," as a means of rebuilding its commercial aircraft industry after the war and as a way of advertising the return of French technological competence to the outside world; in that sense, Caravelle had been a route to industrial modernization as well as a technological ambassador—a visible statement about the renewal of France after the destruction of World War II. But because France lacked strategic bomber programs, its leaders believed they had no hope of maintaining a viable commercial aircraft industry into the supersonic age without an explicit government-supported SST effort. Hence early the following year, the French government invited three aircraft companies to submit proposals for a mid-range supersonic transport that would serve as the Caravelle's successor. Nord Aviation, Sud Aviation, and Dassault all began studies in response.[76]

Despite all these beginnings in the wake of the WS-110A reorientation, however, the biggest impetus toward an American SST program came not from the perception that Britain or France might build an SST first, but from Soviet rocketry. While there was no evidence in the United States that the USSR was planning an SST in 1957, the surprise Soviet launch of the world's first artificial satellite on 3 October 1957 threw the mass media into a frenzy.[77] The tiny satellite, "Sputnik," became a symbol of Soviet technological superiority over the United States. Eisenhower's critics advertised Sputnik as a grievous blow to U.S. prestige abroad. The "Sputnik panic" that ensued as editorialists decried the Eisenhower administration's failure to beat the Soviet Union into space gave the aircraft industry an opportunity to win political support for the next logical step in air transportation, the SST.

The Air Force moved to capitalize on Sputnik immediately by issuing a study requirement (SR-169) on the concept a few weeks after Sputnik's "beep" was heard floating in the heavens.[78] The requirement asked the aircraft companies to present the Air Force with analyses of two different options for pursuit of a commercial SST that could be used as an Air Force tanker. The first option was a "crash" program to produce an SST no later than 1965, while the second option proposed a more measured approach intended to produce a superior SST around 1970. The Air Force typically used these study requirements to prepare the companies for a formal bidding process, hence SR-169 served as notice that the Air Force would be pursuing an SST soon. Although the actual SR was classified, the aircraft manufacturers made sure that their various design efforts appeared in the aeronautical press during 1958 and 1959. The trade journals served as a means to spread enthusiasm for new projects among the industry's legions of supporters, and they reported the design efforts in considerable detail.

The first article to announce the studies appeared in March 1958.[79] It detailed Convair and North American Aviation efforts to adapt their respective bombers into transport configurations. Convair proposed turning its B-58 into an "executive transport" for high-level government officials by extending the fuselage and wing and providing it with more powerful engines.[80] North American proposed a similar treatment for its much larger XB-70. With the bomb bays and some of the fuselage fuel tanks removed, North American's engineers believed they could produce an airliner with eighty to one hundred seats and a range of 4,500 n.m. at its Mach 3 cruising speed.

With the fact of the design studies out, the other aircraft companies responded by inspiring articles on their own programs. On 26 May 1958 *Aviation Week*'s editor, Robert B. Hotz, reported on Douglas and Lockheed SST studies he had been briefed on. "The sages of Santa Monica," Hotz reported, were well along on their Mach 2 transport, and he believed Douglas could have it flying "in about 2 years."[81] Not to be outdone, Lockheed's vice president for engineering, Hall Hibbard, had told Hotz that his company planned to beat Douglas by building a Mach 3 airliner. Lockheed's SST could be in prototype form by late 1961 and in production by 1964. This was, he said, well within the state of the art. Previous estimates of supersonic performance had been far too conservative. Flight research on Lockheed's F-104 had proven that friction drag was much less than had been feared.[82] An eighty- to one-hundred-seat SST could fly over routes as short as a thousand miles and yet have 20 percent lower operating costs than the new 707 and DC-8.[83]

What a transformation! Only five years before, competent engineers had argued that an SST would have operating costs four times those of a subsonic airliner, yet here was a engineering vice president announcing operating costs 20 percent *less* than the new subsonic jets. If such numbers were possible, there was indeed another technological revolution in the offing. Whichever company was first to turn these paper studies into a metal airplane would no doubt crush its competition. No airline could resist buying an airplane with such performance.

The timing of this revolution could not have been worse, however, as the British journal *The Aeroplane* pointed out in June. The launch of the subsonic jet airliner was just beginning, and the new jets were being introduced into a weakening traffic climate.[84] Growth in passenger traffic was slowing, and the massive bills for the new jets, each of which cost five times as much as the old piston-engined airliners, were just coming due. Airline profits were therefore sinking, and forecasts suggested the vastly increased capacity the new jets offered would result in tremendous overcapacity. The jets' higher productivity would not help airline finances if the airlines could not fill planes' seats with paying customers.

Further, the aircraft manufacturers themselves were financially stretched by the development costs of the subsonic jets. In its 1958 annual report, the Douglas Aircraft Company told its stockholders that despite rising sales, its earnings were dropping due to heavy write-offs on the DC-8 program. In 1957 it had written off $55.1 million, followed by $67.7 million in 1958. In 1959 it absorbed a net loss of $33 million, attributing this loss to continued unexpected development costs on DC-8, and only returned to bare profitability in 1961—although it reported that it continued to lose money on the DC-8. Similarly, Boeing reported write-offs on the 707 program of $34 million against net earnings of $61.5 million in 1958 and $58 million against earnings of $25.8 million the following year, although it avoided the net losses Douglas suffered. As late as 1963, Boeing president William Allen believed that neither manufacturer had made a single dollar on the commercial jet revolution.[85]

The financial states of both the aircraft manufacturing industry and of the airline industry, then, were not amenable to an SST development project. Neither the airlines nor the manufacturers could afford it. Yet airline needs did not set the pace for technological revolutions in the aircraft industry. The U.S. Air Force did. If the Air Force went ahead with its supersonic transport, then the winning contractor would be in a position to parlay that plane into a commercial SST just as Boeing had transformed its military aircraft experience into the 707. This new revolution would not reasonably occur before 1970, but *The Aeroplane* made clear that jockeying for the Air Force's contract was well under way.[86]

But because the Air Force's budget was subject to changing political conditions, "progress" in the aircraft industry depended heavily on political factors. In 1959, the embattled President Eisenhower derailed the American industry's charge toward an SST. Eisenhower had used the Sputnik crisis to get amendments to the National Security Act of 1948 passed that removed the Army, Navy, and Air Force secretaries' budget authority and gave it to the secretary of defense.[87] Eisenhower did this to prevent the massive duplication of effort that had allowed each of the three services to maintain its own rocket program. His staff believed that this had hindered progress in rocketry because there were too few competent rocket engineers to support three competing programs, but without a central authority to allocate research and development funding, there was no way to eliminate duplicative programs. The new law allowed Eisenhower to consolidate the rocket programs. They also allowed him to reign in an Air Force budget that the old general believed was out of control. The vast expansion of resources that industry enthusiasts had expected to follow in Sputnik's wake thus did not appear. Instead, Eisenhower had his budget director and his secretary of defense cut the Air Force's aircraft procurement budget.

The first program to feel Eisenhower's budget ax was the nuclear-powered bomber program, WS-125. By 1958, although GE's engine had run successfully, it was clear that whatever airplane came out of the program was not going to be acceptable due to the danger radiation posed to its crew. The airplane part of the program was disbanded in 1958, while the Atomic Energy Commission continued the airborne reactor study through 1963.[88] Eisenhower's secretary of defense, Thomas Gates, limited the money available to

Air Force headquarters for its other bomber programs too, causing headquarters to try to force Strategic Air Command to choose between the B-58 and the XB-70. SAC effectively refused to choose throughout 1958 and 1959, producing a running three-way battle between SAC, Air Force headquarters, and the Defense Department over the B-70.

Advocates of the B-70 tried to convince Eisenhower that the B-70 was the only way the United States could destroy mobile intercontinental ballistic missiles that the Soviet Union was rumored to have and thus had to be kept in the budget. But Eisenhower no longer believed that manned bombers were viable delivery systems for strategic nuclear weapons no matter how advanced they might be. His science advisor, George Kistiakowsky, had made clear to him that the B-70's large radar cross section and its high skin temperature made it "extremely vulnerable to antiaircraft fire" directed either by radar or by infrared sensors.[89] Eisenhower believed that by 1965 the Soviet air defense system would easily be able to destroy high-altitude bombers. And Air Force attempts to justify the bomber as a way to attack mobile ICBMs the old general dismissed as "crazy."[90] Eisenhower had heard that line from the bomber advocates during World War II and had seen them fail to hit, let alone destroy, even fixed and well-known German missile sites. He perceived no reason to believe that Air Force targeting had improved in the interim. Eisenhower instead intended to rely upon ballistic missiles and bombers equipped with "stand-off" missiles that could be launched from outside the Soviet air defense system's range. Such bombers would not need to be fast, and indeed Boeing's "obsolete" B-52 was proving to be an excellent platform for launching such missiles. The B-70 thus lacked support from the top.[91]

In two successive swipes of the budget ax in 1959, therefore, the B-70 was effectively cancelled. In August the High Energy Fuels program was cancelled, and on 1 December Air Force headquarters ordered the Air Development Center to "reorient" the XB-70 program to produce a single experimental airplane.[92] Headquarters intended to preserve the program long enough to get Eisenhower's decision either reversed by his successor or overridden by Congress; in the meantime, it was on life support.

SR-169 was cancelled along with the XB-70 program. No supersonic transport would be forthcoming on his watch, Eisenhower told General Nathan Twining.[93] The Air Force had a new cargo plane going into production, Lockheed's C-141 "Starlifter." It was far too early to propose replacing a brand new aircraft when there were other budget priorities. And it was not the federal government's duty to pay for new commercial airliners. If the aircraft companies believed a supersonic transport was a viable commercial product, Eisenhower told Twining, they could spend their own money on it.[94] He could perceive no reason anyone needed to get anywhere faster than the new 707 could carry them. He would not permit the military budget to be used to fund a commercial program.[95]

Predictably, Eisenhower's effective demolition of the Air Force's only new manned bomber program and its dreams of an SST brought a tremendous political backlash. Air Force and even Army generals publicly decried his decision. Both the mass media and the Democratic Party were happy to entertain their objections. As historian Stephen Ambrose has put it, the Air Force was "the darling of Congressional democrats."[96] Representative Overton Brooks, chair of the House Science and Astronautics committee, called hearings on the cancellation in the wake of a Soviet declaration of an SST project that had appeared in the *Washington Post*.[97] He convened the hearings in May, and Brooks heard from ten enthusiastic witnesses. Brooks asked each of them to spell out what effect a Russian SST would have on American national prestige. His first witness allowed that it would create a "flurry" like the day after Sputnik.[98] The rest of the witnesses agreed. Eisenhower's old friend General Elwood Quesada, recently installed as head of the Federal Aviation Agency, even claimed hysteria would result.[99] Permitting the Soviet Union to be first to an SST would be devastating.

But without the B-70, Quesada told Brooks, a supersonic transport would not occur. As Nathan Twining had told Eisenhower, the B-70 was the technological basis for the SST, and without it the SST was unthinkable. The aircraft manufacturers, which had just gone deeply into debt to launch the first generation of subsonic jets, could not afford to develop an SST on their own. In any case, commercialization of military technology was the traditional route the American aircraft industry had taken to new products; if that route was going to be blocked by the XB-70's cancellation, then "some new approach" would be necessary.[100] The government would have to fund SST development directly, an approach Quesada admitted he did not like. The "heavy hand of Government," he told the committee, should be kept out of "private industry" as much as possible.[101] Sustaining the B-70 was the more appropriate route to the SST because it would protect the virtues of "free enterprise."

The House Science and Astronautics committee had no authority over the

Defense Department or its budget, however; thus these hearings did little but provide ammunition for others. Senate Majority Leader (and presidential candidate) Lyndon Johnson, chair of the special investigating subcommittee of the Senate Armed Services committee, did not miss the opportunity to burnish his aerospace (and defense) credentials. Johnson released his own report on the B-70 program three days before his party's convention, calling for full restoration of the program. The B-70 was vital to national defense, Johnson argued, and in addition to its role in spawning a supersonic transport, the B-70 would serve as the first fully reusable first stage of future rockets. A B-70-based "Recoverable Booster Space System" would also serve as the basis for a satellite interception system to destroy those pesky Russian Sputniks.[102] Thus ending the B-70 program would end progress in not just military aviation but in commercial aviation and in space! Johnson's report made clear that Democrats would restore the supersonic revolution Republicans were trying to terminate. John F. Kennedy's "missile gap" rhetoric reinforced the image of support for aerospace that Johnson had worked to achieve since Sputnik, and his surprise selection of Johnson as his running mate at the Democratic national convention cemented that image. Democrats would deploy the resources that the aerospace industry had been denied by Eisenhower, putting progress back on track. Their victory in the November 1960 election reopened the possibility of an SST program. President-elect Kennedy and Vice President-elect Johnson had built their national campaign on support for a major expansion of the defense budget, and most particularly, on aerospace expansion.[103] Who could doubt that the supersonic revolution was back on?

✦ ✦ ✦

By 1960, the newly minted aero*space* engineering community had convinced itself that a tripling of airline speed was not merely possible, it was inevitable. The United States and the United Kingdom had each made significant breakthroughs in supersonic aerodynamics within months of each other. They were both engaged in constructing supersonic engines. The British leadership's intense desire to sustain their aircraft industry had led them to initiate national SST study programs immediately, lest the United States be unchallenged in the new field. The Soviet leadership, of course, had also begun studying SST possibilities, although this "fact" was mere rumor in the West until 1963. With three nations competing for the prestige (and potential market) of being the first to announce an SST, someone was going to initiate the supersonic age. The only question was which nation would do it first.

Sputnik, the Soviet SST announcement, and the Democratic Party's challenge to President Eisenhower had given the airframers the opening to demand an American SST program to ensure that the United States would be first. These demands were a violation of traditional American business mores, industry leaders recognized, but the alternative was foreign domination of the aircraft market and the destruction of their companies. The XB-70 prototype program was expected to cost about $1 billion, slightly less than twice Boeing's asset value in 1959, or put another way, eighty times its net earnings for the year. This was more risk than a private company could absorb. Hence state financing was the only way the United States could be first to the supersonic future.

Compounding the cost was the risk of an SST. Despite the major advancements in supersonic technologies that the NACA, the U.S. Air Force, the Royal Aircraft Establishment, and the British National Gas Turbine Establishment had made during the 1950s, the most crucial technologies were still untested in 1960. The XB-70 was still years from flying and thus Eggers's "supersonic wedge principle" rested on wind tunnel results that were subject to question. The Maskell and Küchemann "vortex lift" theory was similarly untested in the real world. The Langley lab's researchers, in fact, did not believe either of these theories would work out.[104] And being wrong about new technologies was expensive. The High Energy Fuels program, for example, had hit a technological wall before its cancellation. The Lewis lab had found that the combustion products from the boron-based fuel solidified in the experimental engine's turbine, clogging it. The engine would thus run for a few minutes and then as the turbine became clogged, it would simply stop. In their rush to use the still-untried fuel, however, the Air Force and Navy had spent a quarter billion dollars building full-scale production plants for the fuel.[105] Both plants had to be shuttered. Charging headlong into the future could be expensive.

The SST had any number of such unknown risks. No one knew how long the novel construction materials would last, for example. Military aircraft were only expected to last in the range of six thousand flight hours, but commercial aircraft had to last a minimum of thirty thousand hours or no airline would touch them. And no reputable bank would lend big money to a manufacturer proposing an airplane whose basic construction material was unproven. To make things worse, despite the public face of enthusiasm for the new technology, there were dissidents. Lockheed's brilliant supersonic aircraft designer, Clarence "Kelly" Johnson, did not think much of the SST idea. Douglas's SST team had also quietly warned the Air Force in its response to SR-169 that while an SST was technically feasible, its development would require virtually unlimited resources and the company would still not guarantee its economic viability. The ability to predict supersonic performance was not good enough to base contractual obligations on.[106] Boeing's SR-169 presentation to the Air Force contained a warning that the sonic boom an SST generated might not be acceptable to the public, severely limiting the potential market.[107] Pan American World Airway's engineering vice president, Sanford Kaufman, told the Institute of Aeronautical Sciences the same thing.[108] Boeing, Douglas, and Pan Am were the industry leaders, and their comments on the SST's risks would not be ignored in financial and airline circles.

The combination of high technical risk and high cost made private financing impossible. The supersonic revolution would be state sponsored, one way or another. As historian Walter McDougall has argued, beating the technocratic Soviet Union in the space race meant transforming the United States into what its favorite enemy already was: a state based upon government-directed technological change.[109] Beating Nikita Khrushchev to an SST required the same solution. Thus the Kennedy-Johnson victory in November 1960 put a state-sponsored SST program within the realm of the possible. All the SST enthusiasts had to do was convince the new administration that the SST was important enough to the nation's future to warrant federal support.

The venerable National Advisory Committee for Aeronautics, however, would not be leading the industry's campaign for an American SST program. It too had fallen victim to the Sputnik panic. In the tiny satellite's vast wake, President Eisenhower had decided to put the government's space activities into a single civilian agency. There seemed no better place than the very successful NACA, but it was far too small. It was also badly structured for the magnitude of growth that rocketry was going to require.[110] The National Aeronautics and Space Act of 1958 thus dissolved the old NACA and replaced it with the National Aeronautics and Space Administration. But it was very clear what the new *space* agency's priority was going to be, and the SST enthusiasts—indeed, the entire aeronautics community—were not at all happy about it. The space agency was certainly not going to lead the supersonic revolution.

2
TECHNOLOGICAL RIVALRY
AND THE COLD WAR

In the United States, SST enthusiasm had emanated largely from the airframe manufacturers, which had instituted small SST study programs following the U.S. Air Force's decision to reorient the WS-110A program toward a supersonic cruise aircraft. These efforts were confined to aerodynamic and materials research, and involved fewer than fifty engineers at each of the five interested companies. Much of this effort went toward preparing the briefings on SST feasibility that the Air Force had requested via SR-169, and was funded by the companies themselves, not by the government.

The results of the SR-169 studies had not been particularly encouraging. But the manufacturers' obvious interest in a supersonic transport had ramifications for NASA and for the Federal Aviation Agency.[1] At NASA's Langley Research Center, manufacturers' interest caused the lab to begin researching supersonic configurations that might be more efficient than the XB-70, and would thus improve the SST economic situation. Further, the U.S. manufacturers' interest in an SST caused Great Britain to submit a request to the International Civil Aviation Organization for a study of potential SST impact on international aviation, convincing the American Federal Aviation Agency's head, Elwood Quesada, that Britain intended to produce an SST. In turn, the FAA established a small SST study group. This group's early effort focused on getting informed about the SST issue. The FAA was a new agency, founded the same year NASA had been (1958), and it did not have the technical knowledge to evaluate the SST claims that circulated through the aviation press and the industry technical conferences. Thus for the first couple of years of supersonic enthusiasm, the FAA was attempting to catch up on the state of the art while NASA researchers were trying to improve it.

The 1958 transition from the NACA to NASA had left the Langley Research Center's aeronautical enthusiasts with a severe problem. The NACA had been exclusively devoted to aeronautics and had taken up space-related research only insofar as necessary to provide support to military space programs. Most of this support concerned hypersonic and re-entry aerodynamics, which were, of course, extensions of the agency's aerodynamic specialty. More importantly, the NACA had been barred from developmental work. The agency routinely provided data to other agencies and companies in support of their developmental projects, but it did not design or develop airplanes. It did the "research" part of the "research and development" activity exclusively.

NASA, however, was assigned the mission of researching and developing rockets and other space hardware. Development was by far the more expensive part of R&D, and the publicity surrounding the space race ensured that the agency's rocketry efforts would be high-profile, high-priority projects. The aeronautical specialists at Langley rightly believed that aeronautics would be subordinated to the space mission and starved of the resources to carry on new research by the *space* agency. There was already ample evidence that this would happen. An internal review of the lab's research effort found that research devoted to fighters, bombers, and transports declined from 71 percent of the lab's work to 30 percent during fiscal year 1959, while research on rockets, missiles, and spacecraft exploded from 8.8 to 44 percent.[2] These fears were reinforced by Langley director Floyd Thompson's decision to allow the new Space Task Group to take engineers from the rest of the lab at will, and perhaps more importantly, by the Lewis Research Center's decision to essentially abandon air-breathing propulsion research to better pursue rocketry.[3]

In short, the Langley lab's aeronautics specialists faced the problem of remaining relevant in an agency whose obvious future was going to be much different from its past. They were fortunate that two new major aeronautics efforts coincided with the foundation of NASA: the supersonic transport research and a military fighter program that became known as TFX, for Tactical Fighter, Experimental. Both of these efforts substantially contributed to U.S. supersonic technologies and the eventual SST development program while permitting the Langley and Ames laboratory aeronautical specialists to participate in aeronautical projects that they believed were important to

U.S. national interests. And, because they were well-publicized efforts, they allowed the aeronautics enthusiasts to bask in a little limelight of their own.

But NASA's administrator, T. Keith Glennan, ruled out agency leadership of an SST *development* effort, should one be approved.[4] Glennan believed that the agency's leadership could not effectively manage the space program (its primary mission), the large growth it was undergoing, and an SST development program simultaneously. His decision was widely unpopular at the Langley laboratory, where John Stack was running an SST research program. But it left the way open for the Federal Aviation Agency to take charge of the issue. The FAA had explicitly been given the mission "to undertake or supervise such development work and service testing as tends to the creation of improved aircraft" in its founding legislation, and both its first administrator, Elwood Quesada, and its second, Najeeb Halaby, interpreted that as making the agency responsible for developing new aircraft. Halaby, as historian Mel Horwitch has nicely argued, took the SST as his personal crusade.[5] As the question of whether the United States government should develop a supersonic transport, as opposed to merely conducting research on the concept, was fought out within President Kennedy's cabinet, Halaby played the champion's role.

Halaby knew that both Britain and France were also researching SSTs, and both he and his predecessor had had talks with Britain's Peter Thorneycroft, whose Ministry of Aviation was responsible for sustaining the British aircraft industry. Those contacts convinced Halaby that the United Kingdom would launch an SST development program, and he believed that if Britain were allowed to be first, it would capture the market and become the new leader in commercial aircraft manufacturing. This would be a huge blow to America's already tottering national image. And while the rhetoric Halaby and his supporters deployed to support their program was typically directed at the cold war struggle with the Soviet Union, Halaby, Stack, and others were concerned at least as much with the United States's commercial competitors, Britain and France. Halaby, Stack, and their many supporters could not tolerate the notion that some other nation, even an ally, might beat the United States into the high-tech supersonic future. They thus made use of the national preoccupation with advanced technology to anchor their project in the national psyche.

They were aided by the actions of French President Charles de Gaulle. Elected in 1958, de Gaulle intended to restore French greatness in the world. He did not intend to do it through the pre–World War II route of empire-building but by making France technologically independent from the rest of the world—and especially from the United States. He tackled two primary industries in his drive for national rejuvenation: nuclear power and aerospace. These were the high-technology, high-visibility industries of the late 1950s, and each promised major economic benefits to the nation. They each depended upon precision machining, highly skilled labor, electronics, and substantial educational and research infrastructures. Developing aerospace and nuclear industries would force development of all these other economic realms too, bringing sweeping change to the traditional French economy. There could be no better symbol to the world of French resurgence than an SST, and hence when Britain began looking for a partner with whom to launch a development program, its leaders found France unusually willing. The Anglo-French alliance, ultimately, forced the United States to undertake an SST development program of its own.[6]

The path to that development program, however, was rather tortured. It involved a NASA research program focused on aerodynamics, structures, materials, aircraft performance, and sonic boom studies; the parallel FAA development campaign; and the Anglo-French decision. Examination of each of these is necessary to understand the political nature of the resulting SST programs, for the studies done on both sides of the Atlantic made clear that an SST was at best a highly questionable undertaking. The current state of the art was insufficient to guarantee that economical aircraft could be designed. The programs thus lacked a strong commercial basis, causing the more economically inclined members of President Kennedy's staff to question the value of going ahead. Their arguments, however, were insufficient to overcome the politics of prestige.

U.S. SUPERSONIC TRANSPORT RESEARCH, 1958–1963

Supersonic transport research in the United States during the early 1960s was wide-ranging. NASA's aeronautics research centers sponsored studies of new aerodynamic configurations, engine designs, and sonic booms. These efforts were paralleled by a Federal Aviation Agency program to understand supersonic technologies and a campaign to convince political leaders to finance an SST development program. Despite the extensive research program, however, economically acceptable SST performance eluded aircraft designers.

The basis for NASA's supersonic transport research effort during the late 1950s was a new class of aircraft configurations designed for supersonic flight. These were known as "variable geometry" configurations. The essential idea behind variable geometry was the recognition that by varying the amount of an aircraft's wing sweep, one could optimize an aircraft's aerodynamics for a wide range of speeds. Straight wings were most efficient for low-speed aircraft while swept wings were most efficient for high-speed aircraft. A variable geometry aircraft could have both, by changing the sweep angle in flight.

This was hardly a new idea in the late 1950s. In the rubble of postwar Germany, the United States had found an experimental aircraft, the Messerschmitt P.1101, that could have its wing sweep changed on the ground, thus facilitating flight research on various sweep angles. Beginning in 1945, Langley researchers John Campbell and Charles Donlan had also conducted wind-tunnel investigations of varying degrees of sweep, showing that while the idea was aerodynamically sound, it came with stability problems. In essence, as the wings moved, the changing center of pressure on the wings caused the aircraft's stability characteristics to change, making it difficult to control. Donlan concluded that "wing translation," which would involve moving the normally fixed wing roots to offset the center of pressure movement, would be necessary to achieve suitable control characteristics.[7]

In 1949, Bell Aircraft of Buffalo, New York, proposed a version of Messerschmitt P.1101 modified with a wing translation mechanism to permit changing the sweep angle while airborne. The NACA and Air Force jointly accepted their proposal in July, and the aircraft, named "X-5," was delivered in June 1951.[8] The NACA's High-Speed Flight Station at Muroc flew the X-5 extensively, demonstrating that the variable sweep concept did, in fact, provide the aerodynamic benefits predicted by both theoretical analysis and wind-tunnel testing.[9]

The need for wing translation, however, rendered the idea impractical. The translating pivot was heavy and complex, and while it had not produced substantial mechanical problems for the small X-5, scaling it up to a practical aircraft's size promised to be troublesome. Further, it consumed valuable internal space that aircraft would need for fuel, electronics, or payload. Hence while the variable geometry idea had been intriguing, it was not practical unless engineers found a solution to the stability problem other than wing translation. Variable geometry wings thus were put on the lab's "back burner" until someone had a better idea.

In May 1958, the Langley staff was asked to comment on a British variable sweep design that had been proposed by Dr. Barnes Wallis of the Vickers Aircraft Company. This request came through the U.S./UK Mutual Weapons Development Program (MWDP) steering committee, of which John Stack was a member. Stack had been impressed with the potential of variable sweep, and this request gave Stack and the Langley lab a chance to look at a different solution to the stability problem. Wallis's design was known as the Swallow, and instead of using a single centerline-translating pivot, it used two pivots on the edges of the fuselage. It also placed the engines on the wingtips, mounted on pivoted pylons so that the engine thrust would remain parallel to the fuselage as the wings moved. At a meeting in November 1958, Stack and Wallis agreed to a joint research program on the Swallow.[10]

Working in the Langley lab's seven-by-ten-foot high-speed wind tunnel, project engineer William Alford tested three different variations of the Swallow configuration with relatively disappointing results. Simultaneously, Alford and Edward Polhamus conducted theoretical studies of the idea and concluded that by retaining the outboard location of the pivots, but providing a fixed lifting surface inboard of them (essentially a fixed portion of wing), one could achieve the benefits of variable sweep without the stability problems that had required wing translation in the X-5. They had the Langley shops construct a fourth model along these lines, and in early 1959 they verified that this "outboard pivot" design, as they termed it, worked as predicted.[11] They also determined that placement of the engines on the wingtips impaired stability, and recommended to Vickers that the engines be placed in the fuselage in a more normal configuration.[12]

The new outboard pivot idea happened to occur just as both the Air Force and Navy were looking for a way to design an aircraft capable of very high-speed, low-altitude flight in order to evade the sophisticated Soviet air defenses that threatened the high-speed, high-altitude XB-70. The variable geometry concept was an obvious way to accomplish this. A very high sweep angle would permit low supersonic speeds during the low-altitude portion of a mission, while a low sweep angle would allow efficient high-altitude cruising to and from a combat zone. After a series of briefings held during 1959 at which Stack promoted the benefits of this new technology, both services initiated programs to develop the outboard-pivot variable-sweep concept into working aircraft.[13] The Navy's aircraft program, called "CAP" for Combat Air Patrol, focused on an aircraft that would be capable of supersonic interception while able to "loiter" in the vicinity of an aircraft carrier for long periods of time, while the Air Force wanted a tactical fighter able to fly the high-speed, low-altitude penetration mission. In February 1961, the two programs were merged into one by Secretary of Defense Robert McNamara; this became the TFX program.[14]

Several former members of the Langley staff credit the Swallow effort for resuscitating the variable geometry concept at the laboratory even though the Swallow itself was not a viable configuration.[15] The Swallow effort, which occurred just as the U.S. aircraft companies were becoming increasingly interested in supersonic transport designs and as the supersonic bomber program was winding down, also caused the lab's researchers to begin looking at supersonic cruise aircraft configurations that would use the new variable sweep idea. This effort eventually became known as the "Supersonic Commercial Air Transport" program, or SCAT.[16]

A Supersonic Transport Research Committee established by Stack in 1958 coordinated the SCAT effort. Stack saw in the industry enthusiasm for the SST a means to sustain the aeronautics program as the rest of the agency went into space. This group was co-chaired by Donald Baals and John Swihart. It was primarily composed of aerodynamicists, but also included engineers from Langley's structural research division and its "internal aerodynamics" branch, who specialized in inlet and engine performance. The group's goal was improvement in supersonic cruise performance over and above the XB-70. They did not believe that a supersonic airliner with the XB-70's performance level would be commercially viable, and they thought that further aerodynamic research could improve the situation substantially.

In addition to the new outboard pivot idea, the supersonic transport group had another new idea to incorporate into their prospective SST designs. In 1951, Langley theoretician Clinton Brown had shown analytically that by altering the camber of a wing, one could improve its supersonic efficiency substantially.[17] But when Brown's cambered wing was tested, the gains he had predicted did not materialize. As Dick Whitcomb explained it, the linear theory available at the time required too many simplifying assumptions to accurately model real airflows.[18] While Brown's insight was correct, the theory was useless for designing real airfoils. Engineers still had to have a substantial knowledge of how real air flows behaved in order to design an efficient wing, and the wing would still have to be optimized through a well-informed version of "cut and try" engineering. The supersonic transport research was a

FIG. 2.1. Richard T. Whitcomb's SCAT 4. Its birdlike appearance caused Warner Robbins, one of Whitcomb's colleagues, to tease him about needing newspaper to catch its droppings. [Courtesy NASA, Langley photo #L-59-8567]

chance to revisit Brown's idea and many of the SCAT models laid out by the group reflect attempts to produce more efficient supersonic configurations through the use of camber and twist.

Whitcomb had launched the attempt to find a superior supersonic cruise configuration with a very birdlike model that was eventually labeled SCAT-4. (See figure 2.1.) While SCAT-4 had a fixed geometry, several of the other SCAT models were variable-geometry configurations, which Stack believed would be necessary for a future SST. Whitcomb recalled many years later that his own rejection of variable-geometry configurations got him ejected from Stack's study group. Whitcomb did not believe variable-geometry configurations would result in acceptable aircraft weights and eventually gave up on the SST idea entirely, believing that they could never match the subsonic jet's economic performance.[19]

Stack's activism with respect to the SST concept was not confined to the Langley Research Center. He tried hard to sell the idea to NASA's assistant administrator, Hugh Dryden, who proved to be uninterested. He found more interest in President Eisenhower's aviation advisor, retired general Elwood Quesada, the first administrator of the new Federal Aviation Agency. One of the new agency's tasks was representing the United States at the biannual meetings of the International Civil Aviation Organization (ICAO), which regulated international fares and produced international standards for equipment, procedures, and the like. At the July 1959 meeting in Montreal, the British Civil Aviation Authority had requested that ICAO perform a study of what impact an SST would have on international aviation. Quesada's representatives had tried to block the study, believing that the British would try to use it to block an American SST while working on one of their own, but he failed. After the meeting, Quesada established an interagency SST group to prepare the official U.S. response that included himself, Secretary of Defense Thomas Gates, and NASA administrator T. Keith Glennan.[20] In November, Quesada dispatched two members of his staff to the Langley laboratory to talk to Stack's group, and in mid-December, he formed a small study group inside the FAA to develop detailed knowledge about the state of SST art.[21]

Stack orchestrated a briefing and a report for the FAA describing the SST as his supersonic transport group then saw it.[22] Mark Nichols opened the 11 December briefing with a discussion of the difficulty of achieving adequate cruise efficiency. He reported that during the preceding three years NASA research had improved the aerodynamic efficiency by 50 to 75 percent over previous designs, but this was still not "as high as the designer would like."[23] On the other hand, propulsion system efficiency increased with Mach number, hence gains there tended to offset the reduced aerodynamic efficiency at higher speeds. Thus, he argued, given that an SST could make more trips per day, and thus earn more revenue, an SST could be competitive with the subsonic jets.

Nichols's paper also presented the major problems confronting any program that sought to actually build an SST. The biggest one appeared to be "off-design" performance. "Off design" meant, in essence, whenever the aircraft was not in its supersonic cruise. All aircraft had to take off and land at subsonic speeds, and furthermore they had to be able to function in the same traffic patterns as subsonic aircraft. Because supersonic aircraft configurations were woefully inefficient at subsonic speeds, this problem was a huge one. Nichols stated that a "canard transport" with a design similar to the "canard bomber" that had evolved into the XB-70 would burn one-third of its fuel load before reaching its supersonic cruise point. Further, 55 percent of its gross weight would be fuel, leaving very little for payload. Off-design performance was therefore the biggest direct economic liability of an SST.

Langley noise specialist Harvey Hubbard then addressed the other issue that many airlines were concerned about, the sonic boom problem. Current boom theory indicated that the very high altitudes that an SST would cruise at (above 60,000 feet for the Mach 3 case the report was based upon) would provide enough attenuation so that the boom would not cause damage at ground level. However, SSTs would have to be operated very carefully, he explained, because sudden changes in altitude or direction could cause very intense localized booms on the ground. Acceleration through the sound barrier would also cause an especially intense boom; hence he argued that this needed to be done at the highest possible altitude (which would increase both fuel consumption and the size of the engines). There was, he believed, no chance of eliminating the boom through tailoring of the aircraft's config-uration.[24]

Other Langley researchers discussed relevant structural, materials, and op-erational issues like air traffic control. Eldon Mathauser told the group that aluminum could not be used much above Mach 2, due to loss of strength when exposed to temperatures above 210°F. Materials like titanium and stain-less steel would have to be used on a Mach 3 aircraft. John Swihart argued that a new kind of engine, the afterburning turbofan, would be necessary to make an economically viable aircraft. And Thomas Toll made Stack's argument in favor of variable geometry as a way to help alleviate the off-design perform-ance problem. Such an aircraft could fly with reasonable aerodynamic effi-ciency at subsonic speeds, reducing the fuel burn. There were, in short, a host of problems standing in the way of a viable SST.

Stack concluded the briefing by arguing that all of these problems were solvable with a concentrated research effort. This effort, he told the group, was necessary because the United States was engaged in a "technological cold war," which "made the achievement of supersonic commercial transport of major importance to our survival."[25] Stack believed that "internal communist aggressors" scored their largest gains when U.S. prestige waned, and thus fail-ure to produce an SST would undermine the nation in its conflict with the Soviet Union.[26] The United States could not afford to let someone else develop an SST first. His view, of course, resonated perfectly with the industry's demands for an SST project and with the U.S. Air Force's desire to resurrect the XB-70.

In response to Stack's report, Quesada began promoting a government-financed SST research program, aided publicly by the chairman of the Civil Aeronautics Board, James Durfee, and quietly by the Air Force. Quesada's efforts brought forth the Overton Brooks committee hearings of May 1960 discussed previously. After NASA Administrator Glennan announced in July that NASA would not take charge of an SST program, Quesada drafted a letter for President Eisenhower to sign authorizing an FAA-led development effort costing $450 million.[27] Eisenhower's refusal to back the program, however, derailed Quesada's campaign for a rapid development program. Stymied, Quesada allocated a small sum of money for an SST economic study and got the U.S. Air Force to spend some of its own money on an SST engine investigation.[28] But a major FAA-led SST program was going to have to wait until the next administration.

Quesada's failure to get a program started did not, however, affect the NASA research effort, which was internally funded. During 1960, the Langley laboratory's effort expanded, as the seventeen SCAT configurations entered wind-tunnel testing. The lab's research program included the SCAT aerodynamic studies, structural and materials research, and the sonic boom problem. During the next two years, these studies produced a mixed bag of results. The two most important aspects of the research, the aerodynamic configuration studies and the sonic boom investigation, were split. While the aerodynamic studies produced more efficient configurations, the sonic boom problem appeared to get worse.

The sonic boom phenomenon imposed a number of concerns for both SST manufacturers and their potential customers. The primary issue was how people would respond to the sonic boom. If public reaction was negative, as Pan American World Airways's chief engineer had pointed out in 1959, SSTs might be banned from supersonic overflight of land.[29] In that case, SSTs might find too small a market to be worth the cost. Although the aircraft industry had known about the boom phenomenon since 1950, knowledge about it did not become widespread until 1953, when supersonic aircraft began to enter the U.S. Air Force's inventory in substantial numbers. In 1954, the NACA received a large number of requests for information about the boom, causing Richard Rhode, the assistant director of research at the Washington office, to recommend that the Langley laboratory undertake a study to determine "the noise intensity and the nature of the pressure disturbances at the ground or in the air" necessary to produce damage.[30] The one existing scientific reference to the boom that the Washington office was aware of, by British researcher Geoffrey Lilley, suggested that aircraft would not produce a boom intensity of more than 0.5 pounds per square foot (psf) of overpressure above 20,000 feet. This was not enough force to produce damage to a structure, but Lilley had noted that most observers found the boom frightening.[31] At the Langley laboratory, Lina Lindsay, Dominic Maglieri, and Harvey Hubbard had already started looking into the issue, and had found another British analysis that reported much larger booms could occur from turning while in supersonic flight.[32] Further, booms produced by F-86 fighters during transonic dives had already been measured at up to 2 psf.[33] Some members of the lab staff even believed that the sonic boom might be used as a weapon.[34]

During the next few years, sonic boom studies in the United States and in Great Britain proceeded in very different ways, with British researchers focusing on analytical development of boom theory while the NACA worked with the Air Force and Navy to develop a body of experimental data.[35] Initially, the NACA's researchers believed that boom intensity at ground level was determined primarily by aircraft altitude. According to that theory, an aircraft cruising above 60,000 feet would not produce a noticeable boom, let alone one that would cause damage. By 1958, however, theoreticians in Europe had devised an analytical argument that the dominant determinant of a supersonic airplane's boom signature was the amount of lift it created (which in level flight was equal to the airplane's weight). It this was true, large aircraft would produce booms of substantially greater intensity than a fighter-sized aircraft at the same altitude.

The NACA pursued two approaches to determining whether the lift effect on boom intensity existed. During 1960, using Air Force B-58 bombers stationed in Nevada, Harvey Hubbard and Dominic Maglieri found that the overpressures produced by the aircraft exceeded those predicted by Whitham's original theory.[36] This suggested that the lift effect predicted was real. Their data could not prove the case, however, because the B-58 was not big enough to produce a definitive result. Flight confirmation of the lift effect had to wait for the XB-70, which would be twice the B-58's size. At the Langley laboratory, Harry Carlson also ran wind-tunnel tests on tiny models of the B-58 to determine whether the lift effect could be measured in the aerodynamicist's traditional research tool. Carlson's study also suggested that the lift effect was real.[37] By 1962, then, researchers in the United States and in Britain had realized that boom intensities for a large SST would probably be significantly worse than the manufacturers and airlines had publicized as recently as 1959. This was not a pleasing result, as even those levels had given some airlines pause about the SST's future.

In parallel with the boom prediction studies, researchers at Langley and the High Speed Flight Station had pursued studies to determine human and structural reactions to the boom. The boom was already becoming a political issue for the U.S. Air Force, which was being criticized for its "booming" of cities during training exercises. The Air Force had developed a set of criteria for reimbursing citizens for sonic boom damage in response, but by 1960, its own legal advisors were challenging the validity of those criteria. As of January 1960, the Air Force had processed over five thousand damage claims and had paid on only 2 percent of them, causing its lawyers to argue that "98 percent of the people can't be wrong."[38] Moreover, at a meeting held at Wright Field, Langley's Harvey Hubbard was told by his Air Force counterparts that members of Congress had forwarded four hundred letters complaining about the boom to Air Force headquarters. The group recommended that the payment criteria be somewhat liberalized, and they also agreed that they needed a research program aimed at understanding both boom damage and how people responded to the boom. In short, they needed to find out what boom intensity people and buildings could be subjected to routinely in order to estimate the financial and political cost of the boom.

NASA and the Air Force collaborated to run a series of sonic boom tests, commencing with "Operation Little Boom," run at the Los Vegas Gunnery and Bombing Range about fifty miles from the city of Las Vegas.[39] This test series was run on Air Force personnel and also sought to relate boom intensity to glass breakage levels. Using F-104 fighters and B-58 bombers, the Air Force subjected its subjects to booms up to 120 psf in an attempt to find a damage threshold. Even these very intense booms, however, did not do damage to structures or people, and glass breakage occurred only at very high boom intensities. But both NASA and Air Force researchers recognized that this was not a fair test, since the subjects knew the booms were coming and the structures, being both new and fairly simple, did not represent the "real world" of cities.

The two agencies also planned to bring their tests to that world, and chose

St. Louis, Missouri, as the first city to undergo a "community response" test. Called "Operation Bongo," the test series, consisting of sixty-six booms of under 3 psf intensity, ran from July 1961 through January 1962. Because the researchers were concerned that the public would react differently to booms generated for research purposes than they would to booms produced in the process of national defense, the Air Force told the city's leaders only that the city had been chosen as a Radar Bomb Scoring (RBS) city, to be boomed in the process of training B-58 crews to attack Soviet cities.[40] Harvey Hubbard set up instruments at three federal installations in the St. Louis area to measure the strength of the booms at ground level. To assess reactions to the boom, NASA contracted with the National Opinion Research Center of the University of Chicago to conduct interviews with 1,145 residents of the city during the project.[41] The Langley lab also hired engineering contractors to visit alleged damage sites in company with legal officers from Scott Air Force Base to evaluate damage claims.

The results of the study were not particularly encouraging. Of the residents interviewed, 90 percent felt the booms "interfered" with them in some way, while 35 percent found them annoying. But only 0.6 percent of the interviewees filed formal complaints. The Air Force also received 3,114 complaints and 1,624 damage claims from the city's population during the study, and it approved payments totaling $58,648.23 on 825 of the claims.[42] More significant than the raw numbers, however, was the reaction to increased boom frequency. Immediately after the official end of the Bongo series, a dramatic increase in sonic boom frequency occurred, with seventy-four additional booms taking place over about two months. The increased frequency caused a surge in complaints that the study's authors decided to include. Hubbard and Nixon noted that the increased frequency of booms caused the local newspaper to begin demanding cessation of the booming. This, they believed, helped make complaining about the boom "socially acceptable."[43] But what gave everyone pause was simply that the increased boom frequency was still not realistic in terms of a commercial aircraft schedule at an average of slightly more than one boom per day. A commercial schedule would subject cities like St. Louis to several booms per day. Another set of tests was necessary to evaluate the potential public response to that.

The outcome of the St. Louis tests, while hardly definitive, was sufficiently negative to give airlines and aircraft manufacturers additional reason to question the supersonic future. That unease was reinforced by a publicity stunt run by the U.S. Air Force during 1962. The Air Force agreed to a Convair plan to "market" the troubled B-58 via a series of publicity events intended to boost political support for additional purchases of the aircraft.[44] One of these events, a transcontinental supersonic flight called "Operation Heat-Rise," became infamous. On 5 March 1962, a B-58 flown by Captain Robert Sowers set a new Los Angeles to New York speed record of 2 hours, 57 seconds. He and his crew also set a new damage record, collapsing windows in Anaheim and Riverside (over which the plane had generated an acceleration "focused boom") and causing disturbed citizens to jam police switchboards all along the plane's route—740 calls in New York City alone. The Air Force received over ten thousand complaints from the B-58's flight.

The Air Force, NASA, and the FAA also got a foretaste of the likely public reaction to an SST out of Operation Heat-Rise, although typically the enthusiasts downplayed the public reaction. People would get used to the boom, the FAA's leaders insisted. Air Force and NASA researchers, however, were beginning to doubt that. Booms had become increasingly common as more and more supersonic aircraft entered the Air Force's inventory, and complaints were going up, not down. Further, as one perceptive Air Force office wrote in the wake of the St. Louis and "Heat-Rise" tests, there was a very different question no one had yet considered: *should* the public have to get used to the boom just so businessmen and the wealthy (the only people who could afford to fly in the early 1960s) could get across the country a couple of hours faster?[45] The most boomed cities, after all, would be midcontinent cities that would receive no benefit from the new technology. Hence the sonic boom component of the research program wound up raising an ethical issue for which there was no engineering solution on the horizon.

While the sonic boom research was raising difficult questions, Langley's aerodynamic studies were producing more positive results. By 1962, the SST research committee at the Langley laboratory had gathered enough data to select the best of the configurations and, in concert with the FAA, had decided that the best way to "transfer" them to industry was via study contracts. NASA would pay two manufacturers to conduct evaluations of three (later, four) of the SCAT configurations, and would select the two contractors through a competitive bidding process. In the process of evaluating the configurations, the two manufacturers would learn what NASA's staff already knew, and would, the agency's officials hoped, use that knowledge to improve their own SST designs. The contracts totaled $3 million.[46]

The SCAT configurations that NASA's Supersonic Research Committee chose for the contract studies were Whitcomb's SCAT 4, two swing-wing configurations named SCAT 15 and 16, and a delta-wing configuration from the Ames lab, SCAT 17. SCAT 17 had not originally been part of the program due to the Langley lab's fixation on variable geometry, but had been added because most of the manufacturers were working on delta-wing designs. Lockheed in particular had been unhappy about Langley's promotion of the swing-wing idea and had not been shy in its complaints.[47] And it would not do to entirely freeze out the Ames Research Center's aerodynamicists, whose assistance to the mostly West Coast–based airframe manufacturers was necessary for the program's success regardless of who won.

Of the five companies that submitted proposals in the SCAT bidding process, NASA chose Boeing and Lockheed to perform the studies. This drew a letter from North American Aviation's vice president, Ralph Rudd, to Langley's director, Floyd Thompson, asking for the lab's data so that the company could perform its own studies of the configurations.[48] Rudd hoped that the company would be able to improve its own design and remain competitive with Boeing and Lockheed. Otherwise, he believed Boeing and Lockheed would gain an unbeatable technological "edge." Ultimately, NASA decided to release the SCAT data to all of the domestic airframers to eliminate this potentially damaging issue, but in the meantime North American Aviation's president, Lee Atwood, decided to contact Boeing president Bill Allen about pursuing the SCAT study in a partnership. Allen agreed to a "tie-up," and the two companies split the four SCAT models for testing.

The results of the SCAT study and of NASA's own ongoing research program were presented at a conference at the Langley laboratory organized by Larry Loftin.[49] In essence, this conference represented the state of the art as it existed in the United States. It also reflects the problems that still faced manufacturers interested in pursuing the SST, and hence the research that still needed to be done as the United States made its decision on whether to actually develop an SST.

Lockheed's summary, presented by Richard Heppe and Jim Hong, focused on the issue of weight versus aerodynamic efficiency. Whitcomb's SCAT 4 configuration, they had found, carried a very high structural weight penalty, with the complex wing structure amounting to 19 percent of the aircraft's gross weight. In contrast, the less aerodynamically efficient SCAT 17 had a much lower wing weight, amounting to only 9.2 percent of gross weight.

Since both weight and aerodynamic efficiency affected the aircraft's economic performance, SCAT 17's much lower weight fraction made it a better choice. The same fundamental problem caused Lockheed to drop SCAT 15. In essence, SCAT 15 had two separate wings, the aft fixed portion and the variable-sweep portion. Each of these had to be strong enough to carry the entire aircraft's weight, resulting in a structural weight even worse than SCAT 4's at 21.5 percent. While it was the most aerodynamically efficient concept, its high weight made it uncompetitive with the two less aerodynamically efficient, but more weight efficient, SCAT 16 and 17. Therefore, at the midterm contract review in May, Lockheed, Boeing, and NASA had all agreed to drop SCAT 4 and 15 and focus their efforts on the other two configurations during the remaining study period. That effort did not succeed at producing economical designs, either. Heppe and Hong reported that their efforts demonstrated "configurations substantially improved over that offered by any of the initial SCAT arrangements [were] required for a fully successful United States supersonic transport."[50] In other words, the current state of the art was not good enough.

Lloyd Goodmanson, William T. Hamilton, and Maynard Pennell of the Boeing Company echoed that assessment. They presented a chart comparing the estimated direct operating costs for SCAT 16 and 17 to that of the 707. To make a competitive airplane out of the two configurations, they believed, would take an entirely new high-temperature, low-bypass-ratio turbofan equipped with "duct burners" to provide the thrust necessary for transonic acceleration (when drag was highest) while still meeting the sonic boom limitation NASA had imposed on the study.[51] Primarily, this was due to the importance of minimizing the sonic boom through accelerating at a very high altitude.[52] An improvement in the cruise L/D to 8 (SCAT 16 achieved 6.6) would also likely be necessary to produce an economical aircraft. Perhaps most importantly, however, they pointed out the difficulty of integrating the complex aircraft and the potential pitfalls. The portion of gross weight devoted to payload in all the configurations was not more than 7 percent, while fuel consumed up to 50 percent.[53] Boeing's engineers believed that 7 to 8 percent payload was the lowest acceptable payload fraction; hence missing a weight, propulsion, or aerodynamic target by even a small amount would reduce payload enough to make the airplane uneconomical.[54] (By contrast, the contemporary Boeing 707-320B carried an 11 percent payload fraction to its maximum range.) There was, in short, no margin for error or inefficiency in these designs.

The SCAT study also sealed a long-running argument among SST enthusiasts over the aircraft's speed. John Stack had long argued that the United States should not consider speeds less than Mach 3, because aircraft propulsion systems became more efficient at higher speeds. Further, the sonic boom from a Mach 3 aircraft would be lower due to its higher cruise altitude.[55] However, speeds over Mach 2 required the use of exotic high-temperature materials whose properties were unknown and promised to be very expensive to develop. Some officials at the Federal Aviation Administration therefore believed that a Mach 2 aircraft might be more economical, and during the SCAT program May review they had directed Boeing and Lockheed to evaluate a Mach 2 aluminum version of SCAT 16. The aluminum version otherwise had to meet the same criteria as the Mach 3 SCAT 16. The results were not what the Mach 2 promoters expected, however. Both companies found that the aluminum SCAT 16 would probably weigh some 50,000 pounds more and have higher operating costs than the Mach 3 aircraft. Behind this unexpected outcome was NASA's sonic boom limitation. In order to achieve the specified boom intensity, the aircraft had to fly above 60,000 feet, whereas its most efficient cruising altitude was 50,000 feet.[56] Reduced propulsion efficiency at the higher altitude meant having to carry more fuel to go the same distance as the Mach 3 aircraft. Mach 2 thus made little sense—at least as long as the 1.5 psf boom restriction remained in place.

The SST research program NASA conducted between 1958 and 1963, then, raised the American state of the art in supersonic cruise aircraft, while also demonstrating that there was still a substantial gap between the performance possible with existing technologies and the performance necessary to beat the subsonic jet's economics. From a purely technical standpoint, the United States was not ready to launch a development program if it wished to produce an economical airplane. Development programs had to meet defined schedules and budgets, which meant relying primarily on state-of-the art equipment. It would take, for example, roughly $200 million and seven years to design, build, and test "engine D," and until it had been tested no one could guarantee it would meet the exacting performance required of it. Launching a development program without the certainty promised by the use of existing technology was an extremely high-risk idea that promised to be very expensive and might still be an economic failure. This determination drained what little enthusiasm airlines and many government officials had for the SST idea. But by the time NASA's SCAT conference was held in September 1963, the poor economics of the SST were moot. The British and French governments had already launched a joint SST development program, formally announcing it in November 1962, and in reaction President Kennedy had declared one on 5 June 1963.

THE COMPETITION

The Anglo-French treaty, announced on 5 November 1962 and formally signed on 29 November, committed the two nations to jointly develop and manufacture two Mach 2.2 airliners to be called "Concord" or "Concorde," depending on one's national preference. The first of these aircraft was to serve the transatlantic routes and would go into service in 1970, while the second would be a "medium-range" aircraft intended for routes between European (or, of course, American) cities. Kenneth Owen has recently argued that the project was launched for political reasons. The British Cabinet saw in Concorde a means to persuade French president Charles de Gaulle that Britain was trustworthy enough to be permitted to join the Common Market. De Gaulle was unhappy with the "special relationship" between Britain and the United States, which he contended made Britain insufficiently "European" to be part of an exclusively European organization, and he had made clear that Britain was unlikely to win his approval.[57] The British Cabinet believed that joining with France to produce an SST, even though economic failure of the aircraft was virtually guaranteed, would be a worthwhile act if it got President de Gaulle's blessing on the nation's Common Market application.[58]

In France, the program was also rooted in a political goal. De Gaulle had dedicated his administration to making France diplomatically, militarily, and economically independent of the United States. Robert Gilpin, and more recently Walter McDougall, have shown that de Gaulle believed economic independence could come only from scientific and technological independence, and hence he reorganized the French state to foster high-technology (and primarily defense-related) industries.[59] Primarily, his reforms buttressed the nuclear, aircraft, space, and electronics industries. These received large infusions of capital from the state, designed to spur research and development so that France could eventually compete with the United States with equivalent technologies. Concorde was part of de Gaulle's effort to modernize France through technological revolution.

In Britain, the path toward an SST had started in the early 1950s. Sir Arnold Hall, director of the Royal Aircraft Establishment in Farnborough, had asked Morien Morgan to chair a committee to look at the potential of supersonic transportation. Morgan's committee first met in February 1954, and by mid-1955 had concluded that an SST was not economically feasible. RAE researchers had tried to convert bomber configurations to the transport mission, relying on straight, very thin wings similar to those on the Lockheed F-104 "Starfighter" for the basis of their configurations. They also chose Mach 2 as a reasonable cruising speed, as aluminum could be used for the aircraft's construction up to that speed. "Some horribly large aeroplanes" had resulted from these studies, Morgan recalled.[60] His committee recommended continuing research into making supersonic research more efficient but did not promote an SST development effort.

During 1955, at the same time Morgan's committee was studying existing converted bomber designs, RAE researchers proposed an alternate configuration based upon the "slender delta," a planform like that of the B-58's delta wing but with a shorter span. Reducing the span reduced drag at supersonic speed, but produced an aircraft that needed an impossibly long runway to get off the ground, under the prevailing theory of the day. Aerodynamic theory descending from Sir George Cayley required that aerodynamicists delay separation of the airflow over a wing for as long as possible, ideally producing a flow that did not detach from the wing until it reached the wing's trailing edge. This was the proper route to producing a low-drag airfoil. Historian Kenneth Owen argues that what happened during 1955 to reinvigorate the British SST effort was the recognition that Cayley's theory was not necessarily the only way to produce an efficient airfoil.[61] Dietrich Küchemann, Eric Maskell, and their assistant Johanna Weber had argued from basic fluid dynamics theory that one could produce controlled separation of the airflow through adoption of a sharp leading edge on a slender delta wing. Further, that separated flow would form stable vortices, which in turn would improve the configuration's low-speed lift and stability.[62] In other words, a sharp-edged slender delta might provide the same benefits to an SST that the United States was pursuing through the much more complex variable geometry concept.

On 1 October 1956, the Ministry of Supply, which was then responsible for government aircraft programs, held a conference at which Morien Morgan was asked to form a new committee to study the SST problem. Over the next two years, as the RAE worked to refine the slender delta concept into a efficient planform, Morgan lobbied the various aircraft companies in Britain to take the slender delta configuration, and the SST issue in general, seriously. Andrew Nahum has recently argued that promotion of highly advanced technologies was an unusual role for RAE to play. Typically the agency provided advice to manufacturers and the government and carried out its in-house research program. Morgan's efforts in recruiting manufacturer interest in the SST went well beyond that during the late 1950s, and his committee report reflects his promotional efforts.[63] By 1959, Morgan's second committee had decided that an SST was feasible, and might even have decent economics. Their report therefore recommended that the UK launch development of a transatlantic SST with a capacity of 150 passengers and luggage. The committee estimated that the transatlantic aircraft would cost £75–90 million to develop and put into production. They also recommended development of a shorter-range aircraft carrying 100 passengers be pursued, at a cost of £50–80 million. The long-range aircraft, they believed, should be ready for the airlines by 1970 or so, while the shorter-range aircraft might appear two or three years earlier. To meet this schedule, contracts needed to be awarded in 1960 for the early design phase, with contractors to be selected in 1962. Only this, the committee believed, would permit the UK to achieve market leadership with the SST.

Morgan, in fact, put the issue much more strongly. He argued in his transmittal letter that "a decision not to start detailed work fairly soon on the transatlantic aircraft would be, in effect, a decision to opt altogether out of the long range supersonic transport field—since we would never regain a competitive position."[64] Left unstated was a belief that the United States would transform the B-70 bomber into an SST and destroy the subsonic jet market while conquering the new supersonic market. The enthusiastic public rhetoric of the American aircraft companies had convinced Morgan and his fellow committee members that they were about to launch SSTs of their own, despite the lack of a federal SST development program. That public enthusiasm had been communicated through a variety of channels, but most directly through internal correspondence. Harold Watkinson, the Minister of Transport, for example, had written a memo to Prime Minister Macmillan summarizing the controversy at the 1959 ICAO meeting over whether the organization should write an SST impact study. Watkinson told the PM that the "extreme sensitiveness" the U.S. delegation had shown only made sense if they had already decided to launch their own SST program, and because of this the government had to "make up our minds how we want to face this most difficult and expensive future."[65]

The problem the supersonic future posed for the UK was not simply damage to its aircraft industry. Rather, Morgan's committee had argued, "This country's future will depend on the quality of its technological products and since its scientific manpower and resources are less than those of the USA and USSR it is important that a reasonable proportion of such resources are deployed on products which maintain our technical reputation at a high level. A successful supersonic transport aircraft would not only be a commercial venture of high promise but would also be of immense value to this country as an indication of our technical skill."[66] The SST was necessary not just because it might be a commercial success, but because it was necessary to the sustenance of the nation's image. By implication, Britain's status as a world power would be judged on the basis of its technological accomplishments in the future, not upon the more traditional symbols of armies and navies.

Morgan did not get his development program right away. Instead, for the next two years, as Owen has shown, both British aircraft firms and the government sought partners to help defray the high cost of the proposed development program. After obtaining cabinet approval, the Ministry of Aviation awarded feasibility study contracts to Hawker Siddeley and Bristol Aircraft to examine potential designs based upon the slender-delta idea, and required that the two companies explore the possibilities of collaboration.[67] Lockheed's Hall Hibbard contacted Hawker Siddeley's leadership about a possible collaboration in September 1959; after the British government merged Vickers Aircraft, Bristol Aircraft, and English Electric Aircraft into the British Aircraft Corporation in 1960, the new company immediately contacted Boeing, General Dynamics, Douglas Aircraft, and France's Sud Aviation about potential collaboration on an SST.[68] The American manufacturers were hesitant because they believed the U.S. government might be convinced to fund an SST, making collaboration undesirable. American firms believed they were considerably advanced technologically over the Europeans, and they perceived a substantial risk in giving their competitors valuable knowledge and skills unnecessarily. Further, coordinating a cooperative program with foreign companies, especially companies largely controlled by a foreign government, promised to be extremely difficult.

Sud Aviation, which was owned by the French government, was the most enthusiastic potential partner. Caught up in the general supersonic enthusiasm of the late 1950s, the French government had requested SST proposals from two of its state-owned manufacturers, Sud and Nord Aviation, as well as the private firm Dassault. Nord Aviation responded with a proposal for a ramjet-powered Mach 3 aircraft, while the other two proposed similar Mach 2 aircraft.[69] Most interestingly, however, each had proposed aircraft based upon the slender delta theory. Sud Aviation's proposal won the government's favor, and in April 1960 the company dispatched Pierre Satre, the company's technical director, to Bristol Aircraft's headquarters to discuss an alliance. Their proposed aircraft was to be a medium-range aircraft similar in range and capacity to Sud's own Caravelle. It was, of course, to be called Super Caravelle, and was intentionally sized to avoid competing with American supersonic aircraft, which Sud's leader anticipated would be transatlantic.[70]

In three key areas, the aircraft proposed by Bristol and Sud were very similar. First, France did not yet have the ability to design and manufacture large jet engines, and thus Satre's team planned to use a British engine. Second, the two companies agreed that an SST should be built of aluminum and thus be limited to a speed of around Mach 2. Neither nation possessed the ability to build aircraft in the more exotic high-temperature materials that higher speeds required, and achieving that ability would be time consuming and very expensive. Choosing Mach 2 could get them to market before the United States. Third, their designs were based upon the same general aerodynamic configuration, the wine-glass-shaped slender delta. The proposals differed primarily in size and range, which was not necessarily a roadblock to collaboration. One could, after all, use the same general configuration in two different sizes. Both aircraft could also use different numbers of the same engine. Hence there was a solid technical basis for collaboration.[71] The two companies continued working independently after the April 1960 meeting, but met again during 1961 in May, June, and July. At these later meetings, the two groups discussed the possibility of adopting a single design, an act necessary before a joint development program could be created.

At the national level, meetings between the British minister of aviation, Peter Thorneycroft, and his French and American counterparts revealed that there was a good deal more enthusiasm for a joint program in France than there was in the United States during 1961. The French minister of public works and transport, Robert Buron, met with Thorneycroft in December 1961, and the two agreed to begin constructing a proposal for a joint development project that they could submit to their respective cabinets. That agreement triggered a series of meetings between the two companies aimed at establishing a common design and the outline of a management structure during 1962. By October 1962, the two companies and the two responsible ministries had agreed upon both issues. They would jointly develop two aircraft, each of 100 passenger capacity flying at Mach 2.2, with one sized for 2,400 n.m. range (weighing 220,500 pounds) and one sized for 3,250 n.m. range (reflecting New York to Paris, at 262,500 pounds).[72] The two governments would pay the full development cost, shared on a fifty/fifty basis. And the program would be managed by a committee whose chair would be appointed by the British and French governments, rotating every other year so that neither national effort could dominate the other.

The Ministry of Aviation's drive toward a joint SST development program had its critics, most importantly in the Treasury Ministry. The chief secretary of the Treasury, Henry Brooke, did not believe that the program would be economical for the nation. The aircraft business was not very profitable, he argued, and the high risk involved in this particular project made it particularly unlikely to show a profit. The economic argument that the Aviation Ministry had constructed assumed sales of 150 aircraft worth £200 million, while Britain's share of the development cost, if everything went according to plan, would be £75 million. He saw no way the government would get this development cost back, resulting in a subsidy of 33 percent on the program. Further, "the industry's past record of over-optimistic estimating (including the recent history of the TSR.2) suggests that it would be prudent to consider the £150 million [total program cost] to turn out much too low."[73]

Treasury's antipathy toward the SST proposal caused the cabinet to send the question to a subset of itself, the Committee on Civil Scientific Research and Development, for a thorough airing of the two agency's arguments. At a series of meetings between July and September 1962, the SST supporters and critics presented their cases, with the Bristol Aircraft Company's Sir George Edwards, Bristol Siddeley's Sir Arnold Hall, and Morien Morgan all making their respective attempts to convince the members of the program's economic feasibility. They did not succeed. In summing up the committee's conclusion in October, the chair wrote that if Britain joined France in this project, it would not be on commercial grounds or to protect the industry. Instead, "it will be because we think we can take this in our stride, without interfering with ou[r] chances of success in other major technological projects to which we may attach equal or greater importance; and that in the long run it will not pay to withhold from industry the finance it needs for a natural extension of technological progress, provided that we have the money to see it through."[74] The SST program was not likely to be profitable, the committee concluded. Instead, it was necessary because the supersonic era was simply an expression of technological progress, and Britain had to participate in it somehow.

Thorneycroft, and his successor Julian Amery, had succeeded in making one argument stick in their presentations.[75] They succeeded in convincing the prime minister that failing to develop an SST would mean the end of British aircraft manufacturing. This argument was based upon the assumed inevitability of progress in aviation. Supersonic airliners would displace the subsonic jets just as they had supplanted the propeller-driven aircraft, and any company, or nation, that did not make the technological leap into the supersonic future would be out of the aviation business in a few short years. That prospect could not be faced, but not because of the loss of jobs this eventuality would cause, although that too had been brought up in Thorney-croft's lobbying. The SST was necessary because it was a technologically progressive project.[76] Even if the proposed joint development program were a commercial flop, it would provide the industry's engineers with new skills that they might employ in subsequent (more economical) projects. And, as Morien Morgan had pointed out in the initial STAC report and had been echoed many times in subsequent reports and correspondence, the SST would provide a visible international symbol of Britain's technological competence. It would, in short, be a symbol of national pride, making clear, simply by existing, Britain's continued technological relevance in a world dominated by the relatively vast expenditures of the United States and USSR.

On 6 November 1962, the cabinet approved the joint agreement, citing their hope that President de Gaulle would take the treaty as a sign of Britain's desire to be "European."[77] But despite the optimism of the 29 November signing ceremony in London, de Gaulle summarily rejected Britain's Common Market petition in January, while simultaneously linking Concorde to "European technology." And since the treaty had no escape clause, Britain was quite effectively bound to finish the program.[78] As a ticket to Europe, Concorde was a failure long before it flew. But it could still serve state purposes as a symbol and as a means of modernization. These became Concorde's raisons d'être.

THE FAA AND THE SST: MAKING THE DEVELOPMENT DECISION, 1961–1963

The Anglo-French accord, while hardly a surprise, gave the American proponents of a national SST development effort a new foundation from which to build a base of political support. This was necessary because experience gained on supersonics in the years between the initial burst of SST enthusiasm surrounding the XB-70 program redirection and the Concorde announcement had cast the validity of the SST effort in considerable doubt. Uncertainty about the ability to produce an aircraft with competitive economics had grown in company with increased understanding of the problems involved in supersonic flight. And while the SST's poor economics had ultimately been irrelevant to the French and British governments, this was not so in the American context. American national mythology was based upon technological supremacy *and* economic prowess, and thus the American SST enthusiasts faced a higher hurdle than had Morien Morgan. It was not enough that an SST be possible. In order to receive secure political support the resulting product had to appear profitable, too.

President Kennedy's FAA administrator was a test pilot turned attorney named Najeeb Halaby. After World War II, he served in the Defense and State departments before leaving government for law in 1953. But he did not stray far from aviation, which, he explained in his memoir, had always "evoked the Biblical Genesis" to him.[79] He saw in the SST, as had Morgan and Thorneycroft, "a logical expression of faith in aviation's progress."[80] He thus replaced John Stack, who quit NASA in disgust over the agency's lack of interest in aeronautics in 1962, as the leading American promoter of the SST.

Halaby, as had Quesada before him, believed that the FAA's founding legislation made the agency responsible for the development of civil aviation in the United States.[81] Therefore, any SST development program should belong to his agency, despite its lack of technical and managerial experience in aircraft development.[82] Halaby quickly succeeded in convincing President Kennedy to expand upon NASA's SST research program with a commitment of $12 million for his agency's budget for fiscal year 1962. Congress reduced this to $11 million.[83] The expanded research program was to encompass engine, materials, and structural studies in addition to the ongoing NASA aerodynamic and sonic boom research, and was to be done largely under contracts administered by the Air Force's Wright Air Development Center until the FAA assembled a management staff of its own.

Halaby's promotion of the SST began with the Commission on National Aviation Goals, generally known as Project Horizon. Almost immediately after his inauguration, President Kennedy had directed Halaby to have a general review of national aviation policy done by leading experts in the aviation business. Kennedy was convinced by a series of horrific aviation accidents that commercial aviation was in need of a major infrastructure overhaul and this had been his purpose in commissioning the study. But Halaby also used the report as a tool to promote the SST, arguing that failure to produce an SST would be a "stunning setback" to the United States in the light of Russian advancements in rocketry.[84]

Project Horizon thus downplayed the economic problem to emphasize the importance of prestige, drawing criticism from some of the commission's own members. William Littlewood, for example, American Airline's vice president for engineering and chair of the commission's technical subcommittee, was explicit about the economic question. If the SST could not be proven economical, he wrote, "the bankers will never give us the money."[85] Littlewood was hardly the first airline manager to question the SST. In public, airline presidents like Littlewood's boss C. R. Smith universally toed the patriotic line that the United States had to build an SST for prestige's sake. In private, however, they made clear that they did not want an SST to appear anytime soon. They could not afford it, having overinvested in the new subsonic airliners that had debuted in 1959. Further, detailed route studies done by Pan American World Airways and presented at the 1961 SAE meeting in New York showed that an SST would earn only 29 percent of the net profit its subsonic 707-321 did—hardly an improvement in a generally unprofitable business.[86] At the international level, unease about the economic impact of SSTs was also widespread. The director of the International Air Transport Association contended that national governments would have to pay for the development, then pay their airlines to buy and operate the aircraft. There would not be, he insisted, "enough airlines and enough passengers to foot the bill."[87]

In July 1962, the International Air Transport Association made the job of selling an American SST program still more difficult by releasing a set of "design imperatives" for a future SST. After demanding equal levels of safety and air traffic control compatibility to the subsonic airliner, these design criteria called for no increase in engine noise levels, seat-mile costs at least as good as subsonic jets of the same size and range, and supersonic operation over populated areas, day and night.[88] Neither the cost nor noise requirements were within the state of the art (and the sonic boom issue was highly questionable). Therefore, the IATA's demands stood as a warning to SST promoters. The airline industry was not interested in a prestige vehicle.

The problem of economics worked against Halaby at a meeting of the National Aeronautics and Space Council in August 1962, when his drive to get a development program started was derailed. The council's chair, Vice President Lyndon Johnson, allowed Secretary of Defense Robert McNamara to contest Halaby's economic arguments repeatedly.[89] In short, McNamara believed Halaby's figures were unduly optimistic, and he wanted much more sophisticated economic and market studies performed before agreeing to a development program. But McNamara also did not want the SST program assigned to DOD, so while he questioned FAA's competence he did not advocate replacing the agency.[90] He agreed to let Halaby run the economic studies, allowing FAA to retain control of the research program.

But because McNamara had the vice president's ear, and because President Kennedy had handed the SST issue off to Johnson as part of his private "keep Lyndon busy" campaign, an SST development program was an impossible goal until studies acceptable to McNamara appeared.[91] The vagaries of federal budgeting meant that no program go-ahead could now be had before 30 June 1963, then the end of the fiscal year. In October Halaby let contracts to RAND Corporation and the Stanford Research Institute (SRI) for the studies McNamara had demanded, and he toned down his pro-SST rhetoric. This shift against the SST was dramatic enough to warrant notice by the British ambassador to Washington, who told his government that there was now no substantial political interest in the high-cost, high-risk idea.[92]

The Anglo-French Concorde announcement in November 1962 began to change the political landscape, however, causing Halaby to return to his high-intensity lobbying. Three days after the announcement, he sent an alarmist letter to President Kennedy in which he argued that failing to contest the coalition would allow the French medium-range version to "invade the domestic market" while the British long-range version would replace the Boeing and Douglas jets on the international routes. This would cost the United States its leadership of commercial aeronautics, 50,000 jobs, and $4 billion in export income, and it would force its airlines to spend $3 billion more on imports of the foreign SST.[93] Nevertheless, his tactics did not sway McNamara or Kermit Gordon, head of the Bureau of the Budget, who still wanted to see the results of the economic studies. But Concorde got President Kennedy's attention. Halaby recalled in his memoir that de Gaulle's support of Concorde sparked Kennedy's competitive spirit.[94] That combativeness was exacerbated by de Gaulle's rejection of the British bid to join the Common Market in January 1963. Kennedy thus directed Vice President Johnson to compile a report for him on the SST issue, including recommendations on what the United States should do in response.[95]

Assembling that report, however, reinforced unease about the economic potential of the SST. Halaby's own study committee, the Supersonic Transport Advisory Group, or STAG, chaired by retired General Orval Cook, issued its first report in the wake of the Concorde announcement. The report called for "prompt initiation" of a development program focused on a Mach 3 or faster SST to serve the domestic market (2,400 n.m. range). They called for operating costs better than the existing jets, and proposed that the development would cost $1 billion. The government would finance this, but the group also expected that the government would recoup all or part of the cost through a royalty on sales of the aircraft. Manufacturers, they stated, should provide 10 percent or so of the development costs, as an incentive for good performance.[96]

In essence, STAG was calling for a crash development program to contest Concorde. They had ruled out attempting a transatlantic aircraft because this would take too long to develop, and they wished to get the U.S. Mach 3+ SST onto the market at the same time Concorde was supposed to appear, in 1969. That there was little real time savings to be gained in a domestic-range aircraft from the increase in speed the group ignored, and while they advocated accelerating the sonic boom research program, their decision to forgo a transatlantic aircraft meant that if sonic boom restrictions resulted from that program, their domestic-range airplane would possess no market. They made no attempt to examine the economics of the program other than to explain that the aircraft industry was a major exporter and therefore a substantial contributor to the nation's balance of payments.[97] An SST, by implication, was necessary to sustain this contribution to the economy.

But as the outside economic reports came in, STAG was driven from its original highly promotional position.[98] The Stanford Research Institute's draft report was strikingly negative in its assessment of SST operating costs, giving the airline executives on the committee pause. It also opposed the Mach 3 speed choice that U.S. manufacturers favored, arguing that a Mach 2 aircraft would produce higher annual profits, assuming only first-class fares and coach-density seating. SRI favored the SST program nonetheless, primarily because it projected positive benefits on employment and on the national balance of trade. The chairman of President Kennedy's Council of Economic Advisors, Minnesota economist Walter Heller, attacked the SRI report on a number of grounds, however, including its lack of market data and the quality of its economic model. He also brought up the sonic boom problem, believing that boom-related restrictions could well be the SST's economic doom.[99] In short, Heller believed that SRI's analysts had not adequately defended their pro-SST recommendation, and he advocated retaining the research program until both the economics and the boom were better understood. The Commerce Department's review of the SST situation was even more negative than Heller's, dismissing the economic arguments entirely and contending that the SST development program represented a fundamental alteration in the relationship between government and business. In the Commerce Department's view, government intervention in commercial product development would impair private initiative and create more cries for government support from other business sectors. Finally, Commerce attacked FAA itself for presenting the SST as the only acceptable means to sustain the aircraft industry in the face of the European threat. Instead, Commerce pointed out that there was "a potentially more serious threat presented by an advanced subsonic aircraft for cheaper mass transportation."[100]

Despite the persistent economic questions raised by the SRI, Commerce, and Council of Economic Advisors reports, Johnson's report to the president was very positive, advocating an SST development program be started immediately, with a $100 million appropriation for fiscal year 1964.[101] FAA should manage the program, and the government should require the manufacturers and airlines to pay part of the development costs. Halaby, of course, was happy with the report and expanded upon it in his own memo to Kennedy by proposing that the manufacturers pay 25 percent of the development costs, royalties on sales of the airplane to recoup the development costs, and a limitation of $750 million on the government's share of the cost.[102] In approving of the SST development program, Johnson was reprising his Senate role of aerospace promoter, championing, as he did with the Apollo program, expansive government spending on aerospace technologies.

These reports were not enough by themselves to move Kennedy to approve the program, however. Britain and France had launched Concorde without any orders from airlines, and thus the magnitude of its threat was unclear. Obviously, the state-owned British Overseas Airways Corporation (BOAC) and Air France would operate the aircraft—they would not be given a choice in the matter—but BOAC had publicly opposed the SST idea and could hardly be seen as an enthusiastic "customer." Concorde would become a real threat to the United States only if a large, privately owned international airline chose to buy it. And at the same time all of the economic studies were being done, one airline was pursuing a Concorde purchase—and making certain that Halaby and Kennedy knew it. Pan Am's Juan Trippe, incapable of turning away from the latest technological innovation, told a number of administration officials in early May that Pan Am would place a "protective order" for six Concordes. Trippe was the airline industry's most admired leader and ran the world's largest airline. He preferred to buy American, he told Kennedy's advisors, but there was no U.S. SST to purchase. Pan Am would not forgo the supersonic market on simple nationalism. It would be first to fly an SST, regardless of who built it.

By leaking his intentions to the government in advance of the purchase, Trippe sought to aid Halaby's lobbying campaign for the U.S. SST. Pan Am's purchase agreement bears out that interpretation.[103] But Trippe did not, as was initially reported, order Concordes. Instead, he purchased "options" on them, a then-new device that merely reserved delivery positions but did not represent a contractual obligation to actually buy airplanes.[104] He could therefore drop the Concorde if a superior American SST was built, with no legal penalty. This detail was not reported, however, and Pan Am's "order" thus gave Halaby, Johnson, and Kennedy proof of Concorde's importance.[105] They were well aware that Pan Am's purchase would produce a stampede of orders from all the other transatlantic airlines, transforming the Concorde from curiosity to threat. By the end of May, Kennedy had decided to contest the Concorde, and on 1 June he asked Halaby to head off a Pan Am announcement until after he had had a chance to publicly announce his decision. Halaby's efforts failed, however, and Pan Am's order was unveiled on 4 June, forcing Kennedy to announce his program in Trippe's wake, on 5 June, at the U.S. Air Force Academy.[106]

Kennedy's speech, however, was not the enthusiastic promotion Halaby had wanted. Kennedy did not make an unlimited national commitment as he had for NASA's Apollo program, but capped the government's investment at Halaby's $750 million estimate. He demanded private "risk capital" from the industry to cover the rest of the $1 billion. Finally, he asserted that the United States would not continue the program if the early design phases did not "produce an aircraft capable of transporting people and goods safely, swiftly, and at prices the traveler can afford and the airlines find profitable."[107] Following his public announcement, President Kennedy instructed Halaby to commission an independent review of the SST situation. Kennedy was still not convinced the SST would turn out to be profitable for anyone, and he would not support an SST solely to "out de Gaulle de Gaulle."[108]

✈ ✈ ✈

In the United States, the political value of the SST had not been sufficient to entirely overcome economic considerations, as it had in Britain and France. American leaders perceived little prestige in commercial failure. Yet the strength of Kennedy's belief in American technological prowess was such that he was willing to finance a risk that no U.S. bank had been willing to support, in order to protect America's self-image. The American SST development program was a defensive investment motivated by fear of being supplanted by European competitors and a corresponding fear that the United States would lose prestige in the world.[109] American enthusiasts had deployed an expansive rhetoric of the inevitability of progress to overcome the doubters, pessimists, and economists, just as their European counterparts had. But these arguments had proven insufficient in Britain and the United States, as we have seen. Concorde found approval in British desires to enter the European Common Market, while the American SST development program occurred purely in Concorde's wake. In France, however, the SST was a product not of positive market studies but of de Gaulle's belief in the power of state-directed technological change to modernize the nation. Like John Stack, C. R. Smith, William Allen, Najeeb Halaby, Morien Morgan, Peter Thorneycroft, Juan Trippe, and many others in the aviation industry, de Gaulle equated superior *speed* with technological progress. Hence only an SST would promote what historian

Gabrielle Hecht has aptly described as the "radiance of France"—its image as a modern industrial nation.[110]

Similarly, the equation of speed with progress ensured that the U.S. response to the foreign SST threat was the launch of a program to develop an SST faster than the Concorde, not a Mach 2 SST designed to beat Concorde's economics, and certainly not a "jumbo" subsonic airliner that would permit greatly reduced air fares. Economics and size had not yet been linked to the ideology of technological progress. To maintain its appearance of technological progressiveness in the world, the SST advocates had argued, the United States had to build an SST of its own that would be faster than anyone else's. Anything less, to recall John Stack's words, would be tantamount to surrendering in the technological cold war. But not to the Soviet Union, whose own SST was still only a rumor—to France. In the context of the recent Sputnik disaster, aerospace enthusiasts preferred to promote a socialist (or, perhaps more politely, a Gaullist) model of industrial development rather than accept this loss of face.

Research laboratories in the United States and in Britain played substantial roles in promoting and maintaining the linkage between speed, progress, and national prestige. Morien Morgan at RAE and John Stack at NASA had promoted SST research and development above all other aeronautical projects to any and all potential allies, ensuring that the speed-progress-prestige linkage was always foremost in their arguments. They found many allies in the aeronautical community to whom this argument was second nature. Halaby's elegiac statement of his own belief in the inevitability of aviation progress was only one of many echoes of their arguments. Their rhetoric helped convince politicians that the SST was inevitable, spurring them to support development programs that, like the national space programs in the United States and in Europe, did not pass the usual commercial standards.

The research efforts at NASA and at the RAE had turned a technology that had appeared absurd in 1954 into what even Robert McNamara called a "cinch" in 1963. For both the RAE and for NASA's Langley Research Center, the SST became the next best thing to an Apollo program. The SST gave them a defined and difficult goal to work toward, and because the SST race had become a national priority, it attracted media attention to aeronautical research. And while neither lab ultimately ran their respective development programs, each maintained its traditional role of supporting their industry's efforts. Hence while the space race fundamentally transformed the NACA's relationship with the space part of the aerospace industry, the SST race left NACA's old relationship with the aeronautical part of the industry intact.

With development programs launched, the problem for the SST enthusiasts became one of designing viable aircraft while sustaining the political support necessary to sustain funding. Neither of these efforts proved easy. As engineering work progressed, the economics of the SST designs worsened while political changes magnified the impact of design problems and soaring costs.

3

ENGINEERING THE
NATIONAL CHAMPION

The American SST program began with a government-run design competition patterned after U.S. Air Force procurement
procedures. In the first phase, the manufacturers would work up "brochure"
airplanes that they would submit to the FAA for evaluation by a FAA/NASA/
DOD team. At the end of this "Phase I" competition in March 1964, FAA
would chose one engine and one airframe manufacturer to build a pair of prototype aircraft for flight testing. Then, during "Phase II," the prototype aircraft
would be designed, built, and test flown. Improvements resulting from the
flight test program would be incorporated during a preproduction third
phase. In this way, the agency would select the national champion in the
international SST competition. It would also nearly match Concorde's schedule, with the U.S. SST appearing about eighteen months later. That did not
happen.

President Kennedy's uncertainty about the putative SST's economics led
him to appoint a pair of financial specialists to advise him on the subject in
mid-August. Kennedy gave Eugene Black, a former World Bank director, and
Stanley Osborne, chairman of the board of Olin Mathieson Corporation, a
broad assignment to learn and examine the views of industry, financial, and
government officials toward the SST. In particular, the two men were to advise
the president on the industry's financial capacity. The presidents of the three
airframe manufacturers that intended to participate in the SST program, Bill
Allen of Boeing, Courtland Gross from Lockheed, and Lee Atwood of North
American Aviation, had been unanimous in opposing Halaby's cost-sharing
plan. The companies contended that 25 percent of the SST's development cost
was far more than they could afford, and they publicly campaigned for a

reduction to 10 percent at most.[1] Robert McNamara, on the other hand, would not accept less than 25 percent. He believed that the companies' willingness to accept the cost-sharing arrangement reflected their true belief in the commercial viability of the program.[2] The task Black and Osborne faced was to sort through the conflicting arguments and make a recommendation on the future course of the program. Their presence, and in particular their independent access to Kennedy and Vice President Johnson, immediately reduced Halaby's ability to control the information the two leaders received. Ultimately, they led to Halaby's effective removal from program control.

Henry Lambright has argued that McNamara deliberately slowed the SST development program down in order to ensure that it would result in an economical aircraft.[3] But in his excellent history of the SST program, Mel Horwitch has argued that McNamara was skeptical of the SST's overall viability and used the presidential advisory committee to block it until he became too busy with the Vietnam War to exert the kind of effort necessary to keep the FAA in check. Kenneth Owen has recently reinforced this view, reporting that McNamara retained his skepticism toward it thirty years later. A back-of-the-envelope calculation was all he needed to realize that an SST could not be made to pay.[4] But McNamara went far beyond a "back of the envelope" calculation during the SST controversy. He believed in systems analysis—indeed, he was perhaps the most famous proponent of this new postwar technique. At the Defense Department, he had installed a systems analysis office to provide him with independent assessments of Pentagon programs and implemented the "Planning-Programming-Budgeting System" to enhance civilian control over the uniformed services.[5] He deployed his analysts against the SST, too, producing devastating critiques of the FAA's economic studies.

But while McNamara certainly applied his analytical methods to the SST, his resistance to Halaby's schedule derived not solely from numerical analysis but from the experience—or, more precisely, lack thereof—that the U.S. Air Force and Central Intelligence Agency had gained in the two supersonic cruise aircraft the agencies were building, the XB-70 and the A-12, a supersonic reconnaissance aircraft intended to cruise around Mach 3. Further, he had a much better grasp of the technical history of jet aviation than Halaby did. This knowledge led him to systematically frustrate Halaby's effort to promote a "crash" program to beat the Concorde.

Scheduling is a crucial issue in any complex development program. Schedules drive virtually everything when one is trying to build a complex machine, and they operate at every level, dictating when designs must be finished, when drawings have to be released to the shops, and when money must be raised and allocated. Schedules also fundamentally affect the level of technology one can apply in a new program. One cannot risk an unproven technology in a critical task on a development program (such as a radically new engine) because if that technology does not mature on schedule, it can collapse the entire program. McNamara sought to delay the U.S. SST so that its advanced technologies could mature enough to be relatively low risk when actual construction of the prototypes began. But in the end, even McNamara's delays were not enough to prevent collapse of the FAA's chosen champion, an SST design based upon John Stack's beloved swing-wing.

MCNAMARA, SCHEDULES, AND THE STATE OF THE SUPERSONIC ART

When Najeeb Halaby began organizing his SST program in mid-1963, no aircraft anywhere in the world had spent more than an hour at Mach 2, and no air-breathing craft had ever reached the goal Halaby had in mind for the SST, Mach 3.[6] The XB-70 still had not flown. The CIA's A-12, or "Oxcart," a spy-plane intended to cruise at Mach 3, was still struggling around Mach 2. Technologically speaking, therefore, the relationship between these aircraft and Halaby's SST was not at all similar to that of subsonic bombers to the 707. Two generations of subsonic jet bombers had flown for a decade before Boeing transformed the bomber technology into the 707, and thus the 707 was the product of a set of fairly mature technologies. It was the beginning of a third generation of large jets. On Halaby's intended schedule, the American SST prototype was to be built before either XB-70 or A-12 had completed their flight test regimes. It would have been a contemporary of these first-generation supersonic cruise aircraft, not a second- or third-generation descendant, and would therefore not have the benefit of lessons learned from flying these earlier craft. McNamara first blocked Halaby so that the U.S. SST would benefit from these other programs. The manufacturers unwittingly helped him immeasurably.

A major blow to Halaby's supersonic ambitions came on 26 August 1963, when Donald Douglas, Jr., president of the Douglas Aircraft Company, wrote to Halaby that his company would not be bidding in the supersonic transport program. Douglas felt a detailed explanation of this decision was in order to prevent a backlash against his organization's seemingly unpatriotic decision.

He reminded Halaby of the damaging cost to the company of its DC-8 jetliner program, which had left the company in severe financial straits. The company had also decided to launch a smaller commercial jet, the DC-9, which was intended to serve the short-to-medium-range market in competition with the French Caravelle (which it resembled quite closely) and the British BAC 1-11. The DC-9's cost, combined with the cost of the DC-8's improvement program and the need to perform adequately on various military contracts, did not leave the struggling company with sufficient resources to pursue the SST contract.[7]

What Douglas did not say, either in his letter or in the company's press release the following day, was that neither his engineering staff nor his marketing group thought much of the supersonic transport idea. Richard Shevell, who headed Douglas's advanced design group, tells an instructive story about Douglas's first SST layout, done in the late 1950s: "I'll never forget the first layout done in advanced design of an SST at Douglas. The project guy, who was not highly technical in aerodynamics, but had been given to understand that the L-over-Ds were low, that the fuel requirements were high, drew a picture of an SST which I've never forgotten. It showed a fuselage. You needed so much fuel that the fuselage was full of fuel. So the passengers were sitting in the old-type diving suits with the big metal helmets in the fuel, and all along the top, just above the fuel level over the heads of the people it said, 'No Smoking. No Smoking. No Smoking.'"[8] While the diving-suited passengers had been added as a joke, his point was a serious one. The Douglas design did require that part of the fuselage be used as a fuel tank.[9] As a result, the weight analysis Shevell's group performed on the company's SST in 1963 (known as Douglas model 2229) had turned in a highly negative result, causing him to recommend against the idea. A hundred-passenger transatlantic version would weigh about 420,000 pounds, very close to what Concorde would wind up weighing ten years later. By comparison, the 707-320B carried more passengers and weighed less. Because aircraft economics are fundamentally affected by weight, Douglas's SST design did not appear competitive.

The company's marketing department had also been skeptical of the SST's economics. In a review of the Stanford Research Institute's June 1963 report, E. F. Barnes, the deputy manager of Douglas's market research group, contended that SRI's report was based on highly optimistic assumptions. SRI had assumed that SST's would serve all of the world's air traffic on route segments in excess of one thousand statute miles, leading to a market estimate of 325

Mach 3 aircraft by 1973. But this was unrealistic, according to Barnes. Instead, a more appropriate assumption was that SSTs would serve only those segments "which [had] traffic volumes sufficiently large to support SST operations at a reasonable schedule frequency."[10] This was a much smaller number of routes, leading to a forecast of only 151 aircraft. Given that Douglas had yet to turn a profit on the DC-8 despite sales of more than two hundred aircraft, this was not a promising result, especially since part of that market would go to Concorde. Another member of the marketing staff also argued that SRI had "rather glossed over" the possibility that the sonic boom problem could completely stop the program.[11] The smallness of the market and the risk imposed by the boom made the SST a poor investment of the company's funds.

Douglas's decision caused a small sensation. *The New York Times* reported it as a reflection of "widespread industry caution" toward commercial supersonic flight.[12] Halaby was forced to explain the company's reluctance to President Kennedy, who continued to be uneasy about the SST's commercial prospects.[13] Combined with the industry's resistance to the cost-sharing formula, Douglas's decision underscored the substantial gap between Halaby's enthusiasm and the business world's caution.

President Kennedy's assassination in Dallas on 22 November further weakened Halaby's position as champion of the program by making him a Kennedy appointee in a Johnson administration.[14] And while Johnson supported the SST, shortly after taking office he received the Black-Osborne report. The two men, while supportive of the SST, opposed the FAA's "crash program" to beat the Concorde, wanted an independent authority created to run the program, and agreed with industry that the cost-sharing level should be 10 percent versus 25 percent.[15] They believed tying the U.S. program to Concorde's schedule was dangerous and would almost certainly result in an uneconomical aircraft. Instead, they proposed taking the time necessary to design a truly superior aircraft. Their report therefore constituted a rejection of Halaby's program, while accepting that the national interest demanded an SST.

Several months of bureaucratic warfare ensued as Halaby attempted to protect his program while McNamara, with the assistance of Budget Director Kermit Gordon, sought to supplant him with an advisory board.[16] They succeeded during March 1964, with Johnson's April 1 appointment of a presidential advisory committee (PAC) on the SST, chaired by Robert McNamara.

The committee did not have managerial control over the program but it did have the ability to control the pacing. The committee's task was to make recommendations to President Johnson on how the program should proceed, and Johnson would instruct Halaby on what he was to do. This unusual arrangement gave McNamara effective control of the program without the need to transfer it to the Defense Department.[17]

In January 1964, the FAA received airframe proposals from Boeing, Lockheed, and North American Aviation and engine proposals from Curtiss-Wright, General Electric, and Pratt & Whitney. To evaluate them, Gordon Bain's SST program office put together a two-hundred-person team composed primarily of engineers from NASA's Langley and Ames research centers and the Air Force's Propulsion Laboratory at Wright Patterson Air Force Base. The evaluation took several weeks, but by mid-March the general results were already clear. The three airframe manufacturers had all failed to meet the FAA's specifications.

North American Aviation, Lockheed, and Boeing were the three bidders in the FAA's airframe design competition. They had been asked to produce designs of 4,000-statute-mile range, 30,000-pound payload capacity, with sufficient fuel reserves for a thirty-minute hold at 15,000 feet. Gross takeoff weight was to be limited to about 420,000 pounds, with speed to be Mach 2.2 or greater.[18] Finally, the aircraft were to be competitive with the existing subsonic jets, which the teams interpreted to mean having direct operating costs within 10 percent or so of a 707 Intercontinental.

Of the three competitors, only Boeing came even close to the FAA's criteria. Boeing's design was company model number 733-197, a variable geometry, 150-passenger aircraft designed by a team led by engineers Maynard Pennell and William Cook.[19] Boeing had been exploring SST concepts since 1956, originally by working to improve the company's WS-110 proposal aircraft (a fixed-delta-wing canard design). In 1960, however, the company's SST group had begun to accept John Stack's argument that a swing-wing design was the only one that could meet the difficult challenge of a commercial SST. Pennell's team believed that the swing-wing would provide better "off design" performance than a delta-wing would, would provide a lower noise signature through superior climb performance, and would allow the aircraft to operate off shorter runways. They also believed that the wing pivots would not result in a weight penalty substantial enough to overwhelm these other benefits.

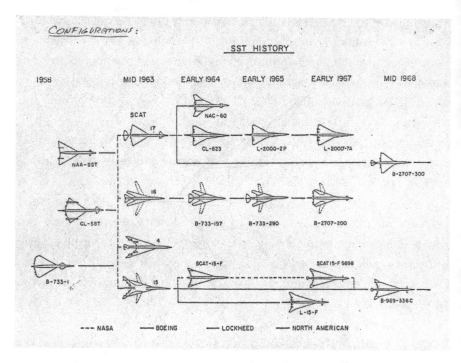

FIG. 3.1. Early SST evolution. The 733-1 (bottom left) was Boeing's entry in the WS-110A competition. The NAA-SST (top left) was a modification of its XB-70, and the CL-SST (center left) was an early Lockheed configuration. The diagram suggests how the four main NASA SCAT configurations influenced each of the company designs. [Courtesy A. Warner Robbins]

There was a good deal of controversy over this, with Bill Cook and some others believing that the swing-wing needed more study than the FAA's timetable allowed.[20] But the swing-wing enthusiasts won out.

The design Boeing submitted in January 1964 had evolved from a model originally laid out in 1960. As figure 3.1 suggests, this early model, 733-101, was refined during and after the SCAT studies into a configuration more closely resembling SCAT 16. The evaluation group, dominated by Langley engineers, favored the swing-wing approach in the assessment, commenting on the design's superior takeoff and landing performance, which also gave the design an edge in community noise.[21] However, they also imposed a 10,400-pound weight penalty on the design, believing that the proposal's weight and aerodynamic performance estimates were optimistic. Because of this, they believed that the aircraft's range was less than required by 660 miles. They also stated that Boeing could recover that distance through "conceivable" aerodynamic refinements and either engine performance improvements or an increase in the aircraft's size to 460,000 pounds (to carry more fuel).[22]

Lockheed's design received a much less favorable review, with the evaluators commenting that the aircraft was "incapable of meeting the range/payload requirements of the [Request for Proposals]." Only major aerodynamic changes could produce a viable aircraft.[23] Lockheed's design team, led by Richard Heppe and Jim Hong, had rejected the swing-wing approach and instead adopted a tailless delta-wing configuration known as a "double delta." This design, company model CL-823, had evolved through a substantial series of models (see figure 3.2), and had been improved through the application of some of SCAT 17's aerodynamics. Like the Concorde's team, Lockheed had relied upon vortex lift to improve the aircraft's low-speed performance, and to keep the aircraft's noise signature within the limitations set by current jets. Heppe's group had chosen a Pratt & Whitney turbofan engine. The evaluators penalized Lockheed for this choice because they believed the Pratt engine had inferior supersonic performance. They also found that the design had substantially inferior sonic boom performance; in order to bring the design within the specified boom level the team had raised its cruise altitude to one at which the propulsion system was inefficient. The evaluators credited Lockheed with a range of only 2,547 statute miles with payload, and with zero payload a range of 3,350 statute miles—far short of that required for a transatlantic aircraft.[24] To Lockheed's credit, Heppe's group had acknowledged most of these problems in their proposal. But the evaluation team, which was dominated by supporters of the swing-wing approach, did not believe Lockheed could fix them within the limitations of the double-delta design.

North American Aviation's proposal received an even more negative review by the team. NAA's submission, company model number NAC-60, was also a delta-wing design, with a fixed, nose-mounted canard. NAA had not had the benefit of the publicly funded SCAT studies, unlike Boeing and Lockheed. While it had attempted to participate in the SCAT studies on its own funding, it had not been able to improve its in-house design using the study results. The evaluators estimated that it, too, could not be brought up to the FAA's requirements without major aerodynamic redesign, and contended that NAA had been optimistic by 46 percent in its transonic drag estimates, by 14 percent in supersonic L/D, and similarly optimistic in its sonic boom estimates.

FIG. 3.2. The Lockheed L-2000, an aerodynamic model being prepared for testing in the NASA Ames 40 × 80-foot wind tunnel. [Courtesy Lockheed-Martin, photo PC026-086]

The evaluation team imposed penalties amounting to 1,530 statute miles on the proposal for these deficiencies.

The engine manufacturers did not fare much better. Three companies, General Electric, Pratt & Whitney, and Curtiss-Wright, had proposed a total of seven engines. GE had proposed two engines, an afterburning turbojet engine based on its YJ-93 engine and an advanced duct-burning turbofan. Both would operate at a turbine inlet temperature of 2,200°F, which the SCAT studies suggested was the minimum temperature necessary to support economical

SST operation. The evaluators believed that the afterburning turbojet had the lowest risk and could be developed within the program's time constraints. The more advanced turbofan, however, could not be. Pratt & Whitney had also proposed two engines, a duct-burning turbofan that relied upon the company's J58 engine core and a more advanced turbofan based on a new engine core. Neither of these met with the evaluators' approval. They were "heavy, poor performing, and deficient in turbine cooling."[25] The performance Pratt proposed was low because Pratt had used a less ambitious turbine inlet temperature of 1,900°F (100 degrees less than the troubled J58 but 300°F higher than any current commercial engine), and the company's engineers steadfastly refused to raise it. None of the propulsion evaluators—or the PAC members—believed that an SST could ever be economical with an engine using less than a 2,200°F turbine inlet temperature, because the higher temperature substantially improved fuel economy.

GE's engine essentially reflected to the evaluators the minimum performance necessary for economic success, while the very high temperature imposed a significant degree of risk. Because the XB-70 had not yet flown with the YJ-93 engine, no one knew whether that engine would perform properly at the temperatures and altitudes expected of it. Pratt & Whitney's experience with the J58, which was beginning to fly on Lockheed's A-12, however, was sobering to the few (like McNamara and CIA Director John McCone) who knew about this "black" program. Pratt had great difficulty keeping the engine operational at its 2,000°F temperature, and the company's adamancy about not raising its proposed SST engine's temperature stemmed from that experience. The company's engineers did not believe that they could produce an engine with commercially acceptable reliability at the temperature the evaluators demanded within the program's schedule. The evaluators perceived Pratt's proposal as excessively conservative.[26]

They held the opposite opinion of Curtiss-Wright's engine, the TJ70. According to company president T. Roland Berner, he had seen the SST program as the company's last chance to remain a viable manufacturer of jet engines. He had given his engineering team substantial resources to work with, and they proposed what NASA engineer Neil Driver remembered thirty years later as the best SST engine proposal he had ever seen.[27] Unfortunately for the company, no one on the evaluation team or on the PAC believed the company could pull it off on the program's schedule.[28] In the evaluation report's words, the "unaugmented turbojet engine incorporates a number of highly advanced, unproven components which result in this engine having an unacceptably high development risk for a commercial transport *within the program timing objectives"* (emphasis added). [29] In fact, the FAA had effectively ejected Curtiss-Wright from the program in November 1963 by canceling all the propulsion research contracts the company held, causing the joint NASA/DOD SST propulsion research team to complain about the FAA's short-sighted attitude.[30] The innovations in the engine were ones that would guarantee an economical SST even if they would not be ready in time for the first-generation SST, and thus the engineers wanted Curtiss-Wright kept in the program.

In addition to the government evaluation group, Halaby had asked his airline committee to examine the SST and engine proposals and submit its own recommendations. This group insisted that because no one had met the economic criteria that FAA had established, two of the three airframe manufacturers (with Boeing and Lockheed, in that order, being preferred) should continue into a new competitive phase. The airlines all believed that competition would force the two chosen manufacturers to put more effort into improving their designs than they would once a monopoly was established, and hence they contended that FAA should delay creation of a monopoly as long as possible. They had the same recommendation for the engine manufacturer, choosing GE and Pratt & Whitney as the two companies that should be kept in the program, although with less solidarity than they had shown with the airframe manufacturers.[31]

The airline evaluation group's report was so negative that it caused Gordon Bain, Halaby's hand-picked SST program director, to produce a review in which he argued that there was no reason *other* than national prestige to continue the program.[32] The SST was not going to be an economical aircraft. Halaby, ever the promoter, submitted an analysis that was considerably more rosy and called for immediate construction of the Boeing SST with GE's engine.[33] But the widely divergent positions of the two caused McNamara to commission his own review of the government evaluation report and the two FAA analyses of it. This was harshly negative, finding Halaby's arguments at best unconvincing and at worst disingenuous. The study's authors implicitly accused him of "cooking" his numbers to produce economical-seeming results.[34] And they stated that his primary analytical tool, an analogy comparing the subsonic jet's evolution to the SST, was an "unabashed evangelistic tract calling for faith in major aerodynamic and engine technology improvements which are nowhere in prospect."[35] These authors contended that Halaby's SST program was misdirected, in that it was biased toward developing the Boeing/GE SST immediately instead of first determining what kind of an SST, in terms of size, range, and weight, would be most economical.

At the first meeting of McNamara's PAC, held at the Pentagon on 13 April 1964, the issue of the schedule dominated the discussion. Halaby had been proclaiming that he would issue his decision on which SST proposal to accept and build, on 1 May, and no one on the committee believed this deadline was feasible any longer. Because none of the manufacturers had met the FAA's criteria, the committee members were heavily biased toward a reexamination of the program's basic direction. McNamara and John McCone believed, for example, that FAA had imposed too many restrictions on the manufacturers. Boeing had actually submitted two designs, one that met the FAA's weight restriction of 430,000 pounds and a 520,000-pound design that appeared more profitable. The evaluation team had not analyzed the larger design because it did not meet the FAA's criteria, but its apparent superiority convinced several members of the committee that a good deal more study of alternative designs needed to be made before a final decision on such a basic criterion as gross weight should be made. McNamara was quite categorical on this point: "The figures that have been presented to date show the unprofitability of the supersonic to be so great that I have no confidence that following down this course of design will ever lead to a profitable airplane."[36] He wanted FAA to take the time to analyze alternatives. This would not necessarily mean a delay in the SST's ultimate delivery, because engines took longer to design, build, and test than airframes did. And he was quite explicit that he wanted development of at least two engines to continue apace.

The committee ended its initial session with a decision to hear from the heads of several airlines. The airline heads were invited to appear on 18 April. During that session, the airline chiefs each told the committee members that they believed a larger SST was the only possible road to an SST, with desired seating capacities running from 200 to 250. Charles Lindbergh, for example, who had accompanied Pan Am president Juan Trippe as member of Pan Am's board of directors, repeatedly emphasized the need for the reduced direct operating costs that came from increased size. The airline presidents were also unanimous in insisting that the SST have a minimum range of four thousand miles. Frank Keck of United, which at this time was primarily a domestic carrier, also made clear that he believed a second SST with more seats but less range (transcontinental, for domestic routes) would eventually be necessary. But the transatlantic airplane needed to come first, and each of the airline presidents told the group that he could wait at least ten years for that airplane, if the result would be truly economical.[37]

The airlines' desire for a delay left Halaby only one leg to stand on in opposing McNamara's desire for a delay: the competition from Concorde. But Pan Am, which had started the whole SST mess by ordering Concordes, demolished that reason, too. Juan Trippe told the committee that Concorde was not going to be a "successful, heavy duty ship for long-distance world operation."[38] Trippe's chief competitor, TWA's Charles Tillinghast, helped Pan Am's case by reporting that Concorde was being "stretched" from 100 to 130 seats in an attempt to reduce its direct operating costs, and that, combined with problems in developing the Olympus engine, would cause the consortium to miss their promised schedule—though they were unlikely to admit that for political reasons. Hence the Concorde's published schedule should not be the basis for an American SST's schedule. TWA thought Concorde was not likely to be serviceable until 1973, and Tillinghast was willing to wait another three years after that for a superior U.S. airplane. There was no reason, therefore, to forge ahead on FAA's original Concorde-based schedule. The U.S. schedule should reflect the need to produce a superior airplane.

The committee agreed to "help" Halaby out of his May 1 deadline by getting Kermit Gordon, the White House budget director, to draft a letter for the president's signature ordering a delay while the PAC heard from the engine and airframe manufacturers and digested the data it gleaned from these hearings.[39] It met again on 1 May to hear from the manufacturers. At this meeting, Pratt & Whitney's representatives made clear that their experience on the A-12 did not augur well for the 2,200°F engines that NASA and the airlines believed necessary for an SST. The J58 engine had achieved only a few thousand hours at 1,900 to 2,000°F on the test stands and far less in flight. This was, its representatives stated, not enough to give "confidence that it will be practical to operate at higher temperatures for extended periods in the initial operation of the SST."[40] The company could deliver a 2,000°F engine reliable enough for commercial operations in the early 1970s, but a higher temperature engine would not be commercially feasible until years later.[41] In effect, Pratt's engineers argued that it was fantasy to believe that one could deliver an engine in the early 1970s that had commercial-level reliability at a temperature that no military engine had achieved in 1964.

With the manufacturers heard from, the committee met again on 8 May. At this meeting, conflict over the schedule, and in particular the scheduling of engine development, dominated the discussion. Halaby and Bain argued that the program should select GE, as the government evaluation team had recommended, and drop Pratt & Whitney and Curtiss-Wright. McNamara, John McCone, and Jim Webb all opposed this decision. Webb continued supporting Curtiss-Wright, pointing out again that the first-generation engine was not going to result in an economical SST. He believed that Curtiss-Wright's technologies offered a 17 percent increase in efficiency over the GE engine, and this second-generation engine offered the key to a truly superior aircraft.[42] He was not forceful in presenting his case, however, and ultimately agreed with the group simply that two engine manufacturers needed to be kept in the program.

That inevitably meant that the other manufacturer would be Pratt & Whitney, which had vastly greater commercial experience and a flying supersonic engine in the J58. McCone and McNamara both believed Pratt was far more likely to produce a workable engine on the program's time scale than was Curtiss-Wright or GE. Webb, Halaby, and Bain all took that as a vote against the SST's future, because Pratt had flatly refused to provide the higher temperatures that economics dictated, while GE would. But neither McCone nor McNamara believed GE would come through, with McNamara telling Webb that he'd bet "$1000 to $100" that GE would never fly on its proposed schedule.[43] The committee finally voted to recommend to President Johnson that both Pratt and GE be kept in the program for another fifteen months, so that each could build a demonstration engine.

This meeting also revealed the primary reason McNamara and McCone were adamant about not immediately approving prototype construction, and it had little to do with economics. McCone, whose agency was responsible for the A-12's development, reported that despite the positive spin that President Johnson and the aeronautical press had given the A-12 after Johnson's public unveiling of the program on 25 July 1964, the agency had been struggling with it for two years and "every notch [we] put it up above Mach 2 and every few thousand feet in altitude [we] run into a whole series of totally unforeseen problems." Because the A-12, which was the only supersonic cruise aircraft anywhere in the world, was so troubled, he and McNamara feared that Halaby's quick schedule would lead to a disaster. McCone stated, "I think this race to see whose plane gets in the air first could very easily lead us into a very tragic situation." He believed that rushing the SST could result in a catastrophe like the De Havilland Comet.[44]

The A-12's problems were not well known, due to the project's classified status, but the aircraft had been a year late on its first flight, had been unable to go supersonic in level flight, and had enormous problems with its inlet control system.[45] The J58 engine did not fly in the airplane until January 1963, nine months after the aircraft's first flight (using J75 engines). Because the aircraft's highly advanced variable geometry inlets had to be programmed for each temperature and altitude combination, getting the plane from Mach 2 up to Mach 3 took sixty-six flights and more than a year.[46] The first time the plane sustained a speed of Mach 3.1 (for only ten minutes), most of the aircraft's wiring disintegrated, and most of the hydraulic fluid that provided the aircraft's control power leaked out because the extreme heat (over 600°F) opened joints in the piping.[47]

The most difficult problem the aircraft's flight test program faced, however, was the "unstart" problem. The engines could not ingest supersonic air, and the engine inlets were designed to slow the incoming air to a high subsonic speed through a set of shockwaves maintained within the inlet. The positioning of those shockwaves had proven to be a very delicate matter. Fairly frequently during the test program, the shockwave would be ejected from one of the inlets, causing what came to be called an "unstart." The unstarted inlet would not feed its engine with enough air, and the unbalanced thrust from the other engine then caused the plane to yaw violently, cracking the pilots' helmeted heads against the canopy. This was obviously unacceptable in a commercial airplane, but it had not been solved, as the PAC was considering the SST's future. The A-12's "black" status protected it from the political backlash that always accompanied seriously troubled aircraft development projects, but no such protection would exist for the very public SST. McNamara and McCone wanted to be certain that there *were* solutions to these problems by fixing them in the "black world" before committing to an SST. The SST was a very high profile project, and they both felt that if the SST prototype experienced these kinds of problems when it began to fly, the project would not survive the resulting political turmoil.

In its 14 May report to President Johnson, the committee underscored the high technical risk inherent in the program. McNamara had told the committee that he was "terribly worried about this program. The technical risks are greater than I have understood. . . . I think they are far greater than the President understands."[48] The emphasis on risk drew protest from Halaby. He had fought to cast the SST program in the most positive light in the report and had been outvoted repeatedly. McNamara allowed his protests to appear in footnotes to ensure that the president knew Halaby did not agree with the committee's decisions, but it did Halaby little good. The committee recommended that Lockheed, Boeing, Pratt & Whitney, and GE all be kept in the SST program for another six months. The two airframe manufacturers were to investigate the most profitable size for their aircraft while refining their proposals, and the two engine manufacturers were to build test components to demonstrate their technologies. The report also called for extensive economic and sonic boom studies. President Johnson approved the committee's report and directed Halaby to execute it on 20 May. By following the committee's recommendations, President Johnson had accepted that postponing prototype construction was the best route to an economical SST.

CHOOSING THE NATIONAL CHAMPION

Whatever his personal feelings about the SST, McNamara could not delay it forever. There was a competition going on, like it or not, so a national champion had to be chosen eventually. NASA and the FAA conducted three more sets of evaluations during McNamara's delays, hoping to find an SST design capable of overcoming his skepticism. Two years elapsed before his committee approved development of a prototype. But it did not approve the prototype because it was convinced of the design's soundness. Instead, it did so because the clock had run out on the competition with Concorde.

In November 1964, the four manufacturers submitted their revised proposals to the FAA. Halaby's SST manager, Gordon Bain, assembled a new NASA/DOD/FAA evaluation team to analyze them, an effort that lasted into December. The report they submitted served as the framework for the next contest between Halaby and McNamara over program timing, which was, once again, won by McNamara. The 126-member team's report ran over a thousand pages. The Boeing/GE SST once again won the group's favor. The new version, model 733-290, was substantially larger than the Phase I model but retained its basic characteristics: a four-engine, turbojet-powered, variable-geometry aircraft. The revised design met the major criteria that FAA had set out, slightly besting the minimum required payload of 40,000 pounds, or about 250 passengers with luggage. The aircraft's larger capacity gave it economics that appeared acceptable in comparison to the 707-320B that had become the program's standard. The 733-290 achieved nearly identical direct operating costs to the 707-320B by having twice the seating capacity—reasonable as long as the 707-320 was the largest subsonic jet flying.

Boeing's close contacts with Langley's researchers gave it a substantial edge in this phase of the competition. Engineers Don Baals, A. Warner Robins, and Roy Harris had devised a new SCAT configuration called SCAT 15F during 1963. As the name suggests, it was a derivative of the swing-wing SCAT 15, but it no longer had the variable-geometry wings. (See figure 3.3.) Instead, it was a highly swept, fixed, "arrow" wing design. SCAT 15F had achieved the highest supersonic cruise efficiency that the NASA engineers had ever seen. Its primary innovation was not in the configuration itself, however. The design contained no new aerodynamic discoveries. Instead, it represented a maturing synthesis of existing supersonic aerodynamic knowledge and the application of a new design tool, the electronic digital computer.

A major component of drag on a supersonic aircraft is "wave drag," which is an artifact of the shockwave that attaches itself to a supersonic aircraft. Engineers mitigated wave drag through the use of the "area rule" that had been devised by NACA researchers during the 1950s. The supersonic area rule

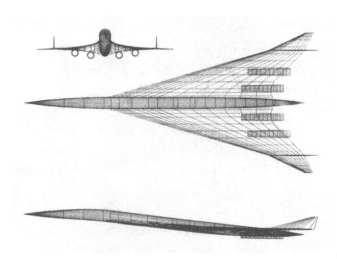

FIG. 3.3.   Computer-generated drawing of SST configuration SCAT 15F.
[Courtesy NASA]

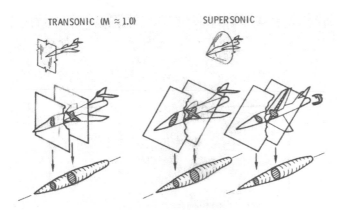

TRANSONIC (M ≈ 1.0)   SUPERSONIC

FIG. 3.4. Illustration of the area rule. The middle portion of the drawing shows the process of passing geometric planes at the Mach cone angle through the vehicle. The tapered cylinders at the bottom illustrate the cross-section of the vehicle encountered by each plane as it passes through the aircraft. The desired outcome is a smooth area transition—no discontinuous changes. On the left is the transonic (Mach 1) case, while the right-hand drawings show the supersonic case. [Courtesy A. Warner Robbins]

was extremely difficult to calculate, however, because it involved having to pass geometric planes through the design at various angles to ensure that there were no sudden changes in the aircraft's area distribution. (See figure 3.4.) Because the calculations involved in computing each plane's intersection with the aircraft were extremely time consuming, human calculators could compute no more than a few such intersections for a given configuration. The area ruling was thus fairly crude.

Langley engineer Roy Harris had been assigned to perform this calculation for a proposed Navy configuration called CAP-2 shortly after arriving at the lab, and recalls that it took him six months. He had been a "little horrified" at the thought of spending his life on six-month-long calculations and had contacted Sarah Bullock, head of the lab's computing section, for assistance in developing a computer program to do this automatically. Bullock assigned programmer Charlotte Creighton to his project, and they computerized the process Harris had used. At the same time, Harris recalls, a delegation from the Boeing Company had visited the lab and mentioned that a company mathematician was working on the same idea. Harris found that they had used a superior calculation method based on one proposed by two British mathematicians.[49] Boeing agreed to give NASA the code if the Langley group would fully validate it with wind-tunnel testing and publish the results. Harris and

Creighton integrated Boeing's code with their own and added to it a program Harry Carlson had designed to calculate optimum amounts of twist and camber in supersonic configurations.[50]

The lab's computer was able to run in a half hour the calculations on the CAP-2 configuration that had taken Harris half a year.[51] More important than the time savings, however, was the improved accuracy. Because the electronic computer was so much faster, it could run more iterations of the calculation than human computers could, and hence its results were much less crude. The computer's speed also meant that it could be used to optimize configurations through parameter variation. One could experiment with the configuration in a way that had not been possible before. Hence, while this research effort did not produce an aerodynamic breakthrough per se, it permitted aerodynamic refinement of supersonic configurations that substantially improved their efficiency. The digital computer thus became a design tool that promised to facilitate substantially better performance.[52] By 1964, the Langley lab had constructed a software complex that could calculate the most important aerodynamic design features of supersonic aircraft.

SCAT 15F was the result of refining the original SCAT 15 with the "lessons learned" from the 1963 SCAT studies and with the computer software complex that NASA had assembled. In Najeeb Halaby's words, it was "an ideal paper airplane."[53] The software had allowed Langley's engineers to make SCAT 15F aerodynamically self-trimming, eliminating the trim drag problem that plagued Lockheed. It optimized the location of the nacelles on the wing's trailing edge, and it allowed "reflexing" of the trailing edge to further optimize the pressure field produced by the interaction of the nacelles and wing.[54] The pressure pattern the nacelles produced in turn allowed the configuration to have lower drag than the bare wing had.

Boeing applied the NASA software complex to optimize the 733-197, while enlarging the aircraft substantially. The company's engineers relocated the engine nacelles and reflexed the trailing edge of the wing for the new configuration, called 733-290. They also made certain NASA and FAA knew they had, prominently placing a discussion of the process in their proposal summary.[55] The new configuration achieved a cruise L/D of 7.95, inferior to SCAT 15F but much better than the earlier configuration—and much better than Lockheed.

For whatever reason, Lockheed's proposal did not benefit from the SCAT 15F or its supporting computer programs during this phase of the competition. Thus their proposal faired badly at the evaluators' hands. It had actually regressed aerodynamically. And Lockheed had chosen the lower-temperature Pratt & Whitney engine to reduce the aircraft's noise signature, further impairing the design's performance. The evaluators judged that the aircraft's range would be only 2,900 statute miles, far less than transatlantic. The primary reason for the shortfall, the evaluators contended, was excessive "trim drag." The aerodynamic surfaces that keep all aircraft level in flight produce drag; in the Lockheed design's case, this component of drag comprised 15 percent of the aircraft's total drag at cruise. The result was so surprising that Larry Loftin, who was responsible for the aeronautics research at Langley, asked for a review of the Lockheed evaluation by four engineers who had not been part of the team. Two of these men were aerodynamicists from the Bureau of Weapons, while the other two were Richard T. Whitcomb of the Langley lab and Robert T. Jones of NASA's Ames lab. In his review, Whitcomb pointed out that the wind tunnel model Lockheed submitted did not match the drawings the company had provided. They were different enough, he believed, that the model did not represent the same aircraft shown in the drawings.[56] The aircraft represented by the drawings appeared to be substantially superior to the one suggested by the model, but there was no way to prove it until Lockheed provided an appropriate model.

The engine evaluation again favored GE's turbojet. Pratt & Whitney had finally bent to NASA's insistence that a 2,200°F engine would be necessary for economics and proposed a version of its turbofan that would obtain that temperature—but not until five years after the SST's 1973 in-service date. It therefore represented a second-generation engine, and because of this, Pratt's higher-temperature engine was not evaluated by the team. Instead, the propulsion evaluators focused on GE's turbojet and the lower-temperature version of Pratt's turbofan. That inevitably meant that GE's proposal came out the winner.

The Boeing/GE combination's clear victory gave Halaby an excuse to push for an immediate decision to authorize construction of the 733-290, but McNamara quickly defeated him.[57] Six months had not been enough time to generate the economic and sonic boom studies that the committee had asked for, and McNamara had President Johnson extend the study contracts into March 1965 so that the committee could receive those before deciding the program's fate.[58]

Because Lockheed had not examined SCAT 15F on its own, Halaby told

President Johnson in a memo that he intended to require Lockheed to examine it in the hope that it would stimulate the company's efforts to improve the delta-wing design.[59] On McNamara's suggestion, however, Johnson instructed Halaby to send the SCAT 15F to both Lockheed and Boeing to avoid giving either company an advantage. The president's letter to Halaby (written for his signature by McNamara) was also quite scathing because Halaby had made the mistake of implicitly criticizing the decision to continue the competition instead of proceeding to immediate construction of the 733-290.[60] The harshness of this letter was a rebuke to Halaby, signaling that he was losing the president's trust.

When the PAC met again in March 1965, after receiving the sonic boom and economic studies, the schedule problem came to the forefront again. The engine problem still worried McNamara, whose staff had done additional research on the subject during the preceding months. The economic studies that the committee had received now showed substantial profitability for the SST, but McNamara rejected them because he believed they were based on faulty assumptions about the SST engine's reliability and maintainability. The economic studies had been based upon the engine's having a "time between overhaul"—the number of operating hours an engine could accumulate before being removed from the aircraft and refurbished—of 2,000 hours.

This was impossible on the SST program's schedule, McNamara argued. The maximum amount of time any single J58 engine had accumulated at Mach 3 was 4 hours, 15 minutes, he told the group, and the A-12/YF-12A/SR-71 series aircraft had accumulated a total of only 294 hours above Mach 2. Neither Pratt nor GE had offered better than 600 hours between overhauls in their proposals, while the existing subsonic engines ranged from 1,600 hours to 5,000 hours. Engines were the most expensive part of an airplane, and their maintenance was a major cost to airlines, both in terms of performing the overhaul and in having the aircraft out of service. The high reliability of the subsonic jets, he and McCone argued, had been achieved through millions of flight hours performed by military aircraft over the dozen-year period between the end of World War II and the 707's debut.[61] By 1958, the subsonic jet engines were considerably more reliable than aircraft piston engines had ever been, allowing the 707 and DC-8 to fly more hours per day than any previous aircraft. The SST would not be as reliable as the subsonic jets were when it debuted, because it would be based on engines that did not have an equivalent body of experience backing them. McNamara was blunt about the issue:

"It is absolutely impossible to take this airplane with these specifications and between 1965 and the end of 1972 develop it to the point of running anything close to nine hours a day in the first year. It just cannot be done. I say that based on three years of struggling with airplanes of this type. . . . I say it not only based on the past three years' experience but based on the clear and unequivocal and unqualified statements of the engine and air frame manufacturers. You can't achieve both a 1965 to 1972 development-production period and the nine-hour per day average thereafter. They are incompatible."[62] If the SST did not achieve the utilization rates upon which the economic studies had been based, then it would be highly unprofitable compared to the subsonic jets. Either the development schedule had to be lengthened, or the SST would be unprofitable for years after its introduction.

The 30 March meeting of the PAC was a disaster for Halaby, because it made clear there was no chance of getting McNamara's (and thus President Johnson's) approval for construction of the 733-290 during 1965. Instead, the committee's May 8 report recommended that FAA give contracts to all four manufacturers to continue their studies through December 1966. The two engine manufacturers were to build demonstrator engines and run them for one hundred hours to verify their performance before submitting their next set of proposals. And to ensure no delay once the winner was chosen, the committee agreed that the White House budget office should request $290 million for fiscal year 1967. But Halaby still believed that the U.S. SST had to arrive on the market at the same time Concorde did in order to succeed, and the delay made that schedule impossible. Frustrated and demoralized over President Johnson's repeated acceptance of McNamara's schedule over his own, Halaby resigned a few days later.

Halaby's departure largely ended the FAA/McNamara conflict. President Johnson appointed a retired Air Force general named William F. McKee to Halaby's post, and McKee was a far better match for McNamara. McKee, who was always called by his nickname, Bozo, had started his career in the Army Air Forces during World War II. But unique among Air Force generals, he was not a pilot. During the war he had been a technical troubleshooter on the Chief of the Army Air Force's staff, and when the Air Force was created, decided to cast his lot with the new organization. This decision was apparently the source of his nickname. His peers told him he was crazy to join the Air Force as a nonpilot because he'd never make rank. Fortunately for Bozo, the Air Force's leadership recognized that they needed people with his skills in their procurement activities, and McKee had spent much of his career at Wright-Patterson Air Force base at various levels of responsibility for aircraft research, development, and procurement. Reflecting his experience, he had been appointed deputy administrator for management development at NASA after he retired. Johnson selected him for FAA after receiving enthusiastic recommendations from both James Webb and Robert McNamara. McKee was exactly the sort of person McNamara could deal with. He had the technical and analytical background Halaby lacked.[63]

McKee recalled in an interview that when Johnson tapped him for the FAA job, the president had told him to get a good deputy to run FAA. His job was to finish the SST.[64] When his appointment finally made it through Congress in late June 1965, he immediately restructured the SST program management. Gordon Bain, unhappy with the new arrangements, resigned in August. McKee appointed another Air Force general, Brigadier General Jewell C. Maxwell, to replace him. Maxwell had previously run the bomber procurement division at Wright Patterson, further reinforcing the SST program management's technical foundation.[65] McKee and Maxwell were going to need their experience to finish the SST, because during the eighteen-month delay McNamara had created, the two airframe designs ran into trouble.

The first signal that McNamara had been right in blocking immediate construction of Boeing's 733-290 arrived at FAA shortly after the "Phase IIC" interim evaluation in December 1965. The design it had submitted, the 733-390, had been refined considerably, but the 733-390's performance was substantially inferior to the earlier proposal. The major change from the 290 was the location of the horizontal tail. One major problem with the 290's configuration had emerged during the refinement process: at the plane's cruising altitude, the hot engine exhaust surrounded the horizontal tail. This was due to the extremely low air density above 60,000 feet, which translated into a very low "q," or air flow momentum. Instead of projecting straight out from the engines as the exhaust would in a high-energy environment, in the low-energy airflow the engine exhaust immediately assumed the angle of the "free stream," the air flowing around the wings. This air curled upwards as it left the wing surface and passed across the tail. The hot exhaust that accompanied the air thus threatened to melt the tail. In the 733-390 proposal, Boeing relocated the horizontal tail and redirected the engine exhaust to pass over it, hoping to fix the problem. But Maynard Pennell's engineers had not had time to fully reintegrate the design before the Phase IIC proposal was due.[66]

If relocation of the tail was the major change to the design, the primary performance degradation came from the sonic boom. By late 1965, NASA's sonic boom specialists had realized that their earlier method of predicting sonic boom intensity was low by 15 to 20 percent, and therefore neither Boeing nor Lockheed met the sonic boom criteria for overland operation. The NASA evaluation team calculated that raising the 733-390's acceleration and cruise altitude to meet the sonic boom requirements would cost 25,000 pounds in additional fuel, decreasing the aircraft's range to 2,819 nautical miles. Lockheed faired much worse, with the L-2000-7 consuming 71,000 pounds of extra fuel, decreasing its range to 2,195 n.m.[67] The L-2000 also had a balance problem that would require it to carry 9,200 pounds of unusable fuel in a forward tank to maintain the aircraft's trim, imposing an 18,000-pound takeoff weight penalty. Despite Lockheed's poor showing, however, this version of the L-2000 was considerably better than its predecessor. In "international" operation, flying over water without the costly sonic boom restrictions, the L-2000's performance was almost exactly equal to the 733-390's. Brief hope within the PAC and at FAA that this assessment would produce a clear winner, allowing the program to be accelerated, was dashed by this result. If one ruled out domestic operation (and several members of the PAC no longer believed that overland operation would be allowed), there was no longer any basis for rating the 733 higher than the L-2000. And neither aircraft possessed intercontinental range in any case.

The decision to delay development of the vehicle for another eighteen months earned McNamara an attack from the industry's leading periodical, *Aviation Week and Space Technology.* In an editorial titled "Flogging with Feathers," editor Robert Hotz castigated McNamara for trying to kill the SST by studying it to death. Hotz argued that the SST's remaining technical problems could be resolved only by building and flying the prototypes, and more paper studies were merely a waste of time, money, and ultimately the loss of the more than one thousand SST sales that he was certain would go to the foreign competition. He dismissed the engine studies as "conveying a false sense of progress," because the critical propulsion problems would not appear until the engines were installed in an airframe. In particular, the nozzles and inlets could not be demonstrated on the ground. Declaring the program to be "penny wise and pound foolish," he called on Congress to either accelerate the program or scrap it.[68]

Hotz wrote from an ideological position that favored action over analysis.

During the 1940s and 1950s, military aviation had fostered a huge number of different aircraft designs without much examination of what the armed services actually needed, a process that led, for example, to the C-141 "Starlifter" being too narrow to carry Army vehicles. To aviation enthusiasts, the plethora of different aircraft types had been exciting, technologically progressive, and symbolic of innovation and creativity. To analysts like McNamara, however, the decade had been wasteful and inefficient. The large variety of aircraft made pilot training and aircraft maintenance more expensive than necessary, and inadequate study had led to aircraft incapable of performing vital missions. To McNamara, getting the design right before building it was worth some delay. Hotz believed that McNamara's analysis was just a sign of "technical timidity" and that whatever problems the vehicle might have could be fixed after building it.[69] The two positions were irreconcilable.

As 1966 wore on, however, the disappointing no-decision decision that Hotz was so scornful of began to look prescient. Both designs got worse. Boeing's relatively poorer showing in Phase IIC caused it to substantially alter the program's status within the company and to reassign its best people to it. Winning this competition was crucial to the company's survival. Without the supersonic design experience this project would provide, the company would soon be surpassed by companies that had won military supersonic contracts. Hence Boeing assigned Thornton "T" Wilson, who was being groomed to replace Bill Allen as president, to oversee the program from headquarters and Holden "Bob" Withington to replace Maynard Pennell, who was promoted.

Withington soon found that the "fix" for the tail problem had in fact imposed substantial stability and control problems on the aircraft and furthermore had not fully resolved the original problem, either. His engineers faced a choice between two possible solutions to the interrelated problems. They could adopt a "T-tail" that would place the horizontal tail atop the vertical stabilizer. Several other aircraft, including the company's new 727, used this configuration. This idea ran afoul of the company's chief aerodynamicist, George Schairer, however. T-tails were susceptible to a particular kind of aerodynamic problem called a "deep stall," which occurred when the horizontal tail became blanketed by turbulent air flowing off the wing. Under this condition, the horizontal tail became ineffective, causing a sudden loss of control. Usually, this kind of stall was unrecoverable, leading to a crash. The "deep stall" occurred only when the aircraft reached a very high angle of attack—something commercial airline pilots never did deliberately—but Schairer was a purist. One did not design an aircraft with a known aerodynamic flaw.[70] He was overridden on the 727, but he was adamant enough that Withington chose not to fight him on the SST. Instead, Withington had the horizontal tail substantially enlarged and mounted the engines directly on it. The engine exhaust would therefore never touch any part of the aircraft.

Lockheed's improved performance in the earlier evaluation, combined with the 733's problems, raised a good deal of concern within Boeing's management. General Maxwell had become increasingly blunt about Boeing's failure to improve the 733-390. His repeated criticisms convinced Howard Phelps, head of the company's Washington office, that Boeing needed to propose a "bell ringer" configuration in September that would be decisively superior to the L-2000.[71] Phelps told T. Wilson that the company had already lost the support of Maxwell's technical advisor, Raymond Bisplinghoff, a former NASA associate administrator and previously a strong ally, and that Maxwell was losing patience with them due to unsatisfactory technical progress and poor communications. In that Maxwell was known to dislike and distrust Lockheed, the suggestion that he might turn to them anyway carried a great deal of weight.[72] A decisively superior proposal (and a strong, positive presentation) would allow Maxwell to take credit for turning Boeing around and win them the Phase III contract.

Because the 733-390 had to be redone anyway to resolve the exhaust impingement problem, Withington's engineers decided to make the swing-wing as much like SCAT 15F as possible. This would demonstrate the company's commitment to using the latest technology in its design. The large, flat tail was sculpted and reflexed to produce beneficial lift interference, and the wing was redesigned to integrate with the tail, forming a single aerodynamic surface much like the "arrow-wing" at supersonic speeds. The team did not adopt SCAT 15F's outboard vertical fins, as these would increase structural weight, nor did they adopt the 15F's blended wing body, which would make manufacturing the aircraft more difficult. This new configuration, designated the 2707-100, was submitted in September 1966 for the final "Phase III" competition.

Nothing quite so dramatic as a complete reworking happened to the L-2000 during 1966, but it too suffered problems. As Lockheed's team, now led by former test pilot William Magruder, worked to refine the aircraft, its range and payload deteriorated. By the middle of the year, the aircraft's gross weight had increased 10 percent, to 550,000 pounds, while its payload at 4,000 statute miles had decreased 25 percent, to 30,000 pounds.[73] The weight increase was largely due to airline demands for greater fuel reserves. The airlines were experiencing the worst year in history for airport delays, and aircraft were being forced to fly holding patterns sometimes for hours while waiting to land. Airline managers were therefore demanding much larger fuel reserves for the SST than the FAA's regulations did. Because the double-delta wing performed very poorly at subsonic speeds, an increase in the time the aircraft had to fly subsonic dramatically increased the amount of fuel it had to carry. The weight increase, in turn, meant that the aircraft climbed more slowly and thus would subject the communities under its flight path to higher noise levels. Further, flight testing of Lockheed's A-12 had demonstrated that small changes in temperature at cruise altitude could substantially affect the aircraft's range. But no one knew how big this effect was, nor did anyone yet know with any accuracy how temperatures in the stratosphere varied over time. This also influenced the airlines' demand for larger fuel reserves. When the company submitted its final version of the L-2000 to FAA in September, its proposal showed that the aircraft would have worse noise characteristics than the existing jet airliners—and these were already unacceptable to many major airports.

Maxwell's SST office assembled a 236-person evaluation team to review these final proposals and also recruited a separate review committee composed of airline executives. The review proceeded generally as the previous ones had. The airline committee was unanimous that the program should advance into the prototype construction phase, but its members agreed on little else. Most of the represented airlines were uncertain enough about the two designs to ask for both to be built and tested, while American chose the Boeing/GE entry for development and Eastern chose the Lockheed entry.[74] The lack of unanimity stemmed from the failure of either company to produce demonstrably superior designs. But while the Phase IIC evaluation had been indecisive because the two aircraft were roughly equal in overall performance, the Phase III evaluation found them to be substantially unequal. The airline managers' problem was that the "winner" of the technical evaluation, Boeing, had won it with an incomplete design.

The design change to the "bell ringer" 2707-100 configuration had occurred late enough in the year that Withington's group at Boeing had not been able to fully integrate the new design. The team had not resolved, for example, the precise location of the engine nacelles on the tail, finalized the sizes of the various control surfaces and the corresponding stability and control system, or done a thorough analysis of the new design's weight. Hence considerable uncertainty over its final configuration, weight, and other characteristics still existed. As proposed, the 2707-100's maximum gross weight was 675,000 pounds, and it could carry a 58,000-pound payload a little over 4,000 statute miles.[75] This exceeded the minimum performance requirements fairly substantially. In a recent interview, Langley engineer Neil Driver recalled that the performance excess gave the evaluators confidence that if the design gained weight during the detailed design phase, it would still be able to meet the performance targets.[76] The design appeared to meet the noise standards that FAA intended to place on all new aircraft, and its high capacity—277 seats in a three-class arrangement—gave it acceptable economics. But because the design was not yet fully integrated, these numbers were almost certain to change.

Lockheed's smaller L-2000-7 did not find favor with the evaluators, despite being fully integrated and more highly refined. The aircraft did not meet the noise standards set for it, largely due to its weight increase (now at 590,000 pounds). And it had no margin at all for weight growth. All aircraft projects the evaluators knew of had experienced weight gain during the detailed design process, and the L-2000-7 could not accept it.[77] The evaluation team believed that any weight growth would force Lockheed to start from scratch on a new design. In an independent review of the evaluation, Arthur E. Raymond, Douglas Aircraft Company's retired vice president for engineering, agreed and told McKee that the design's high degree of refinement meant that it had little chance of further improvement.[78] He saw no chance that Lockheed could both improve its noise signature and provide the larger fuel reserves that the airlines demanded.

The airline committee had therefore faced a choice between an unintegrated design that might still contain major flaws and a well-integrated one that was highly marginal. Faced with such an unappealing choice, the members essentially refused to choose. McKee and Maxwell were therefore free to choose the winning design. They chose the Boeing, citing the swing-wing's superior subsonic performance, its superior airport noise characteristics, and the company's greater experience in jet airliner design.[79]

McNamara's PAC did not question the choice that the two made, but the members still evinced considerable doubt about the SST's future. The committee's last report to President Johnson favored going into prototype construction, but other than that painted a rather skeptical picture. The PAC believed that the airlines were not particularly enthusiastic about the program, and demanded that they prove their support for the program by providing "risk capital" in the form of substantial payments to Boeing. McNamara's committee did not think much of Concorde, which appeared to have very poor economics, and stated that no SST was likely to be allowed to fly supersonic over land. This would, McNamara and John McCone believed, render all SSTs nonviable. They recommended prototype construction only because they believed that further delays to the program would serve no purpose other than to harm its already small chance of success. The Boeing SST would already appear on the market four, and perhaps five, years after Concorde. The committee members believed that if the program was delayed any longer, whatever SST market did exist in the mid-1970s would already have been filled by Concordes. If there was to be an American SST, "cutting metal" had to start soon. The PAC's decision to go ahead with the prototype construction program thus represented not certainty that they had finally reached a superior design but belief that they could no longer wait. It also represented a great deal of faith that Boeing could pull together the highly complex, unintegrated 2707-100.

DEMISE OF THE SWING-WING

Boeing's 2707-100 was technologically ambitious, despite the years of study the company had devoted to it. The company had never built a supersonic aircraft. No one had ever built such a large variable-geometry aircraft. No one, in fact, had ever built an aircraft of this size, weight, or complexity before. In earlier phases of the competition, skeptics like McNamara and McCone had focused on potential propulsion problems, but Withington's engineers faced a different problem entirely: weight, the bugaboo of all commercial aircraft design. The Phase III contract specified a 640,000-pound maximum gross takeoff weight. That goal quickly proved an impossible dream. Within a year after the company's victory, its ambitious swing-wing design collapsed and had to be thrown out.

Boeing and the FAA had agreed that Withington's design team would spend the first half of 1967 integrating the 2707-100 design while incorporating some improvements that the Phase III evaluation team had deemed necessary. The most important of these improvements was to the aircraft's stability and control system. The Dash 100 appeared to be particularly deficient in longitudinal stability characteristics, due to the far aft placement of the heavy engines combined with the short moment arm available to the elevators on the large tail. Because the elevators were relatively close to the aircraft's aerodynamic center, they could not impart enough force to keep the airplane level under certain conditions. They needed to be moved further away to provide sufficient control power. By May, Withington believed this problem could be solved only by the addition of a canard, like that on the XB-70. Essentially a horizontal "tail" mounted near the aircraft's nose, the canard would provide 266 square feet of additional control surface area much further from the aircraft's center. On 26 June, Withington briefed Maxwell on this new configuration, now called the 2707-200. Maxwell was not particularly pleased with the change, believing that the canard would generate new problems of its own, but he had little choice but to approve it.[80]

The evaluation team had considered Boeing's original weight estimate on the Dash 100 to be optimistic, and the addition of the canard made the design's weight problem worse. In addition to the weight of the canard and its hydraulics (about 10,000 pounds), Withington's engineers had to add another 25,000 pounds of structural stiffening to the fuselage to counter the bending moments created in the structure by the large, heavy engines all the way aft and the canard's own weight and lift effects.[81] By early August, the operators' empty weight, which was the weight of the aircraft minus fuel, passengers, and cargo, for the Dash 200 was 360,500 pounds, more than 72,000 pounds over the target.[82] As the payload for the aircraft was 58,000 pounds, this meant that even empty the Dash 200 had insufficient range for transatlantic service. To restore full range and payload to the design, Withington believed that the gross takeoff weight would have to reach 750,000 pounds, far above the contract's specification. George Schairer, who often played the role of internal critic at Boeing, told Withington that the Dash 200 might not achieve an economical ratio of range to payload at less than 1.2 million pounds.[83]

At Withington's June meeting with General Maxwell, Maxwell had been willing to let the plane's weight grow to 675,000 pounds, but not more. Maxwell also had given Withington a deadline of 1 January 1968 to prove the need for the weight increase.[84] Back in Seattle, Withington assigned Fred Maxam, the project's chief of technical staff, to the task of trimming the overweight plane by 75,000 pounds. He was able to remove 23,000 pounds, much of which came out of passenger "amenities," like seat structure, galley size, and so forth. It was not enough to restore the range.

Worse, the weight growth caused other problems for the design. The Dash 200 had already reached its fuel storage capacity limits, and Withington's engineers had to create new space for fuel to carry the additional weight. The only substantial potential volume available was between the wing pivots, which had to be moved further outboard to accommodate the additional fuel. Moving the pivots further out, in turn, reduced the aircraft's effective wingspan, reducing its climb performance and increasing its noise levels. Restoring climb performance meant larger engines or lengthening the wings, which would again increase the plane's weight, starting the entire fuel volume / pivot movement / weight increase / noise increase cycle again.

As the Dash 200's problems became apparent, company president Bill Allen had assembled an "SST Technical Advisory Council" chaired by Boeing's senior engineer, Ed Wells. In mid-October, Withington told Wells that the Dash 200 could not be integrated by FAA's January deadline. Wells later told journalist Charles Murphy that Boeing's experience with the Dash 200 had been strange: "The more we came to know, the less well things worked out for us. Instead of entering into a situation where the problems began to offset one another, the problems were actually compounding. When they should have started to converge, they continued to diverge."[85] In engineering jargon, the Dash 200 design had refused to "close." Its parts could not be integrated into a workable whole.

General Maxwell requested a NASA review of the Dash 200 in January 1968. The analysis he received in early February essentially agreed with Boeing's internal assessment. The Dash 200's canard had resolved one set of maneuvering problems while creating instability in other maneuvers. And the reviewers strongly doubted that Boeing would succeed at reducing the plane's empty weight from 354,300 pounds to the prototype's target, 337,400 pounds.[86] With a 50,000-pound payload, the evaluators gave the Dash 200 a range of only 1,700 n.m., half the range it needed.

At a meeting in early February, Boeing's senior executives hashed out the program's problems and what to do about the project. Wells believed that it was better to take the time to find a better design, even if this risked political cancellation, than to risk the company and its customers on a bad aircraft. Boeing president Bill Allen agreed with Wells a few days later, admitting that they had a design that "couldn't get to the post office. . . . The right thing to do, as a matter of straightforward business ethics, was to tell the FAA that the design could not achieve the contract objectives, and then ask for more time in which to decide on the best configuration."[87] On 14 February, General Maxwell and his chief advisor, Raymond Bisplinghoff, flew to Seattle to discuss what to do to salvage the SST program. Maxwell accepted that the Dash 200 was not going to succeed, and he also accepted Withington's proposal to end the prototype construction effort and return to a study of configurations. But he also demanded that Boeing pay a substantial price for its failure. Allen agreed to spend $45 million to try again.[88]

The previous September, Withington had set up a "Parallel Group" to explore promising new configurations, and after the February decision to shut down the Dash 200 construction effort, this group became the program's focus. Headed by Kenneth Holtby and William T. Hamilton, the group spent some 1.5 million engineering hours investigating twenty new configurations, using wind tunnels at Boeing and at the Langley and Ames labs. They also employed NASA's computer software, flying data back and forth between Seattle and the Langley lab in the bags of United Airlines pilots.[89] By June, three finalists had emerged. One was a swing-wing configuration substantially different from the Dash 200, model 969-404B. Two others were fixed-wing configurations. Boeing model 969-321 was a modified version of NASA's SCAT 15F. And Boeing model 969-302B was a delta-wing design. The two fixed wing configurations seemed superior to the Dash 200, while the swing-wing design retained many of its uncertainties.

The swing-wing design Holtby's group worked up was in substantial ways a return to the earlier 733-390 design. It did not have the tail-mounted engines, or the integrating wing/tail combination, of the Dash 200. This improved the weight problem but restored the exhaust impingement problem, which they still had no better solution to. It also degraded the aircraft's supersonic performance by abandoning the SCAT 15F-like arrow-wing planform. The high weight of the pivot structure gave the 969-404B the highest takeoff weight of the three configurations.

The 969-321 differed from SCAT 15F primarily in the addition of a tail. Its designers had added a tail to improve longitudinal stability. The original SCAT 15F design had been smaller than the 969-321 was intended to be, and like Lockheed's tailless delta, could not be scaled up beyond 600,000 pounds or so without the addition of a separate tail. The outboard vertical fins were deleted, too, because they resulted in a heavier wing structure. Nonetheless, the 969-321 retained SCAT 15F's very high supersonic performance, but its highly swept wing meant that its subsonic performance was quite poor. For the same reason, its noise signature was poor.

The 969-302B had evolved from Boeing's 1958 submission to the WS-110A competition. The company had reviewed it in 1964 and again in 1966, but had not found it compelling. By mid-1968, however, Boeing's engineers had begun to appreciate the simplicity of the 969-302B. It had two advantages, and one disadvantage, compared to the 969-321. Its first advantage was a much simpler, lighter structure, which meant that the 302B would weigh less than either of its competitors at transatlantic ranges. The simple structure was also easier to manufacture than the 321 would be. But the second major advantage was superior subsonic performance, achieved by using a relatively low wing sweep of 50.5 degrees and a full set of leading and trailing edge flaps. These gave the 302B a lower noise signature. More importantly, however, they gave the aircraft a superior subsonic range. By 1968, Boeing's engineers and managers had concluded that the public would not tolerate supersonic overflight of land. Thus part of the evaluation criteria that the company established for the internal competition between the three "finalists" was the addition of an "extra" 350 n.m. subsonic leg to the plane's mission. This extra leg would permit the aircraft to serve many inland cities while still decelerating to subsonic speeds before reaching land. It was with this addition that the 302B showed its superiority. While it was less efficient than the 321 at supersonic speeds, its greater subsonic efficiency outweighed this disadvantage if one assumed no supersonic overflight of land.

In October, Boeing selected the 969-302B as its new SST configuration, renamed the 2707-300, completing the process of abandoning the swing-wing concept. At the Langley lab, Boeing's rejection of the SCAT 15F went over poorly, with many members of the staff believing that they had been "sandbagged."[90] They blamed the "not invented here" syndrome for Boeing's choice. Indeed, within a few months after the decision to adopt the 302B configuration, an improved version of Boeing's SCAT 15F derivative, the 969-336C began to show better performance, although its noise signature remained high.[91]

What Boeing engineer John Swihart recalls, however, was that his peers had become extremely risk averse after having taken a substantial risk in adopting John Stack's swing-wing dream, and then failing to pull it off.[92] When General Maxwell saw the Dash 300 design, he exclaimed that it was "pure vanilla"—and that was exactly the point. It was a lower-risk design. In addition to its superior noise and subsonic performance, it was structurally simpler, and thus the company's engineers could have far more confidence in their weight estimates. The Dash 300 was a plane they *knew* they could build, while SCAT 15F still carried substantial risk.

✦ ✦ ✦

The demise of Boeing's swing-wing design was a result of the pressure to meet a schedule while still attempting a radical advancement in the state of the art. Boeing's variable-sweep aircraft was beyond the state of the art. In that regard, NASA's Langley laboratory had not served either Boeing or the FAA well in its promotion of the swing-wing over the delta wing. Lockheed's simpler approach had proved superior after all, even if Lockheed had done it poorly. McNamara's defense of Lockheed's approach had been correct in retrospect, as had been his belief that the manufacturers had fixed their designs too early. But because McNamara's committee felt pressured to meet Concorde in late 1966, they and the FAA's leadership had gambled on a high-risk design in the hope that it would also be "high payoff."

Fortunately for Boeing and the FAA, Concorde was not on schedule either. Despite the early decision to stick to well-known construction materials and a much simpler design, Concorde's designers faced various development problems. The initial decision to try to make the same airframe serve as the basis of both a medium-range "domestic" version and a transatlantic version proved unworkable, and the "domestic" was dropped. During 1966, the designers increased the aircraft's size substantially, from 260,000 pounds to 335,000 pounds, in order to carry a larger payload.[93] This complete redesign took time. "Unexpected" weight growth of the kind experienced by Boeing eventually increased Concorde's weight to nearly 400,000 pounds. Its economics suffered accordingly.

Cancellation of the TSR.2 bomber also threatened the development schedule, because Concorde's designers had anticipated that the Olympus engine would have flown hundreds of hours on the bomber before Concorde flew. Without the TSR.2's flight experience, Concorde's own flight test schedule had to be extended, and, of course, the program's cost rose. By the time Boeing's 2707-200 design died, the first preproduction prototype's flight had been pushed into early 1969, and the first production prototype would not fly until 1971.

This put Concorde just slightly behind the Soviet SST, which had finally become more than a rumor at the 1965 Paris airshow.[94] Named TU-144, the Soviet entry into the SST sweepstakes bore more than a passing resemblance to Concorde and was often, and unfairly, called "Concordski." This was to be a 286,000-pound aircraft carrying 121 passengers 4,030 miles at Mach 2.3.[95] In the United States, TU-144 took on a life of its own, with proponents of the American SST program using it as a propaganda tool to raise fears in the public of another Sputnik. But it was never taken seriously within McNamara's committee, which had been told by several senior airline officials that no Western airline would buy a Soviet plane. Not only would this cause the purchaser political problems, the uncertainty of receiving spare parts (which could be subject to political embargoes) would make the purchase economically unwise. Hence the Soviet "threat" existed in propaganda only. The TU-144 would fly first, and might go into service first, but it constituted no economic threat at all.

The collapse of Boeing's Dash 200 design created the risk that an unhappy Congress might cancel the program, but this did not happen. In fact, because the redesign and evaluation process was much less expensive than the cancelled prototype production process would have been, the $198 million that FAA had received for the first year of prototype construction was more than sufficient to carry through for the next eighteen months.[96] This provided the Johnson administration with a small political benefit, in that it did not have to request new funding in the FY 1969 budget. It therefore did not have to justify the expense against the mounting costs of the Vietnam War, the Great Society, or the other myriad competitors for federal dollars.

But Johnson's successor *would* have to make a funding decision. Lyndon Johnson's decision not to run for reelection in 1968 cost the SST its chief supporter, and there was no guarantee that Republican Richard Nixon, who defeated Democrat Hubert Humphrey handily, would find the troubled (and Democratic) project worthy. Nixon's first act on taking office in 1969 was indeed not promising to the SST's future. Instead of voicing his support, he placed the program "under review" by a committee, a technique often used to quietly kill undesired programs.

The TU-144 first flew 31 December 1968—the dark horse winner of the SST race. And as Concorde's long-delayed first flight approached, it was not clear that there would ever be an American SST. In addition to the lack of immediate support from the new administration, a movement to prevent any SST from ever reaching U.S. shores was gradually taking shape, something no one had, or could have, foreseen. Not since the increasingly remote nineteenth century had an organized opposition to an entire technology taken place. Najeeb Halaby had commented in 1964 that "liberals and academics" seemed to be opposed to the SST (as, he should have noted, were the conservative *Wall Street Journal* and *The Economist*), but the occasional newspaper and magazine editorial did not make a movement, or pose much of a threat to the aerospace industry's supporters. By the time Nixon took office, however, opponents of the SST, and more specifically its sonic boom, had begun to realize that they had a chance to kill the offensive bird completely by mimicking the other "grassroots" opposition movements that had blossomed during the 1960s. Organized opposition came as an unwelcome shock to the promoters of the supersonic future.

OF NOISE, JUMBOS, AND SSTs

There had always been opponents of the SST. Treasury officials in Britain, Robert McNamara in the United States, and Ian McDonald, head of the International Air Transport Association, had all opposed the race to the SST on primarily economic grounds. Similarly, conservative publications such as New York's *Wall Street Journal* and London's *Economist* had consistently opposed the SST as an inappropriate expenditure of government funds. These believers in the superiority of private over public economic activity adamantly rejected the notion that government should fund commercial ventures. But this was not the critique on which the anti-SST movement was based. Instead, the organized campaign against the SST project started as a campaign against the sonic boom and was transformed when it was adopted by old conservation-oriented "wilderness" organizations, by new "environmental" ones, and by equally new consumer advocacy groups. This coalition succeeded in its efforts to convince Congress to stop funding the SST project in March 1971, in what environmental leaders saw as their first major victory against environmentally destructive technologies.

Historian Scott Hamilton Dewey has recently argued that activism aimed specifically at curbing air pollution, which had long been suspected of having detrimental health effects, finally achieved political potency during the 1960s.[1] President Johnson signed laws aimed at curbing automobile emissions in 1965, water pollution the same year, and air pollution more generally in 1967. He also initiated efforts to curb aircraft noise that were completed under his successor, Richard Nixon. In 1969, President Nixon signed the National Environmental Policy Act, requiring federal agencies to prepare environmental impact statements on their actions (including the SST). While this bill was in many respects a "leftover" from Johnson's term, Nixon also proposed a very broad program of environmental protection, partly enacted in the 1970 Clean Air Act, the creation of the Environmental Protection Agency the same year, and the 1973 Clean Water Act.[2] These legislative successes under presidents of both parties reflected the widespread, mainstream nature of environmental concern.

Various wilderness and conservation organizations had promoted the spread of environmental interest during the 1950s and 1960s through the organization of grassroots campaigns to accomplish specific objectives. The Sierra Club, for example, sponsored campaigns to block dam and power plant construction at sites of aesthetic value. These had mixed results, in that they did not always succeed at blocking construction but always advanced the larger goal of public awareness. The campaign that evolved against the SST was, in one sense, part of that larger goal. Yet it also represented a new approach to tackling environmental issues in the sense that it opposed a new technology, not just the particular siting of an airport, power plant, or dam. In this sense, the anti-SST campaign represented a transitional phase, a period of great controversy within the wilderness organizations as well as with the public at large.

Writing a few years after the defeat of the SST, political scientist Langdon Winner remarked upon the striking spread of a new thought in American politics during the 1960s: that "technology" was out of control. In this view, technology had become autonomous and self-governing, paying no heed to human needs, human values, or to the quality of the human and natural environments. This "autonomous technology," in the eyes of its critics, was inimical to human freedom. It was also anti-life.[3] This critique underlay the environmental movement. Leaders of this movement used the idea that technology was out of control as part of their effort to direct attention to what they perceived as fundamental conflicts between ethics, environmental quality, democracy, and the practice of American technology. This critique presented the SST as a symbol of a particular style of American technology that promoted the "speed-up" and corporatization of life, the concentration of capital, and destruction of the natural (and human) environments. To these groups, the SST represented misdirected technological effort. It represented an old notion of "technological progress" that favored speed over all other criteria. These critics sought a redirection of "progress" along more humanistic and environmentally sustainable lines.[4] They were, in essence, rejecting the notion that technology should be autonomous. Instead, because technology is inherently political, it should be subject to political control.

The anti-SST campaign was what environmental activist David Brower called a "symbolic" campaign, devoted to demonstrating to the public that environmental decline could be reversed by opposition to damaging technologies. He believed he could inculcate an "environmental ethic" into society through these campaigns, producing a society whose vision of progress included things like cleaner air and water, and not merely speed and size for their own sake. Whether he succeeded in this grand, utopian goal of reforming society is highly questionable, but the anti-SST campaign did achieve a lesser goal of convincing the aircraft industry that environmental goals had to be paramount in future aircraft design. By May 1970, it had become clear that the public would no longer tolerate ecologically unfriendly aircraft, and the Boeing SST was one—at least in the public's eye. And in order to make the SST meet the new environmental demands, Bob Withington's engineers had to change the design in ways that made its economics terrible. Hence by the time the anti-SST campaign got organized, the company's leaders had already decided that the two prototypes would not be followed by a production run. Its SST could not be both economical and environmentally sound at the same time.

Yet while Boeing was losing interest in the SST, President Nixon's White House was gaining interest. Late in 1970, after the Republican Party had suffered badly in the midterm election, Nixon realized that the environmental initiatives he had made during his first two years were not only failing to help him but were costing his party votes in the business community. He therefore switched sides and began a quite deliberate antienvironmental campaign. His first act in this new strategy was the highly public campaign to save the SST from the "ecological extremists." Hence the all-out media war between supporters and opponents that swirled around the SST in 1971 had almost nothing to do with the SST or its merits. Instead, the SST conflict was entirely symbolic, with the unhappy bird itself already orphaned.

SONIC BOOMS AND THE CONSTRUCTION OF
AN OPPOSITION MOVEMENT

On 24 March 1970, representatives of the Sierra Club, the Friends of the Earth, and a Baltimore schoolteacher met with staff members from the offices of two congressmen and a senator to establish a group called the Coalition against the SST. They had been brought to this meeting by the efforts of an organization whose representative could not attend, the Citizen's League against the Sonic Boom. Founded by physicist William Shurcliff and biologist John T. Edsall in 1967 and run out of Shurcliff's house in Cambridge, Massachusetts, the league had sought to convince the Sierra Club, the Isaak Walton League, the Wilderness Society, and other conservation groups that the SST was not just another airplane whose noise would be confined to airports and their surroundings. Instead, Shurcliff argued that the SST's sonic boom would endanger the very wilderness these organizations existed to preserve. His success at convincing them to unite against the aircraft was the first victory in the SST endgame and a crucial turning point in these groups' attitudes toward both environmental problems and new technologies.

Shurcliff's opposition to the SST was based upon a series of articles written by Bo Lundberg, who had been director general of the Aeronautical Research Institute of Sweden until his retirement.[5] Lundberg was very well known in aviation circles, and he had been a longtime opponent of the SST, writing against it as early as 1962. Lundberg's early opposition to the SST stemmed from both economic and sonic boom concerns, but by 1966 he had begun to focus exclusively on the boom problem. In a seminal article, Lundberg provided a reinterpretation of the most extensive sonic boom tests to date, the NASA/FAA/DOD test series at Oklahoma City.[6]

Following the sonic boom test series at St. Louis in 1962, NASA, the FAA, and DOD had all concluded that they did not yet know enough to determine where the public's threshold of tolerance of the sonic boom might be, and they agreed to an even more extensive series of tests hoping to find this out. The FAA, which was to coordinate the test series, chose Oklahoma City because about one quarter of the city's economy came from aviation-related sources, including the FAA's own air traffic control training facility. Further, the FAA's primary supporter in the Congress, Senator A. S. "Mike" Monroney, represented the state of Oklahoma and, FAA leaders reasoned, would support the tests as well.[7]

The Oklahoma City test series began in February 1964 and ran through July. Air Force aircraft provided eight booms per day, daylight hours only and on a regular schedule. For the first twelve weeks, the boom intensities ranged from 1.0 to 1.5 psf, and for the last fourteen weeks they were raised to from 1.5 to 2.0 psf. NASA provided instrumentation and contracted with the

National Opinion Research Center at the University of Chicago for determination and evaluation of public response to the boom. The FAA and Air Force also provided a claims office to process and evaluate claims of damage caused by the booms.[8]

Unfortunately for the three agencies, however, despite the local economy's aviation dependence, the tests were no more popular than they had been in St. Louis two years before. For the first few weeks of the tests, as people got used to the booms, complaints decreased. But this did not last. By mid-April, over four thousand claims of damage to buildings had been received by the FAA. Citizens began to form groups opposed to the test series, and in July three members of the city Chamber of Commerce, including the chair of the Chamber's aviation committee, complained to its chief, who in turn complained to Senator Monroney. A local attorney also filed suit to stop the tests with a federal court judge, who refused to intervene. He had better luck with the local district court, which ordered a cessation that was soon overturned on an FAA appeal.[9] Enough complaints reached Monroney about the conduct of the tests that he demanded assurances from FAA that future tests would be assigned to someone else's state.

By mid-May, the political repercussions of the boom tests over Oklahoma City had led McNamara's SST Advisory Committee to recommend to President Johnson that he ask the National Academy of Sciences to handle evaluation of the boom data and to administer future tests.[10] NAS president Frederick Seitz appointed John R. Dunning, dean of the School of Engineering and Applied Sciences at Columbia University, to head a "Committee on SST-Sonic Boom." Dunning's committee began meeting in late July, and FAA's Gordon Bain quickly began a campaign to convince the committee that the Oklahoma City tests had gone well, contending that a public relations campaign could convince the ornery citizens that the boom was acceptable and showing them a movie that he claimed demonstrated that people had "rarely" noted the booms.[11] The committee's members did not accept Bain's conclusions, insisting that the boom and the Oklahoma City data required a much more comprehensive study. They also wanted a set of tests devised specifically to evaluate structural damage. These were carried out in 1966 at Edwards Air Force Base.[12]

In December 1964, the National Opinion Research Center's report on the Oklahoma City tests appeared, documenting that 27 percent of the city's population had stated they would not accept indefinite booms at these levels. In

February 1965, responding to congressional demand, the FAA made the Oklahoma City test results public, hoping that the "truth" would quiet some of the controversy. Instead, further political pressure caused Najeeb Halaby to ask McNamara's presidential advisory committee to find a way to put future boom testing under some "broader jurisdiction" in an attempt to diffuse the blame for this and potential future debacles across several agencies. None of the executive agencies were willing to run future tests, so the committee asked Donald Hornig, President Johnson's science advisor, to accept the responsibility. He agreed, but not without making a statement that would prove prophetic: "I do want to point out that with the best programs and best measurements, after we are all done the Committee will be faced with the major uncertainty of public reaction. There are people, even cabinet members, who feel that the existence of sonic boom at any level with any frequency is an intolerable intrusion on the privacy of the American public."[13] This conversation led the secretary of the treasury to comment that there was a simple way to determine the boom's acceptability—boom Washington, D.C., on a commercial schedule and track Congress's response. The committee did not agree, perhaps recognizing that booming Washington would kill the program. Instead, the 1966 test series planned for Edwards Air Force Base was expanded to include additional human response evaluation.

From a political standpoint, McNamara's committee already had sufficient information to judge the acceptability of the boom. If commercial aviation's biggest supporter in the Senate opposed the booming of his state, it was inconceivable that the other states would acquiesce. The very fact that the FAA had to ask for political cover from its attempt at scientific boom testing should have been the definitive clue in the puzzle over whether the boom would be acceptable to the public. There was no chance that the Boeing SST then being designed would produce booms less intense than the ones Oklahoma City was subjected to, and therefore there was no chance that the American political system would allow permanent booming of population centers. But instead of viewing the boom's acceptability as a political problem, McNamara's advisory committee treated it as one amenable to scientific examination. The majority of the committee continued to believe that an acceptable boom level could be found through scientific methodology, and once found that they could find a way to build the SST to meet it.

The National Academy of Science's committee was much less certain that this was true. The committee's first interim report to the FAA had been very positive, but as the committee began to look more thoroughly at the tests, several members began to believe that what the FAA had concluded about the test results—that an "overwhelming majority" of the city's population had accepted the boom—was not a fair representation of the data. Sociologist Kingsley Davis of the University of California at Berkeley was particularly critical of this conclusion, noting that the tests had omitted the most important question of all, the public's response to night booms.[14] In July 1965, the SST-Sonic Boom Committee endorsed a highly negative report by the Building Research and Advisory Board, which had been formed to examine the data on structural damage. This group had argued that while the boom would cause damage, the more important issue was the public reaction. The full committee issued a new report later in the year that took a somewhat more neutral stance than its preliminary report had, and contended that the sonic boom problem alone was sufficient to defeat the program and that there was little chance of mitigating the boom through design of the aircraft. It also stated that additional public acceptance testing needed to be done before a definitive answer on boom acceptability was possible.[15] These two additional negative reports, however, were not made public until years later. Hence while the National Academy's committee was skeptical of the SST, its public position, expressed in its publicized initial report, was one of support.

In Sweden, Bo Lundberg had reached a much more negative conclusion about the boom than the SST-Sonic Boom Committee had. Using the data from the Oklahoma City tests, Lundberg argued that the boom would never be acceptable. The Oklahoma City tests had showed that boom intensity received on the ground varied greatly from one location to the next, with booms nominally producing 1.5 psf often being recorded in some places as exceeding 3 psf. These higher booms were unavoidable, he argued, and he accused the SST enthusiasts of hiding this reality by referring to the SST program's design boom limit of 1.5 psf as a maximum boom limit, while in fact it would be, at best, an average. This would be true even if the U.S. SST design succeeded in achieving the 1.5 psf limit, while the Oklahoma City testing also showed that the boom prediction methodology that NASA had derived in the late 1950s was flawed. The new data showed that the estimates had been 30 to 40 percent low, and thus the U.S. SST would almost certainly produce booms approaching 2.0 psf on average. Then Lundberg attacked the tests for ignoring the boom's effects on sleep. The data from Oklahoma City showed that 18 percent of the population reported the boom had interfered with their sleep, despite the tests having been confined to daylight hours. And he recounted an anecdote that the FAA had used to put a positive spin on the tests. "Office girls," the FAA had reported, had adapted to the boom by using the first boom of the day as an alarm clock. Lundberg extended the story to ask how the "office girls" would have responded to the boom if it had happened at 2:00 A.M., instead of 7:05 A.M. Inevitably, the boom would happen at night, too, and the question the testers should have asked was this most critical one: how would sleepers be affected? Lundberg believed that a sleeping population would find the boom far more objectionable than the mostly awake Oklahoma City population had, and the city had not exactly welcomed the daytime experiments. He concluded that the SST represented a fundamental conflict between ethics and technology.[16]

William Shurcliff was converted to open hostility toward the SST by Lundberg's argument. In his response to Shurcliff's overture, Lundberg passed on a recommendation from an Elizabeth Borish of Vermont that an anti-SST committee be formed.[17] Lundberg also sent Shurcliff copies of a number of letters he had sent to various conservation societies in the United States as well as a set he had sent to the SST's leading opponent in the Senate, William Proxmire (D-Wis.). In these letters, Lundberg expanded his opposition, arguing that the SST needed to be banned completely, because it would be so uneconomic that simply banning overland supersonic flight would not, in the end, protect the population. In their quest to make the huge SST investment pay off, the airlines and manufacturers would bring so much pressure to bear on politicians that the ban would ultimately be lifted.[18]

Lundberg brought up the economic issue because by late 1966 the National Academy of Science's SST-Sonic Boom Committee and Dr. Hornig's Sonic Boom Coordinating Committee had each concluded that the SST could not be allowed to fly supersonically over land.[19] These conclusions, combined with lobbying by Interior Secretary Stewart Udall and famed aviator Charles Lindbergh, had convinced Robert McNamara that the Boeing and Lockheed SSTs were over-water aircraft only. In his fourth interim report to President Johnson, McNamara's committee advocated an explicit ban on supersonic overflight of populated areas. This did not happen, however, because the FAA did not accept the conclusion and had been able to convince President Johnson not to make any public decisions about the boom.[20] The editors of *The*

Economist, which was stridently anti-Concorde and anti-SST, even accused FAA of doing everything short of actually suppressing the reports to keep this conclusion out of the press.[21]

Nonetheless, Lundberg's many connections within the aviation business had made him aware that the Boeing SST was being designed and marketed to the airlines, to FAA, and to McNamara's PAC as a transatlantic aircraft. This was in keeping with the Boeing Company's own conclusion about the boom, which had always been that the public would not accept being boomed. Lundberg, however, did not believe that the SST proponents were honest in their depiction of the SST's profitability when confined to the over-water routes. Instead, he told Senator Proxmire, the airlines planned to get the aircraft built for over-water flight and then push for unrestricted operations later.[22]

Lundberg did not have to dream up this scenario. Less than two weeks after he had suggested this to Senator Proxmire, an article appeared in the British journal *The Aeroplane* that confirmed that it was exactly what the manufacturers had in mind. Journalist Michael Lumb had attended a Concorde sales policy conference held by the British Aircraft Corporation and described the company's approach to the boom problem. British Aircraft believed that there was an over-water-only market for two hundred Concordes, but this was far too small to break even. Hence it and Sud Aviation had agreed to "produce numbers" showing that the boom was acceptable in order to gain access to overland markets. Shurcliff marked his copy with a well-deserved exclamation point at that paragraph.[23] Outright declaration that the companies intended to fabricate numbers in order to sell Concorde was all the evidence of foreign duplicity he needed. In this light, the FAA's considerable attempts to put a positive spin on the Oklahoma City tests looked like the same thing. Shurcliff now believed that the government was lying about the SST's economics and about the boom.

In addition to his correspondence in January 1967, Lundberg had sent Shurcliff his list of anti-SST friends and asked that he use it to start an anti-SST campaign.[24] In March, Shurcliff and John Edsall agreed to form the Citizen's League against the Sonic Boom, with Shurcliff as director and Edsall as deputy. Shurcliff's sister and his son functioned as unpaid staff members. The league was to recruit members, publish "fact sheets" about the SST, and take out advertisements against it. Shurcliff did not have to try very hard to publicize his tiny group. Less than two weeks into the league's existence, the *Christian*

Science Monitor reported on the its activities and Lundberg's sonic boom analyses.[25] By August, the league had 620 members and had received $14,000 in contributions. It had also begun to generate a deluge of inquiries addressed to the FAA about the sonic boom.[26]

The league received attention so quickly because it did not emerge in a vacuum. In addition to Lundberg's article and Edsall's letter, many other articles questioning the sonic boom and the FAA's economic arguments had appeared during 1966. Kurt Hohenemser, a professor of aeronautics at Washington University in St. Louis, had written against the SST first in the *Los Angeles Times* and then in the *Bulletin of the Atomic Scientists*.[27] John Gibson, dean of engineering at Oakland University in Michigan, had published a very harsh article in the July 1966 issue of *Harper's*, a magazine with a very substantial circulation.[28] The *Wall Street Journal* attacked the SST's economics the same month.[29] Popular singer Woody Guthrie called for public review of the SST's economics in September.[30] Articles like these had already caused the reading public to start questioning the SST program's worth, and Shurcliff sought to expand upon this undermining of public confidence.

For the next two years, Shurcliff made a point of distributing his fact sheets to the editorial boards of major newspapers and magazines. His major success, however, was in convincing the Sierra Club to join the anti–sonic boom crusade. His son Charles had begun writing to and calling the organization's San Francisco office in the middle of 1967, trying to convince whoever would listen that the boom was a threat to wilderness areas. Charles also spent the summer of 1968 working at the Sierra Club's headquarters in San Francisco.[31] In his campaign to gain the club's support, Charles was aided immeasurably by Interior Secretary Stewart Udall, who had complained about the boom's impact on delicate cliff dwellings in the desert West during 1965. Unhappy at President Johnson's continued support of the SST, Udall had established his own committee in late 1967 to review the sonic boom's impact on the "natural resource of tranquility."[32] These events helped convince David Brower, the Sierra Club's executive director, that the SST was a threat worth challenging. Linking the sonic boom directly to destruction of the wilderness, the Sierra Club's board of directors passed an anti–sonic boom resolution on 1 September 1967, and shortly thereafter, so did the Wilderness Society.[33]

The Sierra Club's decision to oppose the sonic boom did not immediately result in an aggressive anti-SST policy, however, because the organization was not yet devoted to challenging "big technology." In fact, that subject was highly controversial within the organization during 1967 and 1968. David Brower believed in the counterculture's critique of American technology and wanted the club to actively oppose technologies that were environmentally destructive. He had also recruited young, environmentally committed staff members from places like the University of California, Berkeley, where the counterculture critique of American technology was particularly strong. Brower and his supporters wanted the club to oppose nuclear power, for its thermal pollution of rivers and lakes, for its potentially destructive radiation and radiation's long-term genetic effects, and for the industry's antidemocratic tendencies. The club's board, however, would oppose only specific sites, not the entire technology. Further, Brower spent lavishly on uncertain campaigns, sometimes without the board's approval or support, alienating board members who might support his causes but would not jeopardize the club's financial standing. This split between Brower and the board of directors became a chasm over the Diablo Canyon reactor project during the mid-1960s, and in the club's 1969 elections Brower's faction was pushed out.[34]

As historian Thomas Wellock has argued, however, Brower's loss and resignation did not mean his ideas had lost.[35] Instead, the club had rejected his financial mismanagement without rejecting his causes. The new executive director, Michael McCloskey, and the new president, Philip Berry, recognized that Brower's ideas had substantial currency within the club and in the larger society. Technology, at least as it was practiced in the United States, was substantially responsible for environmental degradation. The difference between these two and David Brower was simply that Brower had been willing to sacrifice the club in order to save the natural environment, while McCloskey and Berry believed that the club would be more effective if it were managed to survive for the long-term battles that environmental protection would inevitably require.

In his resignation speech, Brower announced the founding of a new organization, the Friends of the Earth, which would be explicitly committed to a morally based environmentalism and to a fight against "undisciplined technology." The new group first met in Aspen, Colorado, in September 1969, and committed itself to "symbolic" antitechnology campaigns as a means of educating the public about the new environmental ethic.[36] And the first of these symbols was to be the death of the SST.

Shurcliff's lobbying had convinced Brower that the SST was a threat to both wilderness areas and the larger environment, and unlike nuclear power, the

SST appeared to provide no environmental benefits at all. It was noisier than the new generation of subsonic jets, it consumed much more fuel per passenger (which implied it would be more, not less, polluting), and it trailed huge sonic boom carpets behind it everywhere it went. Finally, it would serve only a tiny fraction of the nation's population, and the wealthiest segment at that. It was, in short, exactly the sort of "undisciplined technology" Brower despised. Brower thus agreed to finance the publication of a "fact book" Shurcliff had been writing, called *The SST and Sonic Boom Handbook,* which debuted in early 1970 and quickly sold 150,000 copies.[37]

The Sierra Club's new leadership adopted Brower's position on the SST nearly simultaneously. Michael McCloskey remembers this as an easy decision in contrast to the club's turmoil over nuclear power. The SST had no internal constituency, while nuclear power was popular with midwestern chapters because it promised an end to acid rain caused by coal-fired power plants.[38] Even the club's Pacific Northwest chapter based in Seattle did not support the SST program, however. Brock Evans, the chapter head, recalled to an interviewer that he had met surreptitiously with Boeing employees in parking lots to obtain key company documents.[39] Thus even some club members who worked for Boeing opposed the program. With such unanimous opposition within the club, the SST became the Sierra Club's first campaign against "technology-out-of-control," just as it had for the Friends of the Earth.

During September 1969, the same month the Friends of the Earth first met in Aspen, President Nixon had finally announced his support for the SST. This came as a great surprise to the press, which had correctly reported that the majority of his cabinet opposed continuing the program. In fact, four of the five internal review reports on the program had been highly negative, and while the reports were not released, the gist of them had been leaked. The press response to the announcement was therefore quite harsh. Syndicated columnist James J. Kilpatrick, for example, agreed that Nixon had blundered by promoting the SST over mass transit, or modernization of the air traffic control system, which had seen massive summertime delays. Editorials across the nation echoed his argument.[40] But unlike most of his cabinet, Nixon personally supported SST development. He accepted the linkage of "speed" and "technological progress" and perceived supersonic flight in the same quasi-religious terms Najeeb Halaby had, as the next step in aviation's future. He had delayed his announcement of support for purely tactical reasons, not out of uncertainty about its value. The Budget Bureau had recommended waiting until after Congress had raised the nation's debt ceiling, and he had agreed. Further, waiting until September gave him the opportunity to link his support for the SST to the Senate vote on the Sentinel antiballistic missile system. Sentinel was a major policy drive of Nixon's first term in office, but his party was not in control of the Senate. He needed Democratic votes to pass it, and Washington State's two senators, liberal Democrat Warren Magnuson and cold war Democrat Henry M. "Scoop" Jackson, might provide the winning margin.

Nixon's announcement of support led to congressional demands for his administration's review reports, four of which Congressman Henry Reuss (D-Wis.) managed to pry out of the Transportation Department after taking the issue of "executive privilege" before the House Freedom of Information committee and winning. Public release of the four reports Nixon's Transportation Department review committee had generated earlier in the year produced more anti-SST articles. Congressman Sidney Yates, also a prominent critic, inserted the reports into the *Congressional Record* on 31 October, commenting that they were so unfavorable that he was "amazed" Nixon had approved continuing the program.[41] Shurcliff, of course, reprinted this and circulated it to the members of his league.[42] Yet despite all the harsh words about the SST being printed, the plane's supporters had no troubles recruiting the votes necessary to pass an $85 million dollar appropriation for FY 1970 in December. This was about $10 million less than the Transportation Department's request, but it allowed the program to move into the prototype construction stage after seven long years. The press reaction was muted, as there was little left to say. It did not mean that opponents had given up, however. The largest funding request in the program's long history, about $290 million, would go before the House in May. There was plenty of time to mount an organized campaign.

A Baltimore schoolteacher proved to be the catalyst for the opposition movement. Ken Greif, who was independently wealthy, had been appalled at the extreme cynicism toward the political system shown by his students. They believed that the SST was inevitable, because it was typical of "America's misplaced priorities."[43] He thought the SST was product of misinformation leading to bad public policy, and he thought it could be defeated by an effective public education campaign. As an experiment, he decided to finance such a campaign. He hired an attorney named George Liebmann to represent him and dispatched Liebmann to Washington to discover which existing lobby group would be most effective. Liebmann called on the office of Senator

Edmund Muskie, the Senate's leading environmentalist, who pointed him to Laurence Moss.

Moss was a White House Fellow who had gone to the new Department of Transportation as an SST enthusiast in 1969, and had attended the meetings of Nixon's Department of Transportation ad hoc review committee on the SST.[44] He left at the end of his fellowship year believing that the SST was an environmental travesty. Moss had been active in opposing proposed dams in the Grand Canyon during the early 1960s as a member of the Sierra Club, despite having been born in that most unnatural of all places, New York City. He was a nuclear engineer by education and by late 1969 was also a member of the Sierra Club's board of directors. After leaving the Transportation Department, he had taken a job at the National Academy of Engineering in Washington. He responded to Liebmann's overture by setting up a meeting at the Occidental Restaurant, to which he invited Harold Bergan, from Representative Sidney Yates's (D-Ill.) office, James Verdier, representing Congressman Henry Reuss (D-Wis.), Richard Wegman from Senator Proxmire's staff, Gar Kaganowich from the office of Senator Clifford Case (R-N.J.), Lloyd Tupling from the Sierra Club, and Gary Soucie and George Alderson from the Friends of the Earth. The group quickly concluded that the best way to oppose the SST was through an organization dedicated solely to that cause, and they formed the Coalition against the SST. Soucie agreed to be chairman, with Lloyd Tupling as vice chairman and George Alderson as secretary. They also agreed that the group needed a full-time coordinator, and after the meeting Moss and Alderson asked a recent Harvard Law School graduate named Joyce Teitz to take the job.[45]

Shurcliff's campaigning and Ken Greif's experiment had resulted in the formation of a lobby group dedicated to terminating the SST program. But during the same two years, the Federal Aviation Administration, under great pressure from Congress, had begun to talk of banning the boom. So had President Nixon.[46] If a ban on supersonic overflight were put into place, an anti-SST campaign based upon the boom alone would not succeed. Hence the coalition and its partners immediately sought other issues that could be used against the SST. Joyce Teitz did not have to look very hard to find them, of course. The economics problem had clouded the SST for more than a decade, and Shurcliff had not been shy in criticizing it. Furthermore, airport noise had become a major political problem for the FAA in the years since jets had first been introduced, and the SST was even noisier than the first-generation jets had been. Finally, she found that there was a very real alternative future for aviation already being built that promised both lower airfares and much lower noise levels. This permitted the coalition to draw a stark contrast between the noisy, expensive supersonic future and that represented by this new generation of "jumbo jets." This alternative future made it far easier for SST opponents to get their point across during the SST campaign. What the opponents did not yet know, however, was that the jumbos had made the Boeing Company's engineering task on the SST nearly impossible. Hence just as the anti-SST campaign was beginning, the SST's builder was rapidly losing interest in it.

JET NOISE AND JUMBOS

The jumbo jet that would give Boeing's SST program so much grief started within the Boeing Company's own advanced research group as a military cargo aircraft intended to help the Air Force fix a self-inflicted problem: its newest cargo jet, Lockheed's C-141 "Starlifter," was too narrow to carry many of the U.S. Army's vehicles. The project, titled CX-HLS for "experimental cargo, heavy lift system," was based on the existence of a new kind of very powerful, and much quieter, jet engine. Called a high bypass turbofan, the engine was useless for supersonic aircraft but would power comparatively huge subsonic aircraft without generating the earth-shaking roar the earlier J-57 and J-75 produced. As political pressure for airport noise reduction developed during the 1960s, the FAA was effectively forced to adopt a noise standard based on what these new engines could achieve. By 1970, it was clear to Boeing, General Electric, the National Academy of Sciences' SST committee, the Office of Science and Technology—everyone involved in the program, in fact, except the Transportation Department—that the SST's existing engine could never meet this standard.

Rolls-Royce had built the first turbofan engine during the late 1940s.[47] The primary difference between it and its turbojet predecessor was a slightly different airflow path. While all of the air pulled into the compressor of a turbojet engine went through the combustion chambers, was burned, and then exhausted through the turbine, in the turbofan engine part of the air bypassed the combustion chambers and turbine through a set of ducts. In terms of the engine's thermodynamic cycle, of course, this air was "wasted" because it did not contribute to sustaining the engine's combustion process, and wasting air was possible only because compressor efficiency, turbine temperatures, and turbine efficiency had increased markedly. But since the bypass air had also gained energy from the fan, it contributed to the engine's overall thrust.

This new kind of engine, referred to at first as a "bypass engine," had two salutary effects on aircraft. The first was improved propulsion efficiency. Theoretically, engineers knew, they could achieve maximum propulsion efficiency by matching an aircraft's designed cruising speed with the exhaust velocity of its engines.[48] Turbojet engines produced a supersonic exhaust, and they were therefore not particularly efficient at propelling subsonic aircraft. In order to move the cold "bypass" air, however, the turbofan engine's turbine had to extract a great deal more heat energy from the hot air passing through it than did the turbojet engine's turbine, resulting in a colder, slower exhaust. This resulted in a more efficient match between the engine and the airplane.

It also substantially reduced the engine's noise signature. During the early 1950s, at the NACA's Langley laboratory in Virginia, a small group of researchers had conducted experiments using a small compressed-air powered jet that were intended to generate an understanding of how jets produced noise and how that noise propagated through the environment. The Royal Aircraft Establishment in Britain funded nearly identical experiments at the Cranfield College of Aeronautics and at the University of Southampton. It also supported the efforts of a theoretician at the University of Manchester. These efforts proceeded entirely independently, but when the participants all met at a 1953 conference, they found that they had nearly identical results. The primary "cause" of jet engine noise was the impact of the high-speed exhaust on the stationary outside air. Hence reducing the engine's exhaust velocity was the simplest way of reducing its noise signature—which was exactly what the turbofan did.

The first U.S.-built turbofan engine, GE's CJ805-23, first ran in early 1958. Based on the civil version of GE's J-79 engine, this "aft fan engine" generated 40 percent more thrust than its ancestor while lowering specific fuel consumption by 20 percent.[49] In addition to these benefits, the engine promised substantial reductions in noise problems, and the airlines lost no time in forcing the turbofan on Boeing and Douglas. Boeing, in turn, strong-armed Pratt & Whitney into converting the J-57 engine into a turbofan.[50] By 1961, the turbojet had been supplanted by the turbofan on all new commercial jets.

This first generation of turbofan engines had bypass ratios between 1:1 and 2:1, reflecting the maximum amount of air the engines could afford to "waste" and still be self-sustaining.[51] Increasing the compressor and turbine efficiencies, increasing the engine's internal temperature, or any combination of these would result in yet more powerful, more efficient (and quieter) engines. When the Air Force started to pursue Boeing's heavy-lift aircraft idea in 1964, General Electric proposed powering the aircraft with a highly advanced turbofan engine with a 2,300°F maximum core temperature and a bypass ratio of 8:1. The Air Force's propulsion laboratory at Wright Field held a competition between Pratt & Whitney and GE to design and build these new higher-bypass-ratio engines. GE won in 1965, with the TF39.[52]

Boeing, despite having conceived the idea in the first place, did not win the three-way competition between itself, Lockheed, and Douglas for the aircraft, which the Air Force named the C-5A "Galaxy." Lockheed had underbid Boeing by $300 million, and as Bill Allen explained to General Electric's Jack Parker later, at an evening meeting at his home, he, Ed Wells, and T. Wilson had decided that while they could have reduced their bid somewhat, they had not wanted the contract at Lockheed's price.[53] They believed there were better uses for the company's resources.

The widebody, or "jumbo," jet was that better use. Pan Am and Boeing had been discussing a "stretched" version of the 707 to compete with Douglas's DC-8 stretch, which could have up to 259 seats. But Charles Lindbergh, then a member of Pan Am's board of directors, had suggested to Juan Trippe that Pan Am and Boeing consider a new airplane bigger than the stretched DC-8 and using one of the two high-bypass turbofans designed for the C-5.[54] Trippe decided he wanted a 400-seat airplane, and in December 1965 Allen and Trippe signed a letter of intent, with Pan Am to buy twenty-five of the as yet unspecified new giant, which would be the 747.

To build the 747, Boeing needed to build a new plant to build them in, and construction materials were allocated by the Defense Department, due to the Vietnam War. Allen and Trippe had to convince Lyndon Johnson and Robert McNamara to allow them to buy the steel the new plant would need, and Trippe eventually succeeded. But the 747 decision caused McNamara to raise an issue in his PAC during the year that the anti-SST movement would use effectively years later. The 747 was to cost Boeing about a billion dollars, and Pan Am had agreed to make "progress payments"—it would pay 50 percent of the order's cost before the first aircraft was delivered. Pan Am, in essence, was sharing a substantial part of the financial risk of the program, which neither it nor Boeing had been willing to do for the SST. This spoke volumes to McNamara, and to the later anti-SST activists like William Shurcliff, about what the two companies really thought about the SST's future.

Allen and Trippe, however, did not see the SST and 747 as competitors for the same market. John Swihart explained years later that Boeing had a dedicated corps of believers in the commercial cargo market's future, and they had convinced Allen that once the SST debuted and siphoned off all the passengers, proper design would allow the 747 to serve its remaining life as a cargo carrier. Hence when Boeing and Pan Am finally agreed to the new aircraft's approximate size and shape in early 1966, that shape favored efficient cargo carriage. After chief engineer Ed Wells rejected a double-deck layout, the two companies agreed to a fuselage shape created by drawing a circle around two side-by-side cargo containers (each 8 × 8 feet), which had the happy accident of translating into nine-abreast seating with two aisles. Further, instead of placing the flight deck in the nose, they put it atop the fuselage, allowing easy cargo loading through a hinged nose. The passenger interior was designed to be removed easily, so that when the SST arrived, Pan Am could redeploy the 747 in its cargo guise relatively cheaply.[55] They therefore perceived the 747, despite the massive investment it required, as an interim passenger aircraft whose real future was in cargo.

Boeing's public announcement of the 747 produced a wave of concern that spread far beyond McNamara's SST committee. The "jumbo," as it quickly came to be called, produced such an enormous reduction in direct operating costs (some 40 percent) that it threatened the Douglas Aircraft Company's existence. Who would want DC-8s when the 747 would cost less to operate and be much more comfortable? Further, Pan Am's decision to buy such a large number forced many other airlines to buy them. Because regulatory authorities set ticket prices, airlines could compete only on quality of service and levels of comfort. Even airline managers who thought the 747 was far too big for their route structures felt compelled to buy it to offset this advantage.[56] Pan Am's order thus set off a rush to buy the new Boeing giant.

The gaping "hole in the market" between the stretched DC-8 and the 747 caused American Airlines' chief engineer, Frank Kolk, to send a set of specifications for a smaller widebody aircraft to Boeing, Douglas, and Lockheed. This paper airplane, often called "Kolk's machine," would use two of the high-bypass turbofans to propel a 250-seat "airbus" for intercity travel within the United States. Lockheed picked up this "jumbo twin" idea first, but as John Newhouse recounts, was convinced by TWA that the aircraft should have full transcontinental range, which meant the addition of a third engine.[57] Douglas Aircraft, which had merged with McDonnell Corporation during 1966 due to its ongoing financial problems, also allowed itself to be convinced that a three-engine, transcontinental range widebody was what the airlines wanted to replace the DC-8. And when Lockheed lost the SST competition in December 1966, the company reassigned many of its SST people to its widebody project, the L-1011. Hence by 1967, the three major American airplane builders were building three highly efficient jets around the new high-bypass turbofan, of which two were essentially identical.

While the manufacturers were embarking on the "jumbo" revolution, the FAA was starting to respond to a political demand for relief from jet noise. The jet airliners introduced during the late 1950s had generated a great deal more noise than their propeller-driven predecessors, and as the jets gradually spread throughout the national route system, they subjected tens of millions more Americans to high levels of airport noise. The agency had resisted dealing with aircraft noise under Najeeb Halaby, despite a 1962 Supreme Court ruling that held airports financially liable for "taking" the property of people living under airport approaches.[58] Congress had not given the agency authority to regulate aircraft noise, and despite pressure from inside the agency, Halaby was not concerned enough about the issue to try to get Congress to give the agency that authority.[59] His successor, General McKee, was more willing to deal with the problem. Pressure from New York congressman Herbert Tenzer had also caused President Johnson's science advisor, Donald Hornig, to sponsor an aircraft noise panel in late October 1965.[60] As a result of his study, Hornig recommended a high-level federal program to deal with the noise problem.[61] He argued that the government should regulate the amount of noise aircraft could produce, in addition to the FAA's current project to design "noise abating" takeoff and landing procedures.

General McKee's support of Hornig's recommendations had caused President Johnson to add aircraft noise abatement to his Great Society program, and he notified Congress that he wanted FAA granted authority over noise in a special transportation message to Congress in early 1966. He also asked that Congress fund new NASA programs to research noise abating technologies.[62] McKee submitted a proposed bill late in April that would have given the FAA authority to regulate noise via its ability to certify aircraft as airworthy, but the bill languished during 1966. The following year, McKee submitted a revised version that would place this authority in the hands of the new secretary of transportation. Due to the creation of the Department of Transportation and the subsequent reorganization that had to take place, neither DOT nor FAA was able to promote the bill before late in the year. The bill also faced opposition from the Air Transport Association, representing the airlines. The ATA contended that, in the words of FAA historian Richard Kent, "controlling noise at the source was an illusory solution."[63] Instead, the ATA wanted the noise problem resolved by land-use restrictions that banned residential areas from around airports. Neither house of Congress found the ATA's argument compelling, however, and once a dispute over whether noise authority should be vested in the FAA administrator or the secretary of transportation was settled in favor of FAA, the bill passed handily. President Johnson signed it 21 July 1968.

McKee had plenty of information available to aid in establishing the first portion of his noise agenda, a standard applying to new subsonic jets. The agency had contracted with acoustics firm Bolt, Beranek, and Newman to design detailed measurement standards in 1966, including an analytical tool called a Noise Exposure Forecast, to predict the noise "footprint" of jet aircraft designs, and a revision of Karl Kryter's perceived noise level scale, which became known as the "effective perceived noise level in decibels," abbreviated "EPNdB." Members of the Noise Abatement Office that McKee had established in 1966 also kept in touch with research at NASA's Langley and Lewis laboratories, which had both increased their noise research efforts. And McKee had been blunt with the manufacturers, demanding maximum suppression efforts for their new jumbo jets. By late 1968, he found that NASA and the manufacturers had a rough consensus that the three jumbos could meet a standard of 108 EPNdB at the takeoff, landing, and sideline measurement points that had become traditional.[64] The agency released the proposed new rule for comment on 6 January 1969, and the regulation became permanent on 1 December as Part 36 of the Federal Aviation Regulations.

But the FAA and the Transportation Department still had to deal with that important new aircraft the regulation specifically excluded, Boeing's SST. GE's engine at this point generated an unsuppressed sideline noise of 129.5 EPNdB, while Boeing's contract with the government specified that the 2707 achieve a sideline noise of 124 EPNdB. Boeing's internal target was 118.5 EPNdB. All of these were vastly louder than the 108 EPNdB that the agency had just imposed on new subsonic jets. And while the 2707 appeared capable of meeting the community and approach noise limits, that was little comfort, for opposition to the FAA's deliberate omission of a supersonic standard emerged instantaneously. Both SST critics and airport operators perceived that omission of a standard meant that the SST would be noisier than the subsonic jets, and that was unacceptable to them.

The problem that DOT's SST Development Office faced was that General Electric, Boeing, and independent propulsion specialists were united in believing that the SST could not meet the sideline standard with the GE4 engine. In early April 1969, propulsion expert Newell Sanders, writing for the propulsion section of the National Academy of Science's SST-Sonic Boom Committee, was blunt about the possibility of bringing the 2707 into conformance with the jumbo standard, stating that the twenty-decibel reduction necessary to meet 108 EPNdB by 1978 "is highly improbable without dramatic advances in the science and technology of jet noise suppression."[65] Jack Kerrebrock, a gas turbine specialist at MIT, was no less clear in his own final report to John Dunning, the SST committee chair: his committee agreed that GE would have to double the engine's airflow, without increasing the engine's weight substantially, to meet the 108 EPNdB sideline standard.[66] In other words, the SST needed a new engine.

That came as no surprise within either Boeing or GE. Their engineers had already realized that even reaching the lesser goal of 118.5 EPNdB meant an essentially new engine, with an increased internal temperature, a slightly increased airflow, and a substantially reduced afterburner temperature.[67] But in addition to an engineering problem, they faced an institutional one. FAA's own propulsion branch refused to accept what the companies' engineers, like the National Academy of Sciences engineers, saw as a simple matter of fact. In August, for example, in his review of Boeing's noise control document, the SST program office's propulsion chief complained that the document was based on an assumption that "we already know we must have a different engine on the airplane."[68] His superior responded that the noise control document was unacceptable due to its dismissal of the prototype engine, setting the stage for a conflict between Boeing and GE on one hand and the Department of Transportation on the other over the need for a new engine.[69]

Three decades later, Boeing's Bob Withington recalled a meeting he had with the DOT's new SST program director, William M. Magruder, in May 1970 over the engine issue.[70] Late in 1969, the DOT had finally told Boeing and GE that they would have to meet a 112 EPNdB sideline standard with the production aircraft, throwing both companies into a state of shock. They could not do this without a new engine, and Withington had flown to Washington to tell Magruder that the existing GE engine was "absolutely unacceptable for airports." Withington recalled that Magruder exploded at him, telling him that there would be no new engine and that he was jeopardizing the entire program by raising the issue. Magruder called Bill Allen after Withington left to tell him the same thing.[71]

Withington also remembered that, after this meeting, he and many other members of Boeing's management no longer believed that the SST would ever go into production, and the company wanted to complete the prototypes primarily to validate its design and in particular the operation of its highly advanced "fly by wire" electronic control systems.[72] And the records also show that the schism between the two companies and the DOT widened during the year. For a key meeting on the engine issue in September, the SST Development Office prepared a proposed research plan to lower the existing engine's noise signature by the necessary 20 EPNdB through external suppression in the next eighteen months; at the meeting, Withington and GE's John Taylor told them that was impossible. More than ten years of suppressor research by both companies and by NASA had never come close to that level of noise reduction with an add-on suppressor before, and there was no basis for believing that a breakthrough was imminent. Both men told the DOT's staff that, while they accepted the new standard, they had no idea how to achieve it.[73] Into 1971, Boeing, GE, and DOT continued to argue over the need for a new engine.[74]

By the time the organized SST campaign began on Earth Day 1970, then, the aircraft noise problem had already driven a wedge between Boeing and DOT. During the ensuing year, the schism would grow as Boeing and GE struggled to convince DOT that they needed a new engine, while they conducted paper studies of what a new engine would look like (and cost). As 1970 wore on, it became clear that the new engine, even if it were approved, would make the airplane a financial disaster. Had the SST been a commercial effort, both companies would have dropped it. But the SST program had been launched for political purposes in 1963, and it still had a political role to play.

SIC TRANSIT SST

The American SST program lived for fourteen months after the formation of the Coalition against the SST. During that time, the coalition and its supporters engineered a media campaign to generate public antipathy to the SST, a grassroots campaign to produce highly visible protests against it, and a congressional lobbying campaign to turn members of Congress against it. Simultaneously, the Department of Transportation, and after December 1970, the Nixon White House, countered the environmental lobbying with a sales campaign of its own, while Boeing and GE stayed out of the media glare as much as possible. But unknown to the environmental protestors, the SST was already essentially dead. During the year, Boeing had told its DOT managers that its SST was not commercially viable and it would not be put into production. But this decision was never shared with the public or even with SST supporters within Congress. Instead, the administration chose to fight to save the SST prototype program to secure a symbolic victory of its own. Nixon's SST campaign was designed to rejuvenate ties to his business supporters, who did not appreciate his two-year experiment with environmental protection, and to win votes from aerospace workers as part of his "Keep California Green" plan.[75]

The first major act in the coalition's campaign against the SST was part of the first Earth Day, 22 April 1970.[76] A large number of local groups participated in anti-SST protests, teach-ins, and other events during the day-long festival. Senator Walter Mondale declaimed against the SST at the University of Minnesota's Earth Day activities, while in Boston activists held a funeral procession to Logan International Airport and then a mass "die in" to protest it.[77] Schoolchildren signed petitions against the SST. And the Environmental Handbook, written as an information resource for Earth Day activities, included a chapter on the SST that drew heavily from William Shurcliff's "fact sheets."

Earth Day's effect on the SST campaign was to increase public awareness of the SST program and its potential costs, both financial and environmental. It also served to recruit a subset of the youth population into the new environmental movement. Brent Blackwelder, who eventually became the head of the Friends of the Earth, recalled that the anti-SST events during Earth Day at the University of Maryland brought him into what became his life's work. He joined the Coalition against the SST as a volunteer lobbyist.[78] Ultimately, several dozen young people like Blackwelder served as the "footsoldiers" in what Mel Horwitch has called the "war" against the SST, commanded first by Joyce Teitz and later by George Alderson.[79]

The coalition and Senator Proxmire's staff had arranged a set of Senate hearings designed to capitalize on Earth Day for May 7, 11, and 12, and these furthered the anti-SST campaign's efforts to keep the SST's problems before the public eye. Held before Proxmire's Subcommittee on Economy in Government, the hearings proved extremely damaging to Bill Magruder's efforts to keep the SST program alive. The most harmful witnesses were physicist Richard Garwin, and Russell Train, head of the White House Council on Environmental Quality.

Garwin had been chairman of the Office of Science and Technology's ad hoc committee on the SST during Nixon's review of the program the previous year, and he had submitted a harshly negative assessment of the program. In the report he had argued that when an appropriate combination of technology and market demand had materialized, private financing would appear to support development of an SST. The noise and sonic boom problems, combined with the SST's comparatively poor payload fraction, meant that the technology for making an economically and environmentally acceptable SST did not yet exist. He had recommended cancellation of the development program.[80] Lee DuBridge, Nixon's science advisor, seconded this opinion.[81]

Garwin's report, however, was suppressed by the White House, which claimed that executive privilege protected it from public release. This act, and the DOT's continued propagation of what he believed were dishonest statements to the press and to Congress, enraged Garwin. He later wrote that he "felt that it was intolerable for the administration to subvert the democratic process by concealing and misrepresenting relevant facts and reports."[82] Garwin felt strongly enough about this that he had asked Lee DuBridge, Nixon's science advisor, if it would be ethical for him to speak out against the SST as a private citizen. DuBridge believed that it would, as long as he restricted his public statements to information already publicly available and thereby avoided revealing classified or otherwise restricted material, and Garwin passed this on to his committee members.[83]

In his testimony before Proxmire's committee, Garwin argued that the SST's technology would not permit it to be both environmentally acceptable and economical to its operators. Focusing on the noise issue, he contended that while the SST would meet the Part 36 subsonic standards for takeoff and approach noise, its sideline noise would be the equivalent of fifty 747s taking off simultaneously. There was, he believed, no available solution to this that would not cripple the aircraft's economics with substantially increased weight. This was especially problematic in view of the uncertainty in the aircraft's performance. The aircraft's payload fraction was only 7 percent, assuming the companies met all their performance targets. If they did not, he believed, the payload fraction could be as little as 2 percent, which would be highly uneconomical. With the weight of additional noise suppression, the SST would be unusable to airlines.[84]

Garwin's comparison of the SST to fifty 747s got immediate press attention, and Bill Magruder was hard pressed the next day to explain his way out of the mess. Appearing with James Beggs, the deputy secretary of transportation, Magruder claimed that the SST would be a little quieter than the 747 on approach and at the takeoff noise measurement points, but he was forced to admit that it would have a higher sideline noise signature. But Magruder disputed the dramatic "fifty jumbos" claim, arguing that because the decibel was a measure of sound pressure level, not of "noise," Garwin's example was misleading, but his rambling dialogue on the subject convinced no one.[85] His problem, as the head of his office's engineering division had told him that morning, was that Dr. Garwin's testimony was mathematically true and could be proven by anyone with access to the noise data and a high-school-level mathematics education.[86] It was therefore also easy for reporters to check, and the press circulated it very widely. Garwin's claim very quickly became Magruder's biggest headache.

Russell Train's testimony, delivered 12 May, was supposed to help Magruder's sales problem but probably did not. Speaking as the head of Nixon's Council on Environmental Quality, he publicly committed the administration to putting the SST into production only if the environmental problems had been satisfactorily resolved.[87] But he reinforced the critics' claims about the noise problem, stating that the SST's sideline noise would be four to five times louder than the 747's, and it did not appear possible that current technology could meet the 747 standard and still carry a viable payload. Nonetheless, Train told the committee, he had gotten the administration to endorse a proposal that the SST would not be certificated unless it would not degrade the "noise environment around airports."[88]

But Train also introduced a new environmental problem to the public discussion that had not yet appeared. He testified that the SST would place large amounts of water, carbon dioxide, nitrogen oxides, and particulates into the stratosphere, where they might alter the atmosphere's heat balance and cause ozone depletion. The increased humidity might also lead to greater cloud coverage. Any, or all, of these effects could have global impact, and he argued that no nation should introduce large numbers of SSTs until the magnitude of these problems was understood.[89] Gordon MacDonald, an atmospheric scientist who accompanied Train, told Proxmire that the reduction in ozone would cause increased levels of ultraviolet radiation to reach the ground, which might damage leafy plants.

The SST's budget went before the House in late May, and the House Appropriations Committee voted to continue funding the program. Anti-SST congressmen immediately offered an amendment to strip the just-authorized $290 million from the FY 1971 budget, and an intensive lobbying campaign by the Coalition against the SST nearly prevailed. The amendment lost by only seven votes, demonstrating to the activists that they had converted a significant number of congressmen to their side. The SST's funding request went next to the Senate, which ultimately did not take it up until November. Washington State's two senators, Warren Magnuson and Henry M. "Scoop" Jackson, engineered the long delay in order to mount their own campaign to save the program. The delay, however, also allowed new opponents to materialize.

The first of the new critics was one of FAA's major "customers," the Airport Operators Council International, a trade organization composed of the heads of many major airports. In July, the group announced that it was opposed to further funding of the SST unless the Congress required it to meet the FAA's subsonic-jet noise standard.[90] In August, a Massachusetts Institute of Technology study group published a report called the "Study of Critical Environmental Problems," which described the potential hazards in several technologies, including the SST. The study's authors believed that the Transportation Department's putative fleet of five hundred SSTs could generate enough particulate pollution of the stratosphere to cause significant climate change. Similarly, the group believed that water vapor in the stratosphere would increase 10 to 60 percent, but would probably not cause significant ozone depletion. Nonetheless, they concluded, "the projected SSTs can have a clearly measurable effect in a large region of the world and quite possibly on a global scale."[91] To the SST's environmental critics, this study was the smoking gun.

But many opposition leaders believe that the most politically damaging anti-SST arguments emerged in September.[92] In July Joyce Teitz had asked Senator J. William Fulbright to request statements on the SST from sixteen prestigious economists representing the full political spectrum; when they arrived, they were devastating. Only one economist, Henry Wallach of Yale, was willing to support the program. Opposing the Transportation Department's primary defense of the SST, that it was necessary to support the U.S. balance of payments, MIT's Paul Samuelson argued that what the United States needed was exports that could "pay their way—including paying their way in terms of providing their needed capital and the risk-taking inherent to sound projects—not contrived, subsidized additions to our balance of payments." Conservative economist Milton Friedman called the balance of payments argument a "complete red herring" in his response—although he also denied the validity of the environmental claims. Walter Heller of the University of Minnesota, who had been highly skeptical of the SST as chairman of President Kennedy's Council of Economic Advisors, had not improved his opinion of the program in the ensuing years. He did not believe that the SST would be able to compete with the 747, and in any case its funding should not come at the expense of critical social programs. And Milton Friedman's ideological opponent, John Kenneth Galbraith, agreed with him that the balance of payments calculations that DOT promoted were "strictly fraudulent." He considered the SST a frill that came at the expense of more important social needs. DOT's money would be better spent on speeding up baggage handling and customs clearance, he opined.[93]

The Transportation Department had a very difficult time refuting the economists. Magruder's aggressive defense of the project did not help matters. His charge that the economists that Fulbright had queried lacked comprehension of airline economics drew a deeply scornful editorial from the *New York Times,* for example, which pointed out that Paul Samuelson had just won the Nobel Prize for economics—and Magruder was a test pilot.[94] The Transportation Department gave Magruder problems, too, by producing a study that contended the SST might produce climate change after all. A preliminary draft of the report was leaked to Senator Proxmire, who put it in the *Congressional Record*, but it was disowned by DOT, which refused to formally release it until after the funding vote.[95] The anti-SST forces, of course, promptly accused the department of trying to suppress the truth.

Washington State's two senators thus faced a daunting task. It was clear after the election that they did not have enough solid votes to pass the SST's funding bill, and the two used their considerable personal power to try to reverse the tide before the December vote. A *Seattle Times* article detailed the two men's efforts. Senators Goldwater, Tower, Church, and Montoya had gone to Mexico for the inauguration of the new Mexican president, and Jackson and Magnuson convinced SST opponents Church and Montoya to stay in Mexico until after the vote while hurrying Goldwater and Tower, who supported the project, back to Washington. Magnuson blocked funds for expansion of Portland International Airport, prompting SST opponent Mark Hatfield, who had promoted the expansion, to skip the vote. He also shoved through the Senate a bill banning the sonic boom that he had been blocking for five years, hoping this would sway votes his direction. It passed 77 to 0 in two days. The two also tried to convince a number of anti-SST senators to skip the vote.[96] The coalition had expected that tactic, and George Alderson, who had replaced Joyce Teitz as coordinator in August, had filled the Capitol with volunteers detailed to shame the senators into staying.

The vote went badly for the two, who lost 52 to 41 to an odd coalition of environmental liberals and fiscal conservatives. Unsurprisingly, the mainstream press celebrated the vote while the aviation press decried it. The bitterest voice belonged to *Aviation Week*'s Robert Hotz, who called it a "vote against reason in which a wild anthology of environmental mythology was invoked to stifle a scientific exploration."[97] He also castigated President Nixon, and Boeing, for not lobbying to save the plane. Senator Jackson had called the president the day before the vote to ask his help in winning over some recalcitrant Republicans, but Nixon had not done so. Boeing, which had a reputation for poor lobbying skills anyway, had also not turned out to campaign for the program. Hence the environmental movement had won against rather less opposition than there might have been. But the issue was far from over, because the Senate and House versions of the appropriation bills still had to be reconciled.

The funding measure's defeat in the Senate set off a revealing exchange between the Boeing Company and the government that reveals Boeing's growing disenchantment with the program. Thornton "T." Wilson, in a telegram to Magruder, Magnuson, and Jackson, summarized the company's position on the SST. He stated that Boeing could still meet its current contractual obligations on the prototype, but that any reduction in the $290 million appropriation request by the conference committee would not permit the company to meet the contract schedule. Further, Boeing could not assume any further additional costs, including work on noise reduction, unless the government fully reimbursed them—i.e., no more cost sharing. And he questioned whether the program should continue, given the growing emphasis on noise. Wilson went on to state that the company believed that the government would have to participate in production financing, and to make things yet worse, he argued, he expected the airlines to need government financing to buy the SST as well.[98]

Wilson's highly negative assessment of the program and its financial needs reflected Boeing's overall financial troubles, brought on by a virtual collapse of airline orders as the U.S. economy entered the Vietnam recession and by the huge costs of the 747 program, which was approaching $2 billion. But his comments also point to the company's growing unhappiness with the way Bill Magruder's SST office was handling the program. In addition to its continuing refusal to acknowledge the need for a new engine, Magruder was still telling Congress that SST production would take place without government help, and he continued doing so until the very last SST vote in May 1971. This disagreement between DOT's position and Boeing's earned Bill Allen a telephone call from Senator Jackson four days after Wilson dispatched his telegram. Jackson had called Magruder after receiving the telegram to ask about the production financing issue and Magruder had again denied the need for it, so the senator had called Allen to find out why Boeing's position had changed. Allen told him that Boeing had always held the position that production assistance in some form, preferably loan guarantees or government bonds, was going to be necessary and had submitted a financing plan to DOT stating this, as was required in its contract with the government.[99] In a revealing statement, Allen told Jackson that he should call General Maxwell, not Bill Magruder, to confirm that Boeing had submitted such a report. And the senator told Allen that he was very concerned that this production financing issue had come up now. Both he and Magnuson had pushed the program on the Hill based on what DOT had told them—government support would end when the prototypes began flying. If the telegram leaked out, it would imply that the two senators and the Transportation Department were hiding the truth from Congress and the public, in order to sell the program.[100]

That was exactly what the Coalition against the SST had been telling the public through its news releases, fact sheets, and protests, for most of the past year, and Jackson was wise to be worried about the public's potential reaction. But the Transportation Department's dishonesty about the program's true cost was also deepening the rift between itself and both Boeing and GE that the engine noise problem had opened. After the House and Senate conference committee had restored a slightly reduced $210 million SST request to the

Transportation budget, T. Wilson sent a letter to Magruder insisting that the government announce that production financing would be required, and he followed it up with a similar letter to Secretary of Transportation John Volpe.[101] General Electric chimed in with a letter arguing that the government should admit to, and reorient the program toward, the need for and cost of a new engine.[102] Both companies were growing disgusted with the DOT's continued refusal to admit to what they saw as the program's fundamental realities.

As the contractors and DOT argued over the program's future, the SST's congressional supporters and enemies engaged in arcane parliamentary procedures in an attempt to gain an advantage. After the House/Senate conference committee, dominated by SST supporters, had restored the SST's funding, Senator Proxmire announced a filibuster against the Transportation Department's funding bill, knowing that he had more than enough votes to block a cloture attempt. After two attempts by the SST's supporters to shut down debate failed, the Senate's leadership agreed to a deal. The two houses of Congress would pass an interim funding measure to keep the SST program alive until the end of March, allowing time for more hearings and a final, up-or-down vote on the SST appropriation—not on the entire Transportation budget. On 2 January 1971, as part of the agreement with Senator Proxmire, the House and Senate leadership publicly committed themselves to the plan and passed a resolution continuing the program's funding to 31 March 1971.[103]

This arrangement was exactly what Proxmire had wanted, because it forced his colleagues to go on record about the SST in the absence of other popular items, like Coast Guard pay raises, that the larger Transportation budget included. The program would have to succeed or perish on its own merits, notwithstanding the considerable political skills of Jackson and Magnuson. And on the surface, the program's fortunes seemed to improve in February, when a favorable report on the SST's noise problem was released. The previous year, Congress had demanded that DOT commission an independent committee to look into the noise issue, and DOT had asked Leo Beranek to head it. The previous September, the group had told Magruder that the 112 EPNdB goal that the companies had been working toward was still too high and they had recommended requiring the 108 EPNdB standard prompted by the 747, but they had also told him that they did not believe this would be possible.[104] In late January, however, the group had been invited back to Seattle to hear briefings on a potential noise solution. On 4 February GE's representatives had presented a substantially new engine.[105] The engine, as the National Academy of Science's propulsion committee had recommended more than a year before, would be larger in diameter and have a higher compression ratio to increase the engine's airflow to about 870 pounds per second, about a third higher than the prototype's engine. The engine would also have a higher internal temperature. This would allow GE to delete the afterburner, which was the source of much of the sideline noise, and use a retractable multitube noise suppressor that had been tested by NASA recently at the Lewis laboratory. This "quiet engine," Beranek believed, would allow the aircraft to meet the 108 EPNdB subsonic standard by the time the production aircraft was scheduled to arrive in 1978.

Mel Horwitch has argued that the new engine proposal defanged the coalition's best environmental argument against the SST, but in fact it opened a host of new problems in addition to the issue of who was to pay for it. The new engine was larger and heavier than its predecessor, and because it was heavier the SST's wing structure had to be made heavier as well. The need to reduce noise had also caused Bob Withington to select a new larger, thinner wing, which proved to have a flutter problem. Fixing that had further increased the aircraft's weight. *Aviation Week* reported, and testimony from Beranek and Bill Magruder confirmed, that the production SST's weight would grow to somewhere between 800,000 and 850,000 pounds in order to support the increased structural weight and still maintain an "intercontinental" range.[106] They also admitted, however, that even at this weight the aircraft's range would be, at best, 3,300 n.m., barely New York to London, and not even close to the 4,000 n.m. required for what the industry considered true intercontinental range, New York to Rome.[107] This meant that the SST would weigh some 100,000 pounds more than a 747, carry one-third fewer passengers, and do it over a substantially shorter distance, while costing the airlines twice as much to buy. It was not an attractive airplane to airline managers or to its own contractors.

Boeing's program manager, Bob Withington, recalls telling DOT advisor Raymond Bisplinghoff that at this point Boeing no longer had any intention of putting the SST into production, a memory seconded by then–deputy secretary James M. Beggs.[108] They both recall that Boeing and DOT agreed to complete the SST prototypes to validate their design methodologies and in particular the prototype's novel electronic flight control system, which had potential for future aircraft. But DOT had justified national investment in the SST by arguing that sales of the production version would repay the investment, and thus it was trapped by its own rhetoric. It could not admit that the SST would not be manufactured, as they would be admitting the program's critics had been correct all along. There would be no return on the public's investment. Hence for the remainder of the conflict, DOT's leaders continued to insist that the SST could be manufactured profitably and would be sold in large numbers.

Moreover, President Nixon, who had refused to fight for the SST in December, had decided in the interim to begin a campaign of active antienvironmentalism.[109] While no surviving document confirms that Nixon chose the SST as his own first "symbolic campaign," the timing strongly suggests that it was. In his recent study of the Nixon administration's environmental record, historian J. Brooks Flippen has argued that after receiving no credit for two years of environmental initiatives, Nixon had realized that he could not win the votes of environmentalists and instead sought to re-embrace his party's business base.[110] In part, this was in response to his party's loss of nine seats in the House and several governorships in the midterm election, but Nixon also faced a revolt within his party. The governor of California, Ronald Reagan, a strongly pro-business conservative, intended to run against Nixon in the 1972 primaries. A brilliant politician whatever his faults, Nixon understood that federal environmental activism was highly unpopular in the South and West and among the business elite, and so did Reagan. Heading off Reagan's challenge meant reorienting his own administration to undercut Reagan's critique, and the SST was an easy place to start.

Nixon's officials also believed that the anti-SST campaign was part of a larger Democratic attempt to defund science and technology, including aerospace, and divert the money to a "broader voting base" via social welfare programs. Writing to NASA's George Low in early February 1971, Magruder argued that aerospace had not made a case that it represented a large enough voter segment to sustain political support, and worse, it had allowed itself to be separated into "product-oriented camps," each fighting for the same "aerospace dollar." He painted a picture of economic doom should the United States become a net buyer of aerospace products—a massive reduction in gross national product, a brain drain, and "regression from 'industrial' status." To Magruder, defense of the SST was just one fight in promoting the importance of aerospace to the national economy. He urged unity among space and SST advocates, deployment of arguments about the economic benefits of aerospace "spinoffs," and trumpeting the catastrophic job losses he expected to follow the loss of "aerospace leadership."[111]

Finally, during the two years of the SST conflict, the aerospace portion of the economy was in very poor shape, and the SST program was one way the administration could show support for this sector (and its voters). Nixon's decision to terminate NASA's Apollo program, his "Vietnamization" initiative, and an economy-wide recession had triggered mass layoffs in the aerospace sector, with employment dropping from 1.4 million in 1967 to 900,000 in mid-1971, as the anti-SST campaign concluded.[112] It was in this context that Nixon chose to go ahead with the controversial Space Shuttle, which his own staff had rejected repeatedly on economic grounds, and it no doubt influenced his decision to fight for the SST prototypes.

With the Nixon administration finally committed to defending the prototype program, Boeing also began a promotional campaign for its SST. In addition to its technological interests, Boeing's campaign reflected a desire to mitigate problems with its labor unions. Boeing had laid off more than 60,000 people during the previous eighteen months in a desperate attempt to stave off bankruptcy during the deep industry recession, and termination of the SST meant another 8,000 jobs would vanish.[113] Boeing thus contributed $100,000 to a group called "The National Committee for an American SST," which was formed by Donald J. Straight, vice president of SST subcontractor Fairchild-Hiller, and Floyd Smith, president of the International Association of Machinists and Aerospace Workers, the union representing the vast majority of Boeing's employees.[114] The organization took out full-page ads in major newspapers beginning February 23, relying heavily on the cold war imagery of defeat by the Soviet SST and recalling the specter of Sputnik. The ads argued that since the TU-144 and Concorde were already flying, SSTs were inevitable and thus the environmental issues were moot. The United States had to build one or be "left behind."

In the March 1971 hearings that preceded the SST funding votes, Secretary of Transportation John Volpe and Bill Magruder deployed Magruder's expansive rhetoric of economic doom if the SST prototypes were not built. Playing on the deep industry recession, they told the House Appropriations Committee that failure to build the SST would cost 500,000 more jobs, because airlines would stop buying U.S. subsonic aircraft too. Airlines wanted to buy "families" of aircraft from a single manufacturer, and without an SST, Boeing (he ignored McDonnell-Douglas) would not have a full "family." Instead, airlines would buy Soviet aircraft, which did come in a "Red" family, or British aircraft (which, he failed to note, did not).[115] A vote against the SST prototype was thus a vote to dismantle the entire domestic aircraft industry.

Henry Reuss and Sidney Yates in the House and Senators Proxmire, Case, and Percy in the Senate grilled Magruder, Volpe, and the other pro-SST witnesses who accompanied them on the finance issue, and brought out a parade of economists to further their assault. The most scathing testimony came from MIT's Paul Samuelson, who told the Senate that the Securities and Exchange Commission would not accept the kind of briefings being given by the SST enthusiasts because they treated possibilities as probabilities (a comment he also used against some of the environmental claims). Proponents made many such claims: the SST production would return the taxpayer's investment in the prototype; private financing would appear; the prototype investment would provide hundreds of thousands of jobs; Concorde and the Soviet TU-144 would be big sellers, and so forth. In investment circles, such claims were not legally permissible. These things might happen, he told the assembled senators, but they also might not.

Samuelson then attacked the promoters' claims about Concorde, which he reminded the group was being used to justify a defensive investment in the Boeing 2707. He related that he had been privileged over the last six years to dine with MIT's aeronautical engineers at the school's cafeteria, and that they universally regarded Concorde as "the biggest lemon that was ever devised."[116] He suggested that Concorde would be very lucky to sell 75, not the proclaimed 240, and he himself believed that 6 to 8 was the likely market—but only if the airlines who bought them miscalculated.[117] Its operating costs were too high to compete with the jumbo jets, and its range was so marginal that it was capable of only a tiny number of routes. Karl Ruppenthal, a transportation specialist at the University of British Columbia who accompanied Samuelson in the hearings, summed it up rather bluntly: "There is little prestige in a white elephant."[118] There was also little threat in a white elephant, but the DOT's entire campaign was based upon the proclaimed inevitability that Concorde and "Concordski" would displace all of the U.S.'s jetliners. None of the economists who testified before either the House or the Senate believed in the supersonic threat, and a revelation by British Overseas Airways Corporation (BOAC) substantially helped their case. BOAC had told the British government in February that it could not operate Concorde economically.[119] BOAC and Air France had calculated that its direct operating costs were twice that of the 747 and the aircraft would lose money even with a fare surcharge of 20 to 30 percent. If Concorde was such a dog, the economists all felt, the U.S. SST was unnecessary. The jumbos would kill the foreign SSTs anyway and cost the taxpayer nothing.

The House of Representatives, where the SST had never lost before, voted the SST prototypes down by a surprisingly large margin, 215 to 204, two days after the hearings. Thus to win the Senate vote, Nixon, Magnuson, and Jackson faced an uphill battle. The friendly Senate Appropriations Committee restored funding on 19 March, and the Senate leadership scheduled Proxmire's up-or-down vote for the 24th. In the meantime, the White House lobbied hard to save the program, doing itself perhaps more harm than good. White House aides pressured talk-show host Dick Cavett into canceling Senator Proxmire out of a planned live debate with Bill Magruder, for example, so that Magruder could appear alone. Cavett launched into a diatribe against the SST anyway, drawing applause from the audience. Nixon also tried to pressure Senator Margaret Chase Smith, a fellow Republican from Maine, with a promise not to close Portsmouth Naval Shipyard in nearby New Hampshire, but the fiercely independent lady senator published the offer and then voted against the SST anyway. Senator Mark Hatfield, whose airport expansion Magnuson had threatened in the earlier Senate vote, cast further doubt on the wisdom of the "hard sell," announcing that his mail had run heavily against the SST, and many of the letters specifically mentioned outrage against the pro-SST advertising campaign.[120] By the evening of the 23rd, after days of arm twisting, Washington State's two senators knew they had lost, and Magnuson released several of his colleagues from promises to vote in his favor. On the 24th, accordingly, the Senate voted the SST down 51 to 46.

At his victory press conference, Senator Proxmire was asked if he thought the SST was finally dead, and he replied "nothing is ever really dead around here."[121] And he was right. Bill Magruder tried for a few days to recruit private financing to finish the prototypes, but received the response that any independent observer would have expected by now: no. There had never been any serious interest in the SST within financial circles, and the situation was if anything much worse now that the SST was a major public relations liability to whoever adopted it. The White House attempted a restart of the SST program in May, via linking the $86 million that the government owed Boeing to a bill providing a loan guarantee to Lockheed, which was teetering on the brink of bankruptcy. This did not succeed either. After a press conference at which Boeing's Bill Allen estimated the cost of finishing the SST prototypes at over $1 billion, the restart movement foundered.[122]

✈ ✈ ✈

After the March vote against the SST, Nixon characterized its defeat as "the number one technological tragedy of our time."[123] *Aviation Week*'s editor agreed, writing that the SST's demise represented "a fundamental watershed in government that portends a national skid into the hopeless poverty of a technological Appalachia."[124] Nixon and Hotz, like many others in aerospace, saw the environmental campaign against the SST as a campaign against technology in general, not as a campaign against a specific example. To them, the SST was the future. They could not conceive a future in which speed was less important than other considerations, and Nixon soon had other plans to continue working on SST technologies. But partly because the White House took Boeing's action in not supporting the restart attempt as disloyalty, the Boeing SST was truly dead.[125]

Boeing's share of the $86 million the government owed the contractors was about $36 million, which T. Wilson reportedly described as "manna from heaven" to the struggling company.[126] Company legend holds that the reimbursement allowed Boeing to forestall a loan call by Chase Manhattan Bank that would have forced it into bankruptcy. SST cancellation hastened Boeing's, and thus Seattle's recovery. And it ended a public relations catastrophe. The "risk payments" that the airlines had made to Boeing were also eventually returned, having already gained the airlines a significant return in free advertising. Cancellation nonetheless made many people in the company bitter, a fact made very clear to William Shurcliff in a letter by a Boeing executive.[127] Dreams die hard, especially when their failure cost friends' jobs.

The campaign against the SST was successful at winning cancellation due to an unusual confluence of events. First, the SST's own contractor chose not to lobby for it, clearly believing that while the prototypes had technological value, they did not represent a future source of profit for the company. Given the closeness of the Senate vote, strong support from Boeing senior management might well have changed the outcome. Second, the usual process of "logrolling" in Congress—the construction of "iron triangles" of lobbyists, contractors, and congressmen—had not worked, perhaps because the SST contract was simply too small to spread enough "pork" to enough districts. The real money in aerospace contracting flows from large production contracts, not from research and development. As Nick Kotz makes clear in his history of the B-1 Bomber program, it was the expectation of $28 billion in production funding that permitted the aircraft's supporters to keep it alive through several different administrations' attempts to cancel it.[128] By 1971, the Boeing SST contract had only about $200 million left. This may simply have been too little money to satisfactorily grease Congress's bearings. Thirdly, the Coalition against the SST had been very successful at depicting the SST as both an environmental villain and a free-market violator. Using both arguments, they were able to politicize the SST in a way that appealed to both traditional conservatives and the New Left. And finally, the SST campaign occurred at a moment in history when it was particularly easy to convince the nation's citizens that their government was deceiving them. Both the Johnson and Nixon administrations had been caught lying about the costs of and progress in the Vietnam War, creating a deep public cynicism about the quality and honesty of national leadership.

The anti-SST campaign had impacts far beyond the aerospace industry. In October 1972, Congress passed a bill inspired by the anti-SST campaign that established a congressional "Office of Technology Assessment." This agency's purpose was to investigate the potential effects of proposed new technologies before they received public support; in essence, the OTA's function was to provide a source of information about new technologies to Congress. Its mandate reflected a political sense that prospective analysis of technological change might be useful in decision making. More important, perhaps, was that OTA would be a source of information independent of the executive branch.[129] Congressional leaders, even those who had supported the administration's position on the SST, believed they had been deliberately mislead, and only an information source independent of the executive branch could prevent this in the future. The anti-SST campaign, repackaged by SST supporters as an "anti-technology campaign," also became a rallying point for the corporate political mobilization that took place during the 1970s and that financed the Republican Party's resurgence late in the decade.[130] Richard Nixon's strategic decision to abandon his experiment with environmentalism after the 1970 midterm election was a brilliant one in retrospect, clearly aligning his party with the extractive, energy, chemical, automotive, and aerospace sectors of the economy while abandoning the GOP's old support for the far less wealthy conservationists.[131]

Finally, the campaign against the SST was also a moment of transition for the Sierra Club and other "old line" wilderness-oriented organizations like it. During the campaign, their leaders came to recognize the impact that quintessentially urban technologies like the SST could have on the ex-urban world, and after it, they began to tackle other areas of technology. Thomas Wellock has discussed the conversion of the Sierra Club to opposition to all nuclear power after the anti-SST campaign, and while histories of its other post-SST technologically oriented efforts have yet to be written, there are obvious examples: its opposition to dams in general and its support for renewable energy technologies and zero-emission automobiles.[132] These efforts represented not an "antitechnology crusade" as conservative critics claimed, but efforts to redirect the course of technological change in ways the organization's leaders believe to be more appropriate. Beyond the Sierra Club, a "sustainable technology movement" spread in the 1970s, while university-based study of environmentally friendly technologies and the potential to infuse environmental values into capitalist economics also gradually grew.

In the decades since the SST's demise, its supporters have nourished a legend that Congress's termination of the prototype program had been the right decision made for the wrong reasons—the "wrong reasons" being environmental ones. But the environmental activists have always believed that they won the fight with economic, not environmental, arguments. Responding to an earlier draft of this chapter, Boeing's Bob Withington also argued that the SST's own poor economics doomed it, emphasizing the company's decision that whatever happened to the prototypes, the SST would never go into production. The record, and subsequent events, bears this out. From the SST program's very beginning in 1963, the fact that the vehicle had to be publicly financed had drawn widespread skepticism about its commercial viability from business and financial specialists, and the Coalition against the SST had put that skepticism to good use. The coalition had deftly deployed traditional American economic mores against the SST, reminding the public that Americans did not permit their government to develop commercial products—precisely the argument President Eisenhower had leveled against the SST in 1959, and one that had bothered President Kennedy in 1963. American economic ideology is rooted in celebration of free enterprise, and the coalition's depiction of the SST project as free market violator found resonance in the public and political arenas. The American SST died of the "right" causes—rejection of its poor economics. The environmental groups' contribution to the SST's demise thus lay in publicizing, and politicizing, the vehicle's liabilities.

Yet memory has power. The belief that the "wrong" arguments had won lingered through the end of the century, and it fundamentally altered the agendas of future American SST research programs. Since the supersonic future still captivates the imagination of aerospace engineers, NASA research on supersonics adopted the environmental problems as the most fundamental issues to be addressed. Future SSTs would have to satisfy environmentalists. Quieter and less polluting engine design became research priorities, as did stratospheric ozone research and lightweight materials and structures. Hence David Brower's symbolic campaign against the SST, while causing widespread bitterness among enthusiasts, resulted in far more than the death of one airplane. It redirected the course of aeronautical research and forced the industry to redefine "progress," at least in terms of commercial aviation.

OF OZONE, THE CONCORDE, AND SSTs

A few days after the May 1971 SST restart movement ran aground, President Nixon wrote to his principal domestic policy assistant, John Ehrlichman, about other avenues that the United States could pursue toward a new SST development program.[1] Nixon was concerned about the threat posed to American leadership in aeronautical technologies by the Concorde and the TU-144, which had not suffered the same ignominious fate as Boeing's 2707—at least not yet. These two SSTs were still moving toward the airline market despite the anti-SST movement's victory, and just as they had been used by Najeeb Halaby to justify the national SST program in 1963, they would be used throughout the 1970s to justify repeated attempts to launch a new American SST effort.

Ehrlichman responded by having Bill Magruder investigate the potential for private financing of a new SST effort.[2] But Magruder found, just as Eugene Black and Stanley Osborne had in 1963, that investors considered the SST's combination of high technical risk and extremely high capital requirements an unsound investment. Further, the airline industry was deep in one of its periodic downturns—Pan Am had begun hemorrhaging in 1969—and an end to the red ink was not yet in sight. Finally, no sane investor could overlook the obvious public relations problems with being publicly linked to an environmental monstrosity. Mighty General Electric had, after all, been publicly humiliated on Earth Day by a group of University of Minnesota student environmentalists, who managed to force early dissolution of its annual shareholders' meeting in Minneapolis.[3] Unsurprisingly, private investors rejected Magruder's overtures.

On 27 July, Magruder briefed President Nixon on a plan to keep SST research going temporarily.[4] Part of a "New Technology Initiative," Magruder's plan added $20 million to the National Aeronautics and Space Administration's FY 1973 budget for research aimed at resolving the SST's environmental problems.[5] The president apparently agreed to the plan, but in the process of devising the overall budget, the Office of Management and Budget reduced the amount to $11 million.[6] Despite this setback, however, Magruder and John Ehrlichman remained committed to a new SST program. Hence Magruder arranged with Roy P. Jackson, head of NASA's Office of Aeronautics and Space Technology, that the agency begin planning a research program aimed at "technological readiness" for an advanced SST.[7]

Responsibility for designing the program was assigned to William S. Aiken, Jr. He formed an "Advanced Supersonic Transport program (AST)" steering group in late January 1972 to begin planning the new effort. Their stated goal was "to provide the supersonic technology base to permit the U.S. to keep open the option to proceed with the development of an advanced supersonic transport if and when it is determined that it is in the national interest."[8] While the FAA's former program had been explicitly oriented at development and construction of an airplane, this program was what NASA referred to as a "focused technology program," aimed at developing key components upon which an advanced SST could be based—preferably by someone else. The program was to cost a total of $288.5 million through 1978, with $28 million budgeted for FY 1974.[9]

Aiken's group deliberately directed the program's efforts at the 2707's environmental problems. The program's centerpiece was a new kind of engine that propulsion engineers hoped would overcome the noise and weight problems that had rendered Boeing's SST uneconomical. The AST program also included efforts to reduce engine pollutant levels, to reduce airframe weight through the use of new materials and computer-aided design, and to improve aerodynamics, all of which had environmental benefits. Further, NASA was part of a Transportation Department effort to understand the climatic impact of commercial aviation, especially impacts related to supersonic transports. NASA's engineers, and the SST's White House promoters, hoped that this careful attention to environmental problems would permit congressional supporters to deflect the eco-criticism that would inevitably erupt when the White House submitted the plan.

And criticism did indeed erupt when the *Wall Street Journal* exposed it in a rather mocking November 1972 front-page piece.[10] But ongoing anti-SST activism had less to do with the collapse of White House restart hopes than did Concorde. On 31 January 1973, after three weeks of difficult negotiations, Pan Am announced that it was dropping its options on Concorde.[11] Since Pan Am was effectively bankrupt after four years of red ink and was under the control of its creditors, it could not justify a purchase of six Concordes at a cost approaching a billion dollars. Indeed, Pan Am refused a desperate offer by British Aircraft Corporation of a no-charge lease. At this point, the Concorde's direct operating costs were projected to be between two and three times that of the 747's, and it was capable of flying only a small number of routes. Further, Pan Am believed that Concordes would siphon highly profitable first class passengers off its new fleet of 747s, reducing their profitability. None of this made sense to the struggling airline. Even free Concordes appeared a bad deal to Pan Am.

Over the next several months, every other airline that held options on Concorde cancelled as well, save British Airways and Air France. As state-owned airlines, they had little choice but to accept them, although British Airways made clear that it did not want Concorde and could not operate it profitably without a subsidy. Just as Concorde's order book had been filled by Pan Am's initial "order" in 1963, Pan Am's withdrawal emptied it.

SST enthusiasts had long used the existence of the Anglo-French project as prima facie evidence of an imminent threat to American aeronautical leadership, of course, and SST opponents were beside themselves with glee as Concorde's order book evaporated. Concorde's descent validated their own arguments and, in their eyes, Congress's decision to terminate the prototype construction effort.[12] What *Business Week* referred to as "Concorde's bleak arithmetic" destroyed the aura of inevitability that promoters had so carefully crafted around the supersonic future.[13] Despite a major effort by the White House to gain full funding of the Advanced Supersonic Transport program, both House and Senate appropriations committees chose to accede to subcommittee reports recommending that the AST effort be funded at levels "not to exceed $11.7 million" for the indefinite future. Less than half of what the White House and NASA had wanted for FY 1974, let alone following years, this was all Congress was willing to part with for supersonics. With Concorde so obviously a non-threat, the administration's attempt to generate a new national SST development program failed.

SST supporters continued to play the Concorde card during the Nixon years to generate enthusiasm for, at a minimum, an enlarged research program, but they met with no success.[14] NASA headquarters had assembled the AST program at White House request, but Nixon's chief advocate for the SST, John Ehrlichman, resigned on 30 April 1973 due to the gradually widening Watergate scandal, and no more White House effort followed his departure. NASA's administrator, James Fletcher, was a space evangelist who had little interest in aeronautics, a fact that caused one congressman to berate the agency for its disgraceful attitude toward aeronautical problems.[15] Congress in general agreed that NASA did not propose enough money for aeronautical research, and it repeatedly added money to the agency's aeronautics budget during the decade—but mostly for subsonic research aimed at noise abatement and fuel efficiency, not supersonic research.[16] Fletcher accepted the cut to his FY 1974 AST budget, therefore, and after the October 1973 Arab oil embargo began, he decided not to pursue an increase the following year either. There was little point in expending political capital on supersonic research when his own interest lay in advancing the Shuttle program. And, as George Cherry explained under questioning by one congressional SST enthusiast, SST supporters in headquarters had been unable to invent a justification for expanding effort on a technology that used three times as much fuel as subsonic jets in the middle of a fuel crisis.[17] Facing a lack of interest in both Congress and in NASA's front office, promoters of the supersonic future had to settle for a tiny budget and hopes that some future election would dramatically alter the political landscape.

The program's small size did not prevent it from making substantial technological advancements in supersonic cruise technologies during the decade, however. Renamed the "Supersonic Cruise Aircraft Research" program, or SCAR, after the agency's failure to win approval of the full AST plan, this and a related effort, the Variable Cycle Engine program (VCE), generated concepts that suggested a second-generation SST might be feasible sometime in the late 1980s.[18] In a separate set of research programs, NASA demonstrated that an advanced SST might be made ozone-neutral, eliminating the threat to the ozone layer. These technical and scientific advancements led the McDonnell-Douglas Corporation, the most enthusiastic SST promoter during the 1970s, to begin lobbying in 1978 for a greatly expanded research program to validate the new technologies in time for a late 1980s AST development program. But technological progress did not translate into political progress.

Perhaps the most substantial alteration in NASA's research agenda during the 1970s was the agency's construction and institutionalization of a stratospheric research program. The immediate "cause" of what became the Upper Atmosphere Research Program at NASA was an intense controversy over whether SSTs would destroy Earth's thin layer of high-altitude ozone. Beginning in 1970, this controversy sparked a "golden age" of stratospheric research, with substantial federal funds from several agencies transforming what had once been a tiny "cottage science" into a modern, international, large-scale, high-technology scientific enterprise.[19] Because NASA was in need of new problems to solve as the Apollo program wound down, because the stratosphere was, after all, on the edge of space, and because its research centers had developed experimental remote sensing technologies that could help understand the ozone problem, the space agency sought to place itself at the center of this new, very Earth-oriented, research effort.

The existence of Earth's ozone layer was well known within the scientific community, if not among the general public, by the time the SST controversy began. In 1881, Irish chemist W. N. Hartley had hypothesized that high concentrations of ozone existed at high altitudes and served to block ultraviolet radiation, and in 1913 French physicist Charles Fabry proved Hartley correct. In 1926, English physicist Gordon Dobson had designed an optical instrument to precisely measure these high-altitude ozone concentrations, and he established a small but worldwide network of observing stations during the ensuing decade. Finally, in 1931, geophysicist Sydney Chapman had established the mechanism by which ozone was created in the stratosphere.[20] He believed that oxygen molecules, which consist of two oxygen atoms bonded together, would be blasted apart by solar radiation. The resulting highly reactive oxygen atoms would then bond to other oxygen molecules, forming ozone. By 1940, therefore, the small number of scientists interested in the upper atmosphere, who called themselves *aeronomers,* believed they had a basic understanding of the ozone layer.

During the 1960s, stratospheric research received a boost from the Atomic Energy Commission (AEC). The AEC's above-ground nuclear weapons testing injected vast amounts of detritus into the stratosphere, and the agency sought understanding of how this affected the stratosphere. Further, the testing regime had produced an interesting conundrum. The weapons tests, AEC scientists thought, should have been producing a slight, but measurable, decrease in stratospheric ozone concentration, but data from the ground-based Dobson ozone monitoring network suggested that ozone concentration was actually increasing. The increase stopped in 1968 and concentrations stabilized for a few years, but this seemingly strange episode indicated that the stratosphere was a great deal more complex than aeronomers had thought.

The growing public concern with environmental issues caused MIT to host a major study of human environmental impact in July 1970. Released in late 1970, entitled "Man's Impact on the Global Environment: Report of the Study of Critical Environmental Problems," and mercifully abbreviated SCEP, the report contained the first major statement on the stratosphere's health and on the SST's probable impact.[21] A panel on climatic effects, chaired by William Kellogg of the National Center for Atmospheric Research in Colorado, looked at how water vapor from stratospheric aircraft might alter Earth's climate. Water vapor was the second largest combustion product of jet engines after carbon dioxide, and their primary concern was that water vapor from the engine exhaust would cause climate change. The group concluded that stratospheric water vapor concentration would increase 10 to 60 percent in the event a large SST fleet materialized, but that this would not appreciably change surface temperatures. And they recommended that a permanent stratospheric monitoring program be initiated to assess the SST's true impact. But they specifically rejected the notion that either water vapor or nitrogen oxides, the third most significant jet engine combustion product, would cause climate change or ozone depletion.[22] An article by a Boeing scientist accidentally undermined their argument, however.

Almost simultaneously, a scientist at Boeing Scientific Laboratories, Halstead Harrison, responded to the idea that water vapor from SST exhaust might change Earth's climate with an article published in *Science*. Using new reaction constants derived by meteorologist Paul Crutzen and a computer model of the atmosphere he had adapted from one designed for Mars, Harrison argued that the water vapor produced by a fleet of 850 SSTs would deplete the ozone column by 2 to 3.8 percent. Most of this reduction would occur in the Northern Hemisphere due to its high concentration of air routes, producing a temperature rise on Earth's surface of about 0.04°Kelvin.[23] This amount of change was trivial—indeed, it was unmeasurable, and hence it corroborated the SCEP report. One could not separate such a tiny change from Earth's natural variations.

But Harrison's article unexpectedly opened a new and far more controversial issue. Meteorologist James E. McDonald of the University of Arizona, a member of the National Academy of Sciences' Panel on Weather and Climate Modification, read Halstead's article and found his admission that ozone depletion *would* occur startling. McDonald had accepted his own panel's earlier conclusion that significant ozone depletion would not occur, and thus there were no biological effects to worry about. But Harrison's argument caused him to reconsider. By 1970, medical scientists believed that ultraviolet radiation caused certain kinds of skin cancer.[24] If depletion of the protective ozone layer that Harrison predicted did occur, then skin cancer incidence would increase significantly. Indeed, McDonald believed there was a sixfold magnification factor. Each 1 percent reduction in ozone concentration would produce a 6 percent increase in skin cancer occurrence.

McDonald first approached the climate impact study committee that Bill Magruder had formed within the Department of Transportation with his new-found concern, but he was rebuffed. So McDonald made his analysis known to the SST opposition—exactly how is unrecorded—and Representative Henry Reuss invited him to testify in the House hearings on 2 March 1971. The seriousness of McDonald's testimony was undermined by revelation of a long-standing side-interest in UFOs and extraterrestrial visitation. No one in the mainstream press, not even the editors of the anti-SST *New York Times,* was willing to lend credibility to a man with such a bizarre hobby.[25]

The political expediency of destroying McDonald's credibility to undermine his scientific testimony in the public's eyes worked, but only briefly. The scientific community largely accepted that McDonald's assertion of skin cancer risk was correct *if* ozone depletion occurred. The controversy that exploded throughout the Western world in the years following McDonald's testimony was thus over whether the SST, or any other pollutant, would damage the ozone layer. The next round in what Lydia Dotto and Harold Schiff have called the "ozone war" was fired by Berkeley professor Harold Johnston, a specialist in ground-level ozone chemistry and Halstead Harrison's old Ph.D. advisor, after he attended a Transportation Department conference in Boulder, Colorado, in mid-March.[26]

Johnston had been invited by the conference's organizer, University of

Wisconsin physicist Joseph Hirschfelder. A member of DOT's stratospheric impact panel, Hirschfelder had come to believe that the Transportation Department was trying to bias the panel by limiting its data to that provided by Boeing. He demanded access to other sources of data, and eventually the department relented.[27] Hirschfelder was allowed to organize a conference and invite a small number of university scientists, including Jim McDonald. The group convened on 18 March 1971, the same day the House was voting on further SST funding. Johnston rapidly became annoyed at the proceedings. In addition to a very tense atmosphere, the conferees seemed to accept the conclusions of the SCEP study that nitrogen oxides would not be a significant cause of ozone depletion. Johnston's knowledge of ozone chemistry suggested to him that this was wrong.

Johnston spent much of that night working out calculations showing that these gases would be far more potent ozone scavengers in the stratosphere than the group expected, and in the morning he handed out a handwritten paper that estimated NOx-derived depletion of 10 to 90 percent. His effort did not help much, and at an impromptu "workshop" organized by Hirschfelder in the men's washroom and held in a small conference room that afternoon the discussion stayed on water-vapor-induced depletion. Participant Harold Schiff recalled that Johnston lashed out at the group for ignoring the NOx reaction, and this finally prompted the other chemists to grapple with Johnston's idea. Schiff reported that the central question in that afternoon's argument was that no one knew what the stratosphere's "natural" concentration of nitrogen oxides was, as no one had ever measured it. Without that basic piece of information, one could not produce a credible estimate of NOx-induced ozone depletion. If the stratosphere already had a high concentration of NO and NO_2, then the amounts injected by a fleet of SSTs would not matter. If, on the other hand, the stratosphere had none at all, SSTs would be devastating. Johnston tended toward the "devastating" end of the spectrum, while his colleagues were not willing to make that leap. The conference thus ended with a recommendation that more research was necessary to determine whether or not the SST was a threat to the ozone.[28]

Johnston was not satisfied with that result. Back in Berkeley, he turned his calculations into a paper, which he sent out to several colleagues, including Hirschfelder, three of his colleagues at Berkeley, and David Elliot of the National Aeronautics and Space Council. The paper went out on 2 April 1971, and on the 14th, Johnston sent a substantially revised version to *Science*. The journal's editors, in turn, sent it out for peer review, in keeping with its policy, and the reviewers recommended that Johnston rewrite it. They deemed the paper unsatisfactory for two reasons. First, Johnston had not cited an article by Paul Crutzen, who was then working in Sweden, that also suggested the stratosphere's high sensitivity to nitrogen oxides.[29] Second, Johnston's tone was unacceptable. Scientists were supposed to be coldly dispassionate in their writing, in order to appear unbiased and objective. Johnston was not, contending that an SST fleet could cut ozone concentration over the Atlantic corridor in half and allow enough radiation to reach Earth's surface to cause widespread blindness. Publication of Johnston's paper was thus delayed.

But reaction to it was not. Johnston's preliminary 2 April draft had been leaked to a small California newspaper, the *Newhall Signal,* causing the University of California's Public Relations Office to release it. Sensational summaries of Johnston's draft sped east on the wire services and on 17 May, two days before the Senate vote on the SST restart attempt, the story made the *New York Times.* A lengthy follow-up article by the *Times*'s famed science editor Walter Sullivan on 30 May put the issue solidly before the public.[30] According to Johnston, he reported, SSTs, no matter who built, sold, and operated them, could have devastating consequences for humanity. Unsurprisingly, Johnston's draft did not play well in Nixon's White House.

Sullivan's article on Johnston's findings also drew senatorial attention. Senator Clinton Anderson of New Mexico, who chaired the Senate Committee on Aeronautical and Space Sciences, wrote to NASA administrator James Fletcher about Sullivan's piece on 10 June. After summarizing Sullivan's description of the NOx reaction, Anderson commented that "we either need NOx-free engines or a ban on stratospheric flight."[31] Then he laid the ozone issue squarely on the space agency. Paraphrasing the NASA charter, he reminded Fletcher that it was NASA's job to ensure that the aeronautical activities of the United States were carried out so as to "materially contribute to the expansion of human knowledge of atmospheric phenomenon" and to "improve the usefulness, performance, safety, and efficiency of aeronautical vehicles." In short, figuring out the ozone mess was NASA's job, and Anderson "encouraged" Fletcher to establish a program to find out whether or not stratospheric flight was the kind of threat Johnston had made it out to be.

Anderson's letter had little effect, however, in NASA headquarters, which was still struggling to sell the Office of Management and Budget on the Space Shuttle, and the ozone problem temporarily wound up in another agency's hands. *Science* published Johnston's paper in its 6 August issue, producing a resurgence in the public's interest and substantial concern in the White House that the problem would not go away without some concrete action, especially after a National Research Council analysis concluded that the Crutzen/Johnston hypothesis had merit.[32] And because the ozone controversy was framed in terms of human health effects, not mere environmental damage, it promised to have substantial political ramifications.[33] Certain Democrats with presidential ambitions would happily take ozone as a cause, and Democratic senators Birch Bayh of Indiana and Frank Church of Idaho quickly introduced the grandly named "Stratospheric Protection Act of 1971."[34] Senator Henry Jackson also introduced a bill mandating a stratospheric research program, and his passed in late September.[35] The law placed responsibility for a four-year stratospheric research program in the Department of Transportation and provided a budget of $20 million to carry it out. Named the "Climatic Impact Assessment Program"—CIAP to the thousand or so scientists who participated in it—this was the first "golden age" of stratospheric research.

While NASA was not CIAP's lead agency, the agency's research centers nonetheless played important roles in the study. At the Langley Research Center, the ozone question had interested several members of the aerophysics division. Bill Grose, who had spent the 1960s calculating re-entry trajectories for space capsules, recalls that he and others recognized that with the manned space program going largely out of business due to the gap between Apollo's end and the Space Shuttle's expected 1980 first flight, NASA would not need much re-entry work. So Grose's group "bootstrapped" themselves into atmospheric science, initially working on the science necessary to produce sensors for trace gas detection and monitoring, including NOx and ozone.[36] At the Goddard Space Flight Center in Greenbelt, Maryland, Donald Heath had a similar idea, proposing a satellite-based ozone sensor capable of mapping the stratosphere's ozone concentration. At the Ames Research Center in Sunnyvale, California, a group of researchers interested in computer simulation worked to produce a computational model of ozone's complex chemistry. And, of course, because the Transportation Department did not own high-altitude research aircraft and balloons or sounding rockets for sampling the stratosphere, NASA provided these resources to CIAP too. CIAP thus initiated the agency's movement into stratospheric research.

What cemented NASA's role in this new area of science and made it the "five-hundred-pound gorilla" of ozone research, to use one scientist's words, was not CIAP, however, but its own Space Shuttle. The shuttle program office at the Johnson Space Center had to prepare an environmental impact statement, and for the atmospheric portion they contracted with the University of Michigan. Scientists Ralph Cicerone and Richard Stolarski found that the exhaust from the Shuttle's solid rocket boosters would release chlorine, a highly reactive element known to destroy ozone, directly into the stratosphere.[37] Their June 1973 report was initially "buried" by the program office, but NASA headquarters reversed that decision quickly and scheduled a workshop on the problem, held in January 1974.[38] In the meantime, at a conference in Kyoto, Japan, Cicerone and Stolarski presented a paper on chlorine as a potential ozone scavenger—omitting mention of the Shuttle and of NASA's support for their research. Paul Crutzen also presented a paper including chlorine chemistry at this meeting.

The Shuttle ozone problem caused a great deal of concern at NASA headquarters, which was already wrestling with the general issue of how to position the agency within the rapidly growing field of environmental sciences. Within NASA, a general consensus had formed that CIAP was not going to effectively answer the ozone questions because it was too short-term. The program's time limitation inhibited the development of sensors capable of sampling the many chemical species necessary to develop a complete understanding of ozone's complex chemistry because scientists could not design, build, test, and deploy sensors in CIAP's four years and therefore were not bothering to try. Bob Hudson, who briefed chief scientist Homer Newell, deputy administrator George Low, and administrator Jim Fletcher on the Shuttle problem on 13 February 1974, pointed out that of the ten chemical species that were important to the Shuttle's ozone problem, there were accurate sensors for only four.[39] Fletcher and Low left this meeting unhappy with the slow pace of sensor development and a determination to implement a headquarters-directed program to continue stratospheric research once CIAP ended.[40]

The ozone problem became a statutory responsibility for the agency after the British journal *Nature* published a paper by F. Sherwood Rowland and Mario Molina, who argued in mid-1974 that photochemical decomposition of a set of extremely common chemical compounds known as chlorofluorocarbons would release large quantities of chlorine into the stratosphere. Compared to the billions of pounds of CFCs produced every year for use in spray cans, air conditioners, and refrigerators, the Shuttle's exhaust was utterly trivial. Revelation that mundane everyday items like hair spray could be ozone destroyers—and thus cancer risks—quickly produced a media firestorm and caused Congress to move with unaccustomed speed. December 1974 witnessed the first House hearings on the issue, and the following month the Senate Committee on Aeronautical and Space Sciences convened hearings on the ozone problem.

If Administrator Fletcher had let the ozone issue get away from NASA after Clinton Anderson's 1971 overture, he did not do so again. The "Stratosphere Research Program" that had started gelling in headquarters after Hudson's February presentation was initially funded by pulling small amounts of money out of several other programs, and it became the core of a much larger research program once Congress took up the CFC problem.[41] Using this program as a lever, Fletcher and Low were aggressive in pursuing leadership of the CFC problem during the lengthy ozone hearings that followed revelation of the CFC problem. And they were rewarded with a congressional edict handing stratospheric ozone research to NASA, embodied in the agency's FY 1976 authorization bill.

Headquarters created the Upper Atmosphere Research Office, UARO, within the Office of Space Sciences that year, administered by Dr. James King. King reorganized NASA's rather scattered stratospheric research activities, expanding the Langley Research Center's Environmental Quality Office, under James D. Lawrence, into a separate Atmospheric Sciences Division.[42] At Goddard, Nelson Spencer perceived that the stratosphere was "the next big thing" and invited Robert Hudson and Richard Stolarski, then at the Shuttle Environmental Effects Project Office in Houston, to move north and establish a new stratospheric research branch.[43] Previously independent efforts at both the Ames Research Center and at the Jet Propulsion Laboratory were also integrated into the UARO.

The Upper Atmosphere Research Office's short-term goals were to evaluate the potential effects of the space shuttle, fluorocarbons, stratospheric aircraft, and other chemical emissions on the stratosphere. Due to the intense political pressure generated by conflict between environmental groups seeking a ban on CFC use and industrial groups trying to prevent a ban, the Congress expected NASA to produce answers to the CFC questions in time for regulatory action in 1978—a very short time period indeed. Nonetheless, the office did not have money thrown at it. In FY 1976, it received $7.4 million, and in FY 1977 $11.6 million.[44] In addition, supporting efforts in other offices, particularly the *Nimbus G* satellite being developed at Goddard under the auspices of the Office of Applications, effectively doubled that amount.

By 1978, the year NASA launched *Nimbus G* on what turned out to be a fourteen-year mission, the agency's leaders had managed to gain control of a revitalized research field that had sprung unexpectedly out of the SST conflict. When John Stack had started his campaign for an American SST in 1957, no one had thought much about the stratosphere—it was an alien, far-away place that only a handful of scientists understood was vital to the health of nearly all living things. The SST war of 1970–71 had put this strange place on newspapers' front pages, on the agendas of many scientific conferences, and in the minds of young scientists looking for new research areas in which to distinguish themselves. And the research effort that started with CIAP and continued in NASA's Upper Atmosphere Research Office produced news that was not good for the SST enthusiasts. The CIAP study had found that SST-produced nitrogen oxides would indeed cause ozone depletion if the fleet of hundreds of Concordes and TU-144s that the Department of Transportation continued to insist would be manufactured ever appeared. While the exact magnitude of that danger remained controversial well into the 1990s, CIAP forced SST enthusiasts to accept that the politics of ozone could be surmounted only by a technofix that definitively eliminated the NOx problem.

The SST conflict and the subsequent ozone war thus left propulsion engineers at NASA and in the aircraft industry facing two adventures in eco-friendly engine design. They had to devise a supersonic engine that was far quieter than the GE4 engine had been in order to satisfy anti-noise activists, and they had to achieve something like an order of magnitude reduction in NOx emissions to satisfy the defenders of the ozone layer. Further, this new engine had to weigh significantly less than the GE4 had if the "Advanced Supersonic Transport" was to be economical. By the time this supersonic propulsion effort was cancelled in 1981, it seemed to have resolved these problems.

THE SCAR/VCE PROGRAM

The advanced supersonic technology program that had emerged from Bill Aiken's organizational meetings during late 1971 and early 1972 was what the agency referred to as a "focused technology program." It had specific goals for noise and emissions reduction and efficiency improvement in order to establish a coherent research direction, but it would not result in a prototype aircraft. As far as George Low, Roy Jackson, and Bill Aiken were concerned, prototypes were industry's responsibility.[45] At most, NASA would pay for constructing and testing an experimental engine. When Congress refused to fully fund Aiken's program, however, the experimental engine was dropped. The reduced program included only a small amount of component testing.

Aiken had assigned management of the Advanced Supersonic Transport program, quickly renamed the Supersonic Cruise Aircraft Research program (SCAR), to the Langley Research Center. Edgar Cortright, the lab's director, assigned the management task to Neil Driver, who had supervised the SCAT contract with Boeing in 1962–63 and later served as a consultant to the House Appropriations Committee chairman on an investigation of the McDonnell-Douglas F-15.[46] Driver, in turn, assigned the important propulsion effort and about half the program's small budget to the Lewis Research Center, while retaining configuration, aerodynamic, and structural materials research efforts at Langley.

Like the earlier SCAT effort, the SCAR program involved a set of contract studies supplemented by in-house research at the two labs. Because integration of the new technologies promised to be difficult, Driver adopted what he called "the Systems Integration Studies" approach to structure the program.[47] The individual research areas, or disciplines, fed integration studies teams that applied the new concepts to a baseline aircraft. This baseline design, known as AST-100, provided a common reference for the four integration teams—one each from Boeing, McDonnell-Douglas, Lockheed, and NASA. In addition, each company chose a different cruise speed for its design and optimized it accordingly, ultimately generating three slightly different aircraft.

Given the small SCAR budget, the research effort was extraordinarily broad and produced slightly more than a thousand technical publications and articles over its ten-year life. Each disciplinary research area focused on a specific idea, however. In aerodynamics and structures, the research efforts primarily involved the development and validation of analytical design methods, actually computer programs, which promised to make the design process faster, less expensive, and more precise. Richard Heldenfels's materials research group concentrated on improving a technique developed by Rockwell International for making complex titanium parts without machining or joints. Finally, the supersonic propulsion effort at the Lewis Research Center consisted of two separate programs, one funded directly by the Supersonic Cruise Aircraft Research program office at the Langley Research Center and the other financed separately by the Office of Aeronautics and Space Technology at NASA headquarters. The first of these, the Variable Cycle Engine program (VCE), focused on resolving the noise and engine weight problems that the GE4 had presented. The second, the Experimental Clean Combustor program, targeted NOx reductions for both subsonic and supersonic engines. By 1979, Driver's office had spent $86.2 million. Corporate partners had contributed an equal sum.[48]

The large propulsion effort focused on the SST's environmental problems, particularly noise. The fundamental problem Boeing and GE had faced during the national SST program was that while turbojets were the more efficient engine for supersonic flight, they were much less efficient—and a good deal louder—than turbofans at subsonic speeds. Conversely, while a quiet turbofan could power an SST, it would be much less efficient than a turbojet engine at supersonic speeds, probably rendering the aircraft uneconomical. Logically, the perfect SST engine would act like a turbofan at subsonic speeds and then reconfigure itself in flight to perform like a turbojet for the supersonic cruise portion of the flight.

The idea that a real engine could do this was attributed by several NASA engineers to Boeing engineer Garry Klees. Klees recalls that Boeing's propulsion research laboratory, where he worked, had been asked to investigate the 2707's noise problem because the various suppressor designs that Withington's SST design team had already tried had not been satisfactory. Several of his co-workers had the idea of varying the engine's cycle somehow, but Klees was the first to design a means of doing it successfully. With the support of his lab director, Bert Welliver, Klees rebuilt a junked Pratt & Whitney JT8D turbofan into a "variable bypass engine." A set of valves allowed the engine control system to reconfigure the engine's internal flow paths. The engine could therefore run as a low-bypass turbofan or as a high-bypass turbofan. In high-bypass mode, the engine proved to be much quieter, just as NASA and Boeing engineers had hoped. Tests at Boeing's engine facility in Boardman, Oregon, validated the concept.[49]

Propulsion engineers at NASA and in the engine industry understood that Klees's variable bypass engine was just one specific way to vary an engine's thermodynamic cycle, and not necessarily the most efficient. The early focus of the Variable Cycle Engine project at the Lewis lab, then, was evaluation of many possible concepts to determine which ones promised to be the best compromise between noise, efficiency, and weight.

Lewis engineer Larry Fishbach recalls that the VCE effort depended upon the development of computer codes capable of analyzing various combinations of engine cycles. The VCE program started with a set of twenty-five different possible engine concepts, which would have required an enormous number of engineering hours to evaluate. The existing analysis software the lab had was not capable of simulating some of the more unusual engine concepts, and it was not flexible enough to be adapted. He contacted Michael Caddy at the Naval Air Development Center (NADC), who told him about a software package the Navy had developed known as NEPCOMP. This also would not perform the functions Fishbach wanted but he thought it could be modified. He convinced the program's management to contract with NADC to adapt the software so that it could, in Fishbach's words, "simulat[e] any turbine engine the user could conceive."[50] The initial version of the revised code, renamed the Navy-NASA Engine Program, or NNEP, was operational in May 1974, and was used to help reduce the number of engine candidates to four.

During the evaluation process, Pratt engineers made a surprising discovery that impacted the design process. They were working on a duct-burning turbofan engine concept using small air-jet models, and they found that when the bypass air flow was hotter than the core air flow (an "inverted velocity profile"), the model produced less noise.[51] The hotter bypass air provided an unexpected shielding effect, reducing radiated noise by three to five decibels. Both Pratt and GE redirected their design efforts toward concepts that could achieve the inverted velocity profile and its attendant noise reduction during takeoff and landing.

In 1975, Pratt & Whitney and GE each chose one of its engine concepts to carry into a test-bed study. Pratt chose the variable stream control engine (VSCE), while GE's candidate was its double bypass engine. Both engines were designed to produce the inverted velocity profile for noise reduction but accomplished it in different ways. The variable stream-control design, a duct-burning turbofan, achieved it through adjusting the temperatures of the bypass and core streams during takeoff so that the duct burner produced a relatively hotter flow than the core did. GE's double bypass engine achieved the inverted profile by mechanically reversing the two flows using a pair of sliding valves. Both concepts differed from traditional engines only in the presence of a small number of unique parts, and the purpose of the test-bed engine program was to test those unique parts. To save money, each company chose a compatible existing engine and modified it to represent its concept. Pratt chose an F-100 engine for transformation into a test-bed VSCE engine, while GE selected a YJ101. Both engines had been designed for Air Force fighter programs and represented the latest state of the art for supersonic engines.

The Air Force propulsion laboratory at Wright Field supported the first government tests of the variable cycle concept in 1976, using GE's YJ101. This test series investigated the performance of the sliding valves upon which GE's double bypass concept depended. In 1977, the Navy utilized the YJ101 double-bypass test bed to examine a full-authority digital control system, because, as Fishbach explained, the engine's reliance on sophisticated variable geometry components made it the most challenging possible test for the new control system.[52] NASA-sponsored tests on the test-bed engines began in 1978 and ran through 1981, when headquarters terminated the program.

The test-bed program was also designed to validate noise suppressors. The inverted velocity profile concept, combined with the variable cycle engine's inherently lower exhaust velocity, produced approximately enough noise reduction to achieve the 1969 noise standard (known as "Part 36 Stage 2"), but did not produce enough noise reduction to meet the "Stage 3" standard that the Federal Aviation Administration imposed in October 1977 on post-1980 new aircraft.[53] Therefore, Pratt, GE, McDonnell-Douglas, and Boeing all investigated various kinds of noise suppressors to be integrated with the engine exhaust nozzles. The most promising concepts that emerged from the program were the thermal acoustic shield nozzle and the ejector suppressor. The thermal acoustic shield nozzle relied upon the hot-air shielding effect that Pratt had discovered, but achieved it by creating a hot, slow air stream surrounding the bottom half of the exhaust. The other important concept, the ejector suppressor, was a nozzle equipped with doors that could be opened during takeoff. The hot engine exhaust flowing past the doors entrained, or pulled in, outside air, increasing the propulsion system's overall airflow while slowing and cooling the exhaust. (See figure 5.1.) This produced substantial noise reduction at the cost of a fairly heavy, complex nozzle. In a separate, but related, effort, McDonnell-Douglas collaborated with Rolls-Royce in 1978 on a flight test of the concept using a British HS-125 aircraft, to verify that the suppression effect did not disappear on a real aircraft.

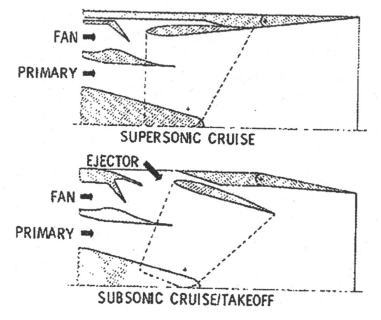

FAN →
PRIMARY →

SUPERSONIC CRUISE

EJECTOR
FAN →
PRIMARY →

SUBSONIC CRUISE/TAKEOFF

FIG. 5.1.   Operation of a mixer-ejector nozzle. In supersonic mode (top), nozzle doors are closed and no outside air is being drawn into the nozzle. In takeoff mode (bottom), ejector doors are open, and the high-speed air inside the nozzle entrains outside air, drawing it in. A mixer downstream (not shown) combines the three airflows. The resulting air mass is colder and has lower velocity, reducing noise. [From L. H. Fishbach et al., "NASA Research in Supersonic Propulsion—A Decade of Progress," NASA TM-82862, 1982, p. 26]

The SST noise problem, then, caused NASA and the manufacturers to investigate a new class of jet engines as well as continue efforts to design more effective noise suppressors. The ozone problem also dictated that the agency investigate low-emission combustors for jet engines, an effort that headquarters incorporated in a separate program. At first, an "Experimental Clean Combustor" program at the Lewis Research Center was initiated in response to the Environmental Protection Agency's intent to establish airport emission standards in 1979.[54] Administered by Richard Niedzwiecki, this program's goal was reduction of engine NOx, carbon monoxide, and unburned fuel emissions by specified amounts. But when the Transportation Department's CIAP study concluded in 1974 that stratospheric aircraft would cause ozone depletion, low NOx emissions at cruise speed was added to the program's goals.

The CIAP study's authors had concluded that for a fleet of five hundred Concorde-type SSTs to cause no discernible change in stratospheric ozone concentrations, propulsion engineers would have to achieve a fourfold reduction in NOx emissions over Concorde technology. A fleet of five hundred Boeing 2707s would have reduced stratospheric ozone concentrations world-wide by an average of 10 to 20 percent, with a much higher depletion over the highest density routes—the North Atlantic, primarily.[55] Unfortunately, because increasing propulsion efficiency required increasing engine combustion temperatures, and higher temperatures meant more NOx production, the study also concluded that a fourfold reduction would permit only one hundred advanced SSTs.[56] Hence if one expected a reasonably profitable production run of five hundred advanced SSTs, elimination of the ozone depletion problem required at least a *sixfold* reduction in emissions. This, then, was the other environmental challenge propulsion engineers faced.

In 1977, Niedzwiecki's low-emissions group spun off an effort called the "Stratospheric Cruise Emission Reduction Program." In the preceding two years, Niedzwiecki had provided several universities with research grants to explore ways of reducing NOx formation without having to reduce engine efficiency through lower turbine temperatures.[57] Two promising ideas emerged from the university research: the use of a catalytic combustor that would reduce NOx concentration via a chemical reaction, and a combustor that vaporized the fuel and mixed the resulting vapor with the incoming air before ignition occurred. The catalytic combustor promised the largest reduction in nitrogen oxides, but also held the highest development risk, so Niedzwiecki chose to pursue the other process, known as lean premixed, prevaporized (LPP), instead.

The LPP combustor concept was premised on a combination of three possible means of reducing NOx production. Because NOx production was directly proportional to flame temperature, a leaner fuel/air mixture (i.e., one in which the ratio of fuel to air was lower than in existing engines) would result in less NOx formation. By ensuring uniform mixing and distribution of fuel and air, local "hot spots" in the combustor could be eliminated, further reducing NOx creation. And finally, because the majority of NOx production resulted from the burning of droplets in the fuel/air mixture, NOx could be reduced through prevaporization of the fuel so that only gas-phase combustion occurred.[58] Niedzwiecki had a test rig built in order to demonstrate this approach's validity, but the program was cancelled before it could be used.

The parallel airframe technology research carried out at the Langley Research Center supported a series of study contracts with Boeing, Lockheed, and McDonnell-Douglas aimed at improving the performance of arrow-wing planforms like SCAT 15F, while also coordinating in-house structural and aerodynamics research. For the first three years, the four organizations focused on improving computer design tools, especially structural analysis programs like ATLAS and NASTRAN, to permit superior optimization of their designs. After a conference held in November 1976 to discuss the three manufacturers' results, the program shifted its emphasis from developing analytical tools to validation of the results of the computer simulations.[59] The most successful of the three companies during the SCAR effort was McDonnell-Douglas, which by 1980 was convinced that it had resolved all the economic problems that had undermined the Boeing SST and could build an economical transpacific AST.

One of the earliest decisions made in the SCAR program was to drop the Mach 2.7 speed that NASA, FAA, and Boeing had all promoted during the national SST program in favor of Concorde's Mach 2.2 speed. The program's leaders thought it was crazy to pick a speed that required an all-titanium airframe when titanium weighed (and cost) a great deal more than aluminum. Choosing the slower speed made the aircraft lighter and much less expensive to build. And without the overland sonic boom intensity limitation that NASA had imposed on the designs during the 1963 SCAT study, the lower cruise speed and altitude did not produce the high weight penalty that the aluminum SCAT 16 had incurred.[60] In their opinion, the SST's higher speed had not been worth its cost. Instead, economical operation suggested that slower was a superior choice.

The Mach 2.2 baseline configuration that the SCAR researchers chose was based upon a modification of the SCAT 15F "arrow wing" configuration from the 1960s program. Re-optimized for the slower speed, it became the "AST-100" configuration. (See figure 5.2.) All of the airframe contractors then developed their own versions of the AST-100, to which they then applied the results of their research, allowing them to assess the impact of various new technologies on vehicle performance. The AST McDonnell-Douglas defined was a 350-seat, 5,500 n.m. range that met the Stage 2 airport noise standards and that the company's engineers estimated would have total operating costs slightly lower than the company's DC-10-30.[61]

Douglas engineers R. L. Radkey, Bob Welge, and Robert Roensch described the evolution of the Douglas AST design in a 1976 paper. Beginning with the basic Langley arrow-wing design, Douglas's engineers had refined its aerodynamics with the aid of two computer programs. One of these, the Woodward

FIG. 5.2. AST model in the Langley Research Center Full-Scale Tunnel, 8 February 1977. [Courtesy NASA, Langley photo #L77-1140]

program developed by the Boeing Company under a contract from the Ames Research Center, permitted analysis and optimization of the wing, including camber, thickness, and nacelle integration.[62] Once the team had settled on an optimum wing planform, they used an in-house wave drag program to help optimize the fuselage. The resulting area-ruled fuselage would have higher construction costs than a simple constant cross-section (e.g., cylindrical) fuselage would, however, so the group also examined a variety of constant cross-section fuselages. These, they found, produced enough extra wave drag to overwhelm the construction cost benefit, and they dropped the idea in favor of the more complex area-ruled design.

Under a SCAR program contract during late 1975, McDonnell-Douglas validated the resulting design in the Ames Research Center's Unitary Plan tunnels, finding that it achieved a trimmed cruise L/D around 20 percent better than Boeing's 2707-300 but about 5 percent lower than predicted by computer simulation.[63] Using the data from these tests, the Douglas aerodynamics group corrected the Woodward program and then used it to re-optimize the arrow wing.[64] The new configuration, described at the 1979 SCAR conference by Roensch and Page, appeared to meet their goal, a 38 percent improvement in cruise aerodynamic efficiency over the Boeing 2707.[65]

In addition to the supersonic cruise efficiency improvement effort, the Douglas AST team participated in a cooperative effort with the Langley Research Center's Full Scale Tunnel staff to design and evaluate high-lift devices—flaps and slats. Boeing had not adopted the SCAT 15F–like arrow wing in 1968 due to its poor takeoff, subsonic performance, and noise level, and these issues had to be fixed in a way that preserved the supersonic qualities of the configuration. The highly swept thin wing had poor subsonic performance because the incoming airflow tended to "separate," or become turbulent, as soon as it encountered the sharp leading edge. Theoretically, the way to fix the problem was to make the wing appear more like a thicker, straighter subsonic wing to the incoming air without actually changing the wing. This might be accomplished through the use of a set of leading edge flaps that could be deployed at subsonic speeds and retracted before accelerating to supersonic speed. The Langley Research Center thus built a 10 percent scale model of the Douglas 1975 configuration for an extensive series of tests on various flap designs.[66] These were complemented by another series of tests on a leading-edge "vortex flap" devised by Boeing.[67]

Another factor in the economic improvement evinced by McDonnell-Douglas's AST was its use of a new construction material, superplastically formed diffusion-bonded titanium sandwich. This differed from the titanium construction used during the Boeing SST effort in how the titanium parts were made. In traditional aircraft manufacture, both aluminum and titanium parts had to be relatively simple in shape, and complex shapes had to be built by bonding simpler shapes to each other with glues, welding, or fasteners. Glues generally would not withstand high temperatures; welding and brazing introduced weaknesses to the metals and were very difficult to apply to complex parts; fasteners added a great deal of weight. Under an Air Force contract related to the B-1 bomber, however, Rockwell International had found that titanium could be formed into complex parts by heating it up and blowing it into molds, somewhat like glass. Langley's SCAR office contracted with McDonnell-Douglas to refine the process, and Douglas pursued the idea of manufacturing sheets of sandwich material. Previous methods of making aluminum and titanium sandwich had been both expensive and unreliable, and the new method appeared to be superior. Using the AST wing as an example,

Douglas engineers estimated that replacing the older construction methods could reduce the wing weight 16.3 percent, and because it eliminated a large amount of finishing labor, could reduce the wing cost by 63.7 percent. Overall, the savings from the new method would reduce the AST's direct operating costs by 6 to 7 percent.[68] The Douglas group thus dropped the original all-aluminum structure for one that would consist of 78 percent titanium and 22 percent aluminum or composite materials.

Substantial cost savings also accrued from the use of computer-assisted structural design. During the late 1960s, the aerospace industry had begun designing a wide variety of computer software complexes aimed at the design and analysis of complex structures. Under NASA contracts, Boeing designed ATLAS and FLEXSTAB, while NASA built NASTRAN and SAVES internally.[69] These programs were primarily intended to save time and manpower in the design process, reducing overall program cost. Boeing, for example, found that an early, incomplete version of ATLAS reduced the manpower needed to analyze its SST's structural design by slightly more than half.[70] After a decade's refinement of computer-aided design methods, Douglas's William Rowe stated that a complete SST structural design cycle could be done in one-sixth the time manual methods required.[71]

In a 1981 paper, the Douglas engineers presented a summary of their design's economic potential. They contended that the incorporation of the SCAR/VCE–derived technologies into Douglas's AST promised to bring the 350-seat aircraft's direct operating costs to within 3 percent of its 270-seat DC-10-30.[72] In addition to the AST's improved economic performance, they contended that the AST would carry only full-fare passengers, effectively splitting the market into a full-fare supersonic luxury market and a discount-fare subsonic market.[73] The resulting AST's higher productivity and higher average airfares would produce nearly fourfold higher annual revenue.[74] And they discussed a computer-based market study conducted by Douglas's marketing department. Based on a route analysis of 191 city pairs, the study suggested a market for 294 Douglas ASTs by the year 2000. A former member of Douglas's marketing department has challenged this result, however, remembering that the real number of potential sales calculated during the SCAR program was thirty. Such low numbers would not suffice to keep the program going and had been altered to reflect better performance.[75]

Whatever the truth happened to be regarding Douglas's marketing study, McDonnell-Douglas management began an advocacy campaign to win NASA a substantial increase in the SCAR/VCE budget. At this point in the program's life, most engineers in and out of NASA believed that SCAR was becoming technologically stagnant, because there was not enough money to transform the new concept technologies into hardware and thoroughly test them. Without the measure of certainty provided by a validation program, the SCAR/VCE technologies were too risky for a private company to base a hugely expensive aircraft program on. And there was little point in continuing SCAR if it could not validate the technologies it had devised. Thus McDonnell-Douglas argued that NASA should finance a technology validation phase to preserve the nation's "supersonic option" as Concorde finally went into service. Unfortunately for both McDonnell-Douglas and for NASA, McDonnell-Douglas's lobbying campaign failed.

CONCORDE, THE AST, AND THE BUDGET CRISIS

The SCAR program's technical progress encouraged SST enthusiasts to promote a new national SST development effort. Advocates tried to convince political leaders to launch a new SST effort to compete with Concorde. They won some sympathy from members of the House Committee on Science and Astronautics, and from NASA's new administrator, Robert Frosch. But even with this help, they could not overcome the SST's fundamental political liabilities. In addition to lingering anti-SST sentiment, the national economy had entered a period of high inflation that the Carter administration, elected in 1976, sought to mitigate by cutting government spending. Late 1978 saw the return of fuel shortages and resulting dramatic increases in prices. Finally, NASA was experiencing huge cost overruns in its highest priority program, the Space Shuttle. In this context, the SST enthusiasts not only were unable to make substantial progress toward a new development effort; they witnessed cancellation of the SCAR/VCE program. Instead, Congress funneled money into NASA's subsonic jet noise reduction and fuel economy research.

The campaign to start a new SST development program coincided with Concorde's introduction to the United States. In mid-1974, the British and French governments had notified the FAA that their two national airlines, British Airways and Air France, wanted to begin scheduled Concorde service to the United States in 1976. Each airline sought two flights per day to John F. Kennedy International Airport and one per day to Dulles International Airport in northern Virginia. In May 1976, Concorde began flying to Dulles after

President Carter's secretary of transportation, William T. Coleman, reviewed the CIAP study and a draft environmental impact statement and decided that the small number of Concordes—only fourteen were actually built—did not constitute an significant environmental threat.[76] He approved a sixteen-month trial period during which the FAA would monitor community reaction to Concorde's noise signature. He also hoped that the trial period would determine whether or not Concorde, or future SSTs, would be economically viable.[77] Because Dulles was owned and operated by the FAA, Concorde service started immediately.

New York proved not so easy. In March 1976, the Port of New York Authority had acceded to local anti-noise groups and banned Concorde due to its high noise signature. The DOT's environmental impact statement on Concorde had documented that the aircraft produced a noise signature much like that of the first 707s and well in excess of the Part 36 standards for new subsonic aircraft. In 1975, Concorde's two manufacturers had retained public relations firms to try to "sell" Concorde to the American public (and the U.S. Congress); after the New York ban, they also hired law firms. Mere lobbying would not succeed in overturning the ruling, the two governments eventually concluded, because no politician in New York could be induced to vote for "foreign noise."[78] On 17 March 1976, the two airlines filed a complaint against New York in federal district court, arguing that states could not impose a stricter noise standard for aircraft than the federal government had, but more than a year elapsed before the case proceeded. When it did, however, both the district court and the federal appeals court ruled for the airlines. New York did not have the right to impose stricter noise standards, and the Transportation Department had specifically exempted supersonic aircraft from the Part 36 rules.[79] Hence Concorde was legal, and Concorde service to New York finally began on 22 November 1977.

As the Anglo-French lawsuit percolated its way through the courts, McDonnell-Douglas launched a campaign to get a new U.S. SST development program going. Despite Concorde's 20 percent surcharge over standard first-class fares (which were also heavily subsidized by the two nations), Concorde's Dulles service was very popular with first-class and business passengers.[80] Both British Airways and Air France averaged load factors of over 90 percent during their first year, effectively "skimming" high-fare passengers off Pan Am and TWA, the two major domestic transatlantic carriers. Because first- and business-class passengers were the source of the majority of these airlines' transatlantic revenue, the uneconomical Concorde was indeed a threat to their profitability. It was this threat, combined with the belief that the SCAR technologies could produce a substantially more economical airplane than Concorde, that motivated Douglas's campaign to begin a new SST effort in 1977.

Douglas's head of advanced SST research penned an article for *Acta Astronautica* in which he laid out the argument for a new start. "The supersonic era is a reality," he argued. "For the first time in twenty-five years on the major airline routes of the world, the domination by U.S. manufactured aircraft is seriously being challenged."[81] McDonnell-Douglas believed a "near term U.S. AST program" was necessary to protect the U.S. aircraft industry if the traveling public continued to demand supersonic flight, as Concorde's high load factors appeared to promise it would. McDonnell-Douglas's AST would be profitable in competition with the current jumbo jets by pulling first-class and "full-fare economy" passengers off the jumbos, although discount fares would not be feasible. Forty-two percent of the transatlantic fliers paid either first-class or full-fare-economy rates, and this was enough to support an AST manufacturing effort.

Thus in the Douglas team's vision of the supersonic future, the aviation market would be segregated into a supersonic "first class" and subsonic "tourist class" market. This would result in substantial fare increases for both classes of fliers. The removal of high-fare, first-class traffic from the subsonic jets would render them unprofitable, and airlines would have little choice but attempt to replace that loss by increasing yields. And, of course, the lack of discount fares on the supersonics amounted to a fare increase by another name. The supersonic future would be more expensive for everyone.

In June 1977, McDonnell-Douglas president John C. Brizendine wrote to NASA administrator Robert Frosch to complain that NASA was not putting sufficient resources into its AST effort. After describing the accomplishments NASA had sponsored under the SCAR/VCE program, Brizendine argued that the SCAR technical advancements would not be transformed into a viable commercial product without much more substantial government investment. The problem, he contended, was that there was still far too much risk involved to justify private investment, especially in the critical noise area. Brizendine believed that NASA needed to fund at least half-scale tests of noise suppressors, variable-cycle engine components including nozzles and inlets, and high-temperature structural tests of large sections of wing and fuselage.[82]

Brizendine's position, in short, was that American manufacturers would not commit the huge capital resources necessary to put an SST into production until NASA had answered the crucial environmental question, noise. While the other manufacturers' inputs to the NASA study are not preserved, the plan the agency finally submitted to the House Committee on Science and Technology in October strongly reflected that argument. All three airframe manufacturers contended that the engine and associated propulsion components had to be demonstrated at subscale before they would commit to a production aircraft. Each also noted that large-scale structural components using the SCAR-derived manufacturing techniques needed to be built and tested before production commitment. The primary difference in the three airframers' plans was that Boeing wanted to build a research airplane that could become a supersonic executive jet, while Lockheed and McDonnell-Douglas did not believe one was necessary.[83]

NASA's report to the committee, however, included an unexpected shock. The agency estimated the cost of sustaining the "technological readiness" program to the point at which the airframe manufacturers would commit to the airplane to be between $900 million and $1.7 billion, depending on how many of the three airframe and two engine manufacturers were sustained through the program. Keeping all five companies in the program would minimize technical risk but maximize program cost.[84] This was much more money than the committee members had in mind. Furthermore, the committee had asked NASA for a complete plan, including discussion of financing issues for production and social impact. Deputy Administrator Alan Lovelace believed that such issues were not in NASA's purview, and had refused to respond to these issues. The committee therefore declared the report "unresponsive" to its request and kicked it back to NASA for retooling at a lower cost.[85]

Unwillingness to support an expanded SCR effort did not stem from a general lack of interest in aeronautical research, however. In response to the energy shortages of the early 1970s, NASA had formulated the Aircraft Energy Efficiency program. The major component of this program was the Energy Efficient Engine project, or E^3. The project's goal was to reduce fuel consumption by at least 12 percent relative to the latest generation of turbofan engines while meeting upcoming EPA emission standards and dramatically lowering maintenance costs.[86] A ten-year, $90 million program, E[3] promised to do the inverse of an AST program—lower airlines' operating costs by making possible a set of still more efficient subsonic airliners for the 1990s.[87] In 1978, the House Science committee attempted to take $12 million away from this project and transfer it to SCAR/VCE; the Senate committees blocked their effort and protected the program through its completion in 1985.[88] Improving the environmental and economic qualities of subsonic airliners took precedence over supersonics, a political reality that NASA administrator Robert Frosch made clear in a letter to McDonnell-Douglas leaders the same year.[89]

NASA's new supersonic technology plan, submitted in September 1978 for the FY 1980 budget, was to cost $561 million for a "technology validation" effort that stopped short of the "technological readiness" level that the manufacturers sought, and would only support one airframe and one engine manufacturer for the eight-year program. Again, the House showed more interest in supersonics than did the Senate, but neither supported anywhere near the funding levels the NASA plan entailed.[90] Instead, the conference committee agreed to a $9 million increase in the VCE program's hardware demonstration phase, and demanded—again—that NASA come up with a new plan. The new plan never emerged, however, a victim of two major traumas: Concorde's final disgrace, and the Great Inflation that took hold in the late 1970s.

SST supporters had long relied on Concorde as proof of the inevitability of the supersonic future, but in 1978 the first data on Concorde's financial performance from airline service was revealed. After winning permission to serve New York in 1977, British Airways and Air France, along with Concorde's two manufacturers, had embarked on a massive sales campaign to both sell Concorde's existing service to the public and to sell more Concordes to airlines. Both efforts failed. While *Aviation Week* editor Robert Hotz argued that a 72 percent British Airways New York-to-London load factor and 65 percent Air France New York-to-Paris load factor proved the existence of a supersonic market, he ignored the fact that both companies lost a great deal of money on Concorde's overall service.[91] Writing in *Fortune* magazine, Frank Melville reported that while the two airlines were able to make money on the lucrative New York and Washington, D.C., routes, they lost money on all their other routes. Air France's once-per-week Paris-to-Caracas run managed only a 38 percent load factor, for example, far less than the 65 percent break-even load factor. Furthermore, to break even overall, the airlines had to achieve a 65 percent average load factor for 7.5 flying hours per day, or 2,750 hours per year.

But the aircraft averaged only 4.5 hours a day in the air, because, he said, the two airlines lacked enough profitable routes to yield another 1,000 flying hours per year.[92]

Pan Am, TWA, and the rest of the world's airlines did not ignore that unhappy fact, and despite the manufacturers' sales campaign, they continued to refuse to buy Concordes. Finally, after spending almost a half billion dollars on the sales campaign, the British and French governments gave up. In September 1979, they shut down the assembly lines, destroyed the tooling, and ended all public support for a follow-on "advanced Concorde," but they continued subsidizing British Airways and Air France Concorde service.[93] All fourteen Concordes were given to the two airlines, thus ending the twenty-three-year-long, $4 billion adventure. Concorde's final collapse gave the lie to SST supporters' "inevitability" argument, demonstrating that airlines would not buy into the supersonic future if that future included increased costs.

Similarly, the Soviet TU-144 had earned no customers, not even the state-owned airline Aeroflot. The 1976–80 Five Year Plan for Aeroflot had contained no TU-144 service, reflecting the reality that the aircraft had been unable to sustain the light-duty cargo service it had been assigned for evaluation purposes in December 1975. It was supposed to fly twice weekly carrying airmail but could not carry much of it; in June 1976 this was reduced to one flight per week and was then cancelled altogether in December. The following year, the eight production TU-144s were placed in "service" on a once-per-week domestic route as a face-saving gesture or perhaps, as Howard Moon put it, a "triumph of hope over experience."[94] One hundred and two passenger flights took place before supersonic service was cancelled again in June 1978 following an engine fire that destroyed one aircraft. After some improvements, the TU-144s had gone back to cargo service in June 1979 but never carried passengers again. Even the limited cargo service was cancelled in 1983, reflecting the beginning of recognition even in the Soviet Union that operating efficiency mattered.

The striking feature of the late 1970s promotional campaign for a new American SST effort, in fact, was the complete absence of airline interest. Only Pan Am, now a shadow of its former self, participated in an Office of Technology Assessment study of advanced supersonic airliner potential, and it provided SST skeptic John Borger.[95] No one from an airline testified before Congress in support of the SCAR/VCE program during the 1970s. And the promotional articles that appeared in the trade press during the decade relied upon interviews with manufacturers' representatives and NASA engineers, not airline officials. During the national SST program of the 1960s, the FAA had demanded public expressions of faith in the supersonic future from the airlines, proclaiming them to be "proof" of a hugely profitable "market." Believing that the SST was nothing but an aerospace welfare program, Robert McNamara had further forced the airlines to commit themselves via deposits—a tactic Boeing picked up and used for the 747, demanding pre-payments to ensure customers did not backpedal. McNamara, Najeeb Halaby, Bozo McKee, and Bill Magruder had all understood that Congress would not keep funding such an expensive program without explicit airline support. No such expressions of support from the AST's future customers accompanied the OTA report. Without airline support, with Concorde's highly visible failure fresh in congressional minds, and in the face of substantial budgetary constraints, the attempt to generate a new post-SCAR technology validation program went nowhere. A major shift in political attitudes was going to be necessary to bring forth an American SST.

Unfortunately for NASA and for SST supporters, the political debacle that would sweep President Carter out of office in 1980 did not sweep in a pro-NASA, pro-aeronautics regime. The late 1970s had witnessed a painful round of inflation. Investors saw the value of their assets eroding, while the middle class witnessed rapidly escalating property values being transformed into high property taxes. This had enormous political consequences, some of which had been obvious to the Carter administration. Inflation, at least by one definition, was the result of demand exceeding supply, and the federal government was the nation's largest single consumer. Hence Carter had resolved to reduce federal spending as a means of combating inflation. Unfortunately for both Carter and NASA, most of the largest federal programs were indexed to inflation and thus rose in lock-step with it. The agencies that were not indexed to inflation had to bear the brunt of the cutting.

NASA was one of the agencies unlucky enough not to be indexed to inflation, and the fiscal restraints imposed on it kept its budget growing at a rate less than inflation, steadily eroding its ability to pay for its Space Shuttle. In 1978, NASA headquarters began transferring funds from other programs to the Shuttle development account to cover shortfalls without notifying Congress. In 1979, it was caught at this game by the appropriations committee staffs. It received a verbal slap on the wrist, which failed to change its behavior. In 1980, it was caught again, and this time the two committees agreed to impose new reporting requirements on the agency, which they made a matter of federal law. But they also agreed to a $285 million supplemental budget request to help cover the Shuttle's soaring cost.[96]

The 1980 election changed the status quo dramatically, however. The landslide election of Ronald Reagan installed a director at the Office of Management and Budget who was ideologically opposed to federal spending on civilian R&D, which was, after all, NASA's purpose. Reagan had promised to balance the federal budget during his campaign, which meant further, deeper cuts, and his instructions to his transition teams had included identification of potential rescissions.[97] Reagan also ushered in the first Republican Senate since Dwight Eisenhower's first two years. In this environment, a supplemental request to cover the Shuttle overruns was out of the question. Instead, in March 1981 headquarters notified the House appropriations committee staff that it intended to transfer $190 million from the Office of Manned Space Flight to the Shuttle, and another $60 million from the Office of Aeronautics and Space Technology, including the SCAR program's entire budget.[98] Neither House nor Senate committee chose to come to SCAR's rescue, and they approved the reprogramming on 3 June. Headquarters instantly terminated SCAR/VCE, not even permitting the management office resources to prepare final reports.

✦ ✦ ✦

The 1970s had not been kind to supersonic enthusiasts. The anti-SST campaigns and the ozone problem had left the public deeply suspicious of the SST's environmental consequences and equally suspicious of the motives of SST supporters. Concorde's long-predicted failure destroyed the promoters' chief argument, that the Europeans would gain technological superiority over the United States through their dogged persistence in getting the highly advanced airplane to market. Ongoing energy shortages made even second-generation SST economics appear questionable, and airlines were not interested—especially since NASA was funding a program to substantially reduce subsonic aircraft fuel consumption, thus improving subsonic jets' economics vis-à-vis prospective SSTs. Whatever the merits of an advanced supersonic transport, finally, neither Congress nor the Reagan administration would pay for it, given the tremendous budget pressures produced by the inflationary economy. For several reasons, then, the AST could not make political progress.

Despite its lack of political success, however, the advanced SST research effort was not wasted. By far the most important outcome of the 1970s research was the accidental discovery of the ozone problem. By 1980, research funded by DOT and NASA had demonstrated that SSTs were less threatening to the ozone layer than Johnston and Crutzen had initially thought, leading to a relatively well-accepted belief that future combustor technology could render an AST ozone-neutral. Instead, SST-inspired research had shown the major threat to Earth's ultraviolet shield was chlorofluorocarbons, which led to a ban on CFC-powered aerosol sprays in the United States in 1978, a worldwide ban on CFC manufacture in 1992, the institutionalization of stratospheric research at NASA, and a permanent NASA program to monitor the stratosphere's composition.

The SCAR/VCE program held its own successes. The Navy-NASA Engine Program was still the dominant engine cycle analysis software in the year 2000, and the successor that was evolving relied heavily on its code. ATLAS, NASTRAN, and FLEXSTAB became the cores of more sophisticated software and thus continued to evolve as well. These tools, in turn, reduced the cost of designing new aircraft—even subsonic ones—substantially. Hence while SCAR/VCE did not succeed at generating a second-generation supersonic transport development program, the effort still increased the manufacturers' future capabilities. The program was also highly productive. Over its lifetime, its researchers produced more than a thousand technical publications on subjects ranging from supersonic aerodynamics to advanced materials, structural design, software complexes, and emissions and noise reduction.[99] Viewed solely on its merits as a technology program, SCAR/VCE was very successful.

The desire to make a supersonic future possible, finally, also helped transform the kinds of aeronautical research NASA performed. In previous decades, the agency had focused on aerodynamic and structural research. Due to the space race, it had ceased air-breathing propulsion research and had had to restart it to support both the SCAR program and the noise abatement research that airlines, aircraft manufacturers, the Congress, and urban residents all demanded. As NASA official Charles W. Harper had noted in 1969, the agency's traditional aerodynamics focus was no longer sufficient to address the SST's fundamental problems.[100] To meet Bill Aiken's stated goal of an "environmentally acceptable, economically viable" SST during the early 1980s, the agency had had to address noise reduction and pollution control in addition to its more traditional activities.

6

THE AIRBUS, THE ORIENT
EXPRESS, AND THE RENAISSANCE
OF SPEED

The cancellation of the SCAR program in 1981 was a reflection of the general state of NASA's aeronautics program in the early 1980s. The Shuttle's overruns had led to termination of several other aircraft and rotorcraft-related projects in 1981, and several surviving projects, such as the Aircraft Energy Efficiency project, were scheduled to terminate upon reaching their goals in mid-decade. And it was not clear that there would be any successor projects. President Ronald Reagan's new administration was openly hostile to federal support of civilian research and development, making NASA's aeronautics program a prime target for elimination.

The agency's problem in 1981 stemmed from the unhappy reality that the very ideals upon which it had been founded had been driven out of the American political mainstream. NASA's aeronautics program, like its predecessor, the NACA, was based upon a belief that government should play an active role in the development of new industries and new business opportunities. Of the several possible ways government could do this, the American political system had settled upon federal support for scientific and technological research.[1] The major move toward this "science state" occurred in the wake of World War II and public revelation of the stunning impact new knowledge had had on the prosecution of the war, although before the war a handful of specific scientific disciplines had benefited from public support—geodesy via the Coast Survey, geology via the United States Geological Survey, and of course aeronautics via the NACA. War's end, however, brought with it advocacy for permanent, state-funded research institutions. Many of these were attached to the armed services—the Office of Naval Research and the Defense Department's Advanced Research Projects Agency are the best known among this group. Their defense ties made these institutions largely uncontroversial. Federally funded civilian science was a good deal more controversial after the war, finding opposition among conservatives wedded to the concept of a laissez-faire state. These politicians had only been able to curb the movement of government into civilian science, not stop it, however. Postwar belief in an activist government, coupled with the government's obvious wartime success at fostering technological progress—always a good thing in Americans' minds—made the spread of the science state impossible to resist. The notion of activist government reached its postwar peak with the 1960 election of John F. Kennedy and Lyndon B. Johnson.

But simultaneously, a strident form of antigovernment ideology was gaining currency, particularly in the South. During the Eisenhower administration, the federal government had taken the first tentative steps toward ending racial segregation in the southern states, reviving a latent but deep hatred of federal power in the region. Westerners had also always resisted federal activism, wishing the (eastern) government would simply keep sending them agricultural (and especially water) subsidies without annoying strings. Because the nation's population was clearly shifting south and west, certain Republicans perceived an opportunity to fundamentally restructure the nation's political culture.[2] The first expression of this transformation was Barry Goldwater's 1964 campaign. Goldwater fashioned a quixotic campaign out of western antigovernment, antiprogressive rhetoric and strident anticommunism and, partly because George Wallace had siphoned off the anti–civil rights vote in the South, went down to a spectacular defeat. But Richard Nixon ran on the same basic ideas in 1968 and found the western antigovernment rhetoric played well in the South, too. Both these men were heroes to Californian Ronald Reagan, who achieved the White House in 1980 using an antigovernment, anticommunist campaign.[3] And unlike Nixon, who had substantially expanded government's role in society despite his campaign rhetoric, Reagan, and especially the man he appointed to head the Office of Management and Budget, David Stockman, meant to carry out a substantial dismantling of the civilian government.[4]

Reagan's election also marked the ascendancy of what multibillionaire investor George Soros has called "market fundamentalism."[5] Market fundamentalism preaches that all good things descend from the unfettered operation of free markets—government, in this view, "distorts" markets and "slows" progress. Like religious fundamentalism, market fundamentalism has its bases in belief, not in historical evidence.[6] Government had always played a variety of roles in the nation's development. The U.S. Army had developed the most important manufacturing technology of the nineteenth century, interchangeable parts.[7] The government had provided a large portion of the financing for the nation's rail network. It had made the arid regions of the American West habitable through vast public water projects—wealth transfers, as it were, from East to West.[8] It had set high import tariffs to encourage domestic manufacturing, imposed regulation on the financial system, and protected wilderness areas from development. Yet despite the federal government's history of active participation in the nation's economic development, the idea that government should remain separate from the nation's economic life continued to dominate the nation's politics until the Great Depression. That long national trauma convinced the majority of the electorate, and the nation's governing elite, that free markets were no longer to be trusted with the nation's economic health. Instead, they adopted the idea that government should actively manage the economy. This idea became policy in the United States with Franklin Delano Roosevelt's appointment of Marriner Eccles to the Federal Reserve Board in 1934, and was cast into rigorous theory by British economist John Maynard Keynes. Keynesian economics dominated the American government for nearly two generations.

The Great Inflation of the 1970s saw the Keynesian consensus fracture and break apart.[9] Because ideas never die, free markets had had disciples in the United States throughout the postwar era, but as long as the nation's economy functioned well, those critics of Keynesian intervention received little attention. But the Great Inflation represented a sickness in the economy, and free marketeers had an explanation that resonated with the New Right's antigovernment ideology. Government itself was responsible for the economy's ills, through its overstimulus of the economy via deficit spending and its distortion of free markets through regulation and spending patterns. The solution was the same one sought by the New Right's leaders for other reasons: deconstruction of the federal civil government. President James Earl Carter, a conservative Democrat, had been the first postwar president to place a significant number of free market economists in positions of authority within his administration, most notably Alfred E. Kahn, who led the administration's airline deregulation effort in 1977 and 1978 and then became his "Inflation Czar" in 1979.[10] Carter had faced a great deal of opposition from his own party over this rejection of now-traditional economics, however, preventing the kind of sweeping reduction in the regulatory state that free marketeers sought. President Reagan faced no such internal opposition. Reagan's staff set out in 1981 to disassemble as much of the federal civilian government as it could get away with, including NASA.

The space agency's dilemma was that it was an egregious violator of free markets. When NASA embarked on a new project, the agency chose the goals, the technologies, and the contractors. These decisions gave NASA influence over which technologies succeeded or failed, because the agency's decisions influenced companies' own internal research and development funding (called IRAD in the jargon). If, for example, the agency's known preference was for hydrogen-fueled rockets, companies were unlikely to spend their own money on petroleum-fueled rockets, even if petroleum-fueled rockets promised to be less expensive. Companies would orient their internal research and development efforts to NASA's, in order to be better candidates for the agency's dollars. To free marketeers, therefore, the agency suppressed innovation, reduced technological diversity, and denied the nation the progress that free markets would have provided. To these NASA critics, it was a dispenser of corporate welfare and a distortion of the aerospace market.

The most vulnerable portion of NASA to this criticism was its aeronautics program. While aeronautics was only a very small portion of the agency's budget in 1980, free marketeers perceived it as the biggest violator of their principles. In 1980, there was only a single purchaser of launch vehicles—the federal government—and a market consisting of a single customer is unfree by definition. In time, Reagan administration officials forged a policy to force the agency out of the expendable rocket business in a privatization effort.[11] But the aeronautics program, which explicitly existed to help an aviation business that already dominated the world market, was an easy target for the "corporate welfare" label. Bolstered by the conservative Heritage Foundation's arguments against NASA, Stockman sought to eliminate its civilian aeronautics research effort entirely by reducing its budget to a level that would permit bare sustenance of the agency's vital aeronautics facilities—some of the wind tunnels and laboratories—and its military-related programs, but no more.[12]

Stockman's effort failed, and in a bizarre twist, led to a new supersonic transport research program late in the decade. President Reagan's science advisor, Edward Teller protégé George A. Keyworth II, began a counterattack by forming a policy review committee that resulted in a series of reports specifying the conditions under which NASA should support new technologies and that laid out a "road map" for the nation's aeronautical research through the end of the century. One element of that road map, a Mach 25 hypersonic transport that could reach space or, perhaps, Tokyo in two hours, inflamed President Reagan's love of political symbol and became the "Orient Express" in his 1986 State of the Union address. Under the cold light of economics, however, the hypersonic transport rapidly devolved back into a merely supersonic one, to be called the High Speed Civil Transport. Backed by wide political support and bolstered by European hints of a future "Concorde II," the HSCT became the space agency's top aeronautics priority in 1989.

"EUROGAULLISM" AND THE AIRBUS PROBLEM

The free marketeers' attack on NASA's aeronautics program actually began in the final years of the Carter administration, and came at a very bad time for the American aircraft industry. For the first time in the postwar era, it was beginning to lose market share to foreign competition. The threat, however, was not the one that a generation of supersonic enthusiasts had used to demand government investment in supersonic transport research, the unfortunate Concorde. Instead, it came from the airplane American Airlines' old chief engineer, Frank Kolk, had tried to get the U.S. manufacturers to build in the late 1960s, a widebody, twin-engine aircraft for short-range intercity flying. Unfortunately, this "airbus" was not being assembled in Boeing's vast Seattle complex or McDonnell-Douglas's Long Beach works. It was taking shape in Toulouse, France, a product of an industrial consortium jointly owned and financed by France, West Germany, and Britain known as Airbus Industrie. This state-financed competition represented the antithesis of American free market ideology, and it drove the U.S. manufacturers and their supporters to campaign against the Office of Management and Budget's cuts in NASA's aeronautics program.

Airbus Industrie, like Concorde, was a product of French ambitions to achieve technological and economic independence from the United States. French leaders believed that the fragmented European market could not sustain each individual nation's existing aircraft industry because no one nation was able to sell enough airplanes to make back the investment new aircraft designs required. The total world market was simply not large enough to sustain one French, two British, and three American aircraft manufacturers. At best, the world market of the 1970s might sustain three manufacturers—total. If one of these survivors was to be "French," it would have to be a French-led *European* consortium. Once the largely state-owned European manufacturers had been united in some way by their national governments, each of the member states would have a strong incentive to "buy European." In this way, the existence of the consortium would help defragment the European market, perhaps resulting in enough sales to eventually achieve profitability (and knock one or more American manufacturers out of the commercial business). It would also contribute to the resurgence of France as a respected world power, the central goal of Gaullist policy.[13]

In 1965, years before Concorde would actually fly but after it was already clear that the "short range" version would not be worthwhile, an Anglo-French committee had explored the idea of a large airliner suited to European intercity distances, 800 to 1,200 n.m. The American widebody jets being launched the same year were too big and were better suited to North American distances, which would make them less efficient for European use. These publicized efforts drew German attention, and at the 1965 Paris air show, a group of West German manufacturers formed "Studiengruppe Airbus" specifically to discuss technical standards with their French counterparts. Later that year, a group of European airlines also discussed their needs for an economical intercity airliner. The design the airline committee converged on was the "jumbo twin," Frank Kolk's ideal airliner.[14]

The attraction of the jumbo twin for airlines was simple economics. The wide fuselage would permit a much greater passenger load than a narrow body of similar length, without substantially increasing drag (and thus fuel costs). And because engines were the most expensive part of an airplane in terms of purchase and maintenance, reducing the number of engines to two, the minimum allowed under existing safety regulations, would produce lower costs. The widebody twin was thus the perfect airliner, assuming its manufacturer did a competent design job. McDonnell-Douglas and Lockheed had chosen tri-jets instead, leaving a "hole in the market" that the airbus might fill.

The Airbus twinjet first flew in October 1972, received commercial certification in March 1974, and went into service with Air France in May 1974. But Airbus Industrie only managed to sell thirty-eight aircraft to four customers by 1978. Between late 1975 and the middle of 1977, Airbus sold only two more. As writer John Newhouse eloquently put it, the plane seemed to be just like its other European forebears—"well designed, well built, and a commercial flop."[15] Airlines' lack of interest in the A300 changed dramatically, however, after Eastern Airlines placed an unexpected, large order in April 1978. Run by former astronaut Frank Borman, Eastern had leased four A300s to try out on its routes, and after several months Borman was satisfied with the aircraft. Eastern placed orders for twenty-three A300s and for twenty-five of a slightly smaller version, the A310, which Airbus Industrie was planning to launch later that year. Airbus Industrie had offered substantial inducements to Eastern, including loans at interest rates well below those prevailing in the United States, which was suffering from high inflation. But Eastern was known in the industry as a savvy buyer of aircraft, and this sale gave the A300, and Airbus, the credibility that its only other large sale (to state-owned Air France) had not. The Eastern sale, combined with surging fuel prices, caused the A300 to become a sudden hot seller. By the end of 1979, Airbus Industrie had sold more than three hundred A300s.

Airbus Industrie, therefore, had suddenly become the threat to American aeronautical leadership that Concorde was supposed to have been. With the reentry of Great Britain to the consortium in 1978 as a partner in the A310, it was a truly multinational entity with access to the treasuries of the three richest European nations. Its products were technically competent and economically competitive and evinced an attention to airline needs that previous European aircraft projects had not. And in perhaps the biggest blow of all to the American manufacturers' pride, it was evidence that the much-derided European "state socialism" model of economic development might actually work.

The sudden Airbus threat, combined with the rapid shrinkage of NASA's aeronautics research budget under the combined assault of Shuttle overruns and Office of Management and Budget cutbacks, caused the American aircraft industry to scramble to protect its own rather limited subsidies, as contained in NASA's applied technology programs. These efforts, like the SCAR/VCE and Energy Efficient Engine projects, were intended not only to develop new technologies but to make the new technologies' transition into products more affordable—and thus more likely. "Technology" is essentially a form of knowledge. If NASA did all of its research and development internally, its technologies would disseminate only very slowly, if at all, and at great expense as companies would effectively have to relearn them. Performing R&D and developing knowledge jointly made the conversion of that knowledge into new products less expensive. This was why the manufacturers were intimately involved in the projects, and, of course, it was also why free marketeers perceived them as corporate welfare.

With its aeronautics research budget shrinking, NASA headquarters had asked the Aeronautics and Space Engineering Board (ASEB) of the National Research Council to study the agency's future role in aeronautics.[16] The ASEB was composed of engineers drawn from industry and universities, and its function was to serve as a relatively independent source of expertise for policymakers. Its parent organization, the National Academy of Sciences, maintained a study center in Woods Hole, Massachusetts, and this was where the ASEB's panel met in late July and early August 1980 to construct its response to NASA's request.

The panel's chair, H. Guyford Stever of Stanford University, divided the attendees into four groups, to consider the agency's potential role in military, transport, general aviation, and rotary-winged aircraft. The transport group's chair was Boeing vice president John Steiner, and included representatives from General Motors, Douglas Aircraft, TWA, Pratt & Whitney, Lockheed, MIT, and Pan Am. The group was tasked to deal with the central question raised by the free marketeers' criticism of NASA's aeronautics research: what parts of the research and development spectrum belonged to NASA, and which should belong to the manufacturers?[17] All but the most dogmatic free marketeers accepted that the government should play some role in advanced research, after all. But because there was no obvious demarcation line along the research-development-new product spectrum, the group's challenge was to create one that could be conveniently defined for policymakers.

In their report, the transport panel laid out an argument that NASA had to remain involved in aeronautical technology development because state-funded foreign competitors had been able to adopt advanced technologies that the U.S. manufactures could not afford. The group used French support of Dassault as a primary example. Dassault was building a complete composite wing to demonstrate the commercial viability and weight savings of new structural materials using state funding.[18] This was the same idea that the American manufacturers had been trying to get funded via NASA's Primary Composite Structures research effort that was being denied by OMB. The

weight savings promised to reduce fuel consumption, thus making future foreign aircraft more economical than U.S. airplanes. This loss of technological
leadership, the group argued, would have profound economic consequences.
The American aircraft industry had generated $35 billion in export sales in the
past decade, accounting for hundreds of thousands of jobs and a substantial
portion of the nation's balance of trade. Continued erosion of U.S. market
share to Airbus would cost 100,000 jobs in the near future, an impact that
would be felt in all fifty states through the companies' network of suppliers.[19]

In essence, therefore, the transport panel was arguing that NASA had to
remain involved in aeronautical research and development because other
governments were following the U.S. model of aircraft development—state
funding of new aerospace technologies. They were doing it in a somewhat
different way, to be certain, but it made little difference in the end whether
aviation technologies were developed by DOD and NASA first under the
"national security" rubric and then transferred to the private sector (707, DC-
8, 747) or whether the state directly funded new commercial aircraft technologies (Caravelle, Concorde). Aviation simply was not a classical "free
market." If the United States did not continue its own subsidization of technological progress, it would cease to be the world's leader in aeronautical
technologies and forfeit the economic advantages that it currently had. Hence
Steiner's panel was arguing that cessation of NASA's applied technology
research amounted to unilateral disarmament in an industrial cold war. How,
then, to explain to the free market enthusiasts that NASA's "corporate welfare" functions should continue?

NASA headquarters had divided the research and development spectrum
into eight sections for the committee, and in order to address the issue of
where NASA's responsibilities should end and industry's begin Steiner's group
addressed each one. Four of them, the maintenance of national aeronautics
facilities, the conduct of basic research, and the evolution of generic and
"vehicle class" technologies, were far enough distant from product development to be uncontroversial, the group thought.[20] One section, prototype
development, had never been within NASA's purview and the agency did not
want the role. Another, operations feasibility, required government involvement because operations feasibility research involved integration of aircraft
into the nation's air navigation and traffic control infrastructure, which was
government-owned and operated. Hence the primary issue of state support

devolved upon two categories of research, in the committee's opinion: *technology demonstration* and *technology validation*.

The difference between the two was rather subtle. A technology demonstration was necessary when evaluation of a new concept required full, or at least, large-scale testing. The Energy Efficient Engine program had required a technology demonstration phase because after each individual component had been built and tested, engineers would still not know whether they had been successful at meeting the project's goals until they saw how the components functioned within an engine. This was because in an engine the components interacted with each other in complex ways that could not be predicted in advance. Without the confidence produced by an engine test, the companies involved had a very strong disincentive to applying the new technologies—high risk of an expensive product development failure. Steiner's group therefore argued that NASA should undertake technology demonstration when the benefits of the technology were "judged to be high."[21]

Technology validation was more controversial, even within the group. Validation was really the final step before actual product development, and included large-scale ground and flight testing. Its purpose was "to make possible, with reduced risk and without prohibitive development costs to industry, the practical utilization of high-benefit, high-risk conceptual, component, or subsystem technological advances."[22] Steiner's committee agreed that this should be undertaken only when the technology is clearly in the national interest. But a member of the group who did not work in the aviation industry objected to the use of public money for this purpose. The unnamed member pointed out that the only arguments the other members had made to support this were based on the industry's size and the presence of state-funded foreign competition—both of which were conditions that obtained for other industries too, such as automobiles, which did not receive government aid. Why, this perceptive critic seemed to be asking, should aviation receive special treatment?

Why indeed? The answer the committee gave was a historical exegesis of the NACA and NASA, hoping to convince unbelievers that because the government had always supported aviation development it should continue to do so. The problem the committee faced in convincing this and many other skeptics of the need to continue subsidizing aviation was that the answer to the critic's question wasn't economic, technical, or historical. It was political.

The United States government had given civil aviation a special place in the nation's economic life decades before, and it had done so for expressly political reasons. Those reasons, the desire to improve internal communications through greater speed, to improve national defense through better military aircraft, and to create new business opportunities, were part of the activist government ideology that was rapidly becoming extinct. Military aircraft development belonged to the Defense Department now, not NASA, and the United States' internal communications network was highly effective. In any case, free market ideology placed internal communications in the hands of private enterprise—except, of course, for roads. Free marketeers also insisted that new business opportunities should come from entrepreneurs, not from government. Steiner's transport group, and the larger effort by the ASEB, could not respond to the free market critique adequately because they were not equipped to address an issue of political ideology.

The ASEB's report, therefore, had no real effect. Released in January 1981, it seemed moot, at first, because the president elect, Ronald Reagan, was widely perceived to be more supportive of aerospace than President Carter had been. He had taken pains to ensure that perception, for example asserting in a letter to Clifton von Kann, then head of the Air Transport Association, that he would be a friend of the aircraft and air transport industries.[23] Further, he was a former governor and long-time resident of California, which contained a major portion of the aerospace industry. Hence the industry's leaders were shocked when his budget director, David Stockman, launched his assault on NASA's aeronautics program. President Carter's FY 1982 budget had requested $469 million for NASA's Office of Aeronautics and Space Technology, with $323.6 million for aeronautics programs. After taking office, President Reagan launched a major drive to reduce that budget, and the amended budget the OMB sent to Congress reduced the aeronautics portion of the budget to $264.8 million. The approved amount, $284.2 million, was higher than the White House wanted, but was still a significant decrease from FY 1981.[24] The FY 1981 budget had contained a 10 percent decrease, and as discussed in the preceding chapter, a substantial fraction of the approved FY 1981 aeronautics budget was actually diverted to the Shuttle.

Due to the structure of the budget process, the first true Reagan budget was FY 1983, and it was this budget that represented Stockman's primary efforts to eliminate NASA's subsidies to the aircraft industry. In "Special Analysis K," a review of federal research and development policy, Stockman had specified that in FY 1983, federal support for aeronautics would be focused on fundamental research, facilities maintenance, and technology activities "critical to the Nation's defense needs." But "technology development and demonstration projects with relatively near term commercial applications [would] be curtailed as an inappropriate Federal subsidy."[25] The budget submitted to Congress in March 1982 therefore essentially eliminated the aeronautics-related "systems technology" studies. The depth of the cuts caused outrage in the House Science and Technology committee. The amount requested for aeronautics was $232 million, an 18.6 percent decrease from the previous years' budget. The committee's members were particularly unhappy with the way Reagan's White House had gone about achieving these cuts. OMB had presented a revised FY 1982 operating plan that eliminated all technology validation activities except those that were clearly related to military aircraft—without consulting the committees. In response, the chairs of the two independent agencies' appropriation subcommittees, Edward Boland in the House and Jake Garn in the Senate, requested that NASA administrator James Beggs have the budget reviewed by the National Research Council's Committee on NASA Program Reviews. Their response was to be due in time for the July budget hearings.[26]

The letter caused a stir in the White House, where the Office of Science and Technology Policy, led by physicist George A. Keyworth II, had also been concerned with the impact of Stockman's "Special Analysis K." Senator Garn was one of President Reagan's oldest friends and closest political allies, and the implicit rejection of the cuts contained in the letter did not bode well for the idea's chances. Hence as the NRC's Committee on NASA Program Reviews began its study of the cuts, the OSTP put together its own policy review committee. Called the Aeronautics R&T Steering Group, this committee was tasked with examining U.S. aeronautical policy and composing a new policy that would cause private industry to "assume research for near-term commercial application."[27] That's not quite how things worked out, however. In the near term, Congress reversed most of the White House's cuts without the need of new policy, restoring the aeronautics budget to roughly the previous year's level.[28] When it finally emerged, the new policy called on the United States to embark on a series of revolutionary new commercial aircraft technologies to ensure continued U.S. aeronautical superiority. In short, the new policy enshrined state-financed technological change in commercial aviation—

Gaullism, without the provocative name. It also helped launch the United States on a most improbable quest for a *hypersonic* transport.

POLICY ADVOCACY AND SYMBOLIC POLITICS: ON THE ROAD TO NASP

In his State of the Union message of 4 February 1986, President Reagan announced the launch of a new aerospace research program whose ultimate result he named the Orient Express. This vehicle, he said, would allow passengers to fly from New York to Tokyo in two hours, perhaps by the end of the century. Whereas Boeing's 2707 would have traveled the world at a mere Mach 2.7 had it been built, this transatmospheric vehicle would have a top speed of Mach 25. It was, in short, the ultimate dream of speed. Numerous critics at the time assumed that Reagan's announcement had been a defensive act precipitated by the space shuttle Challenger's explosion, which took place on the originally scheduled date of the State of the Union address, but in fact the address had been long prepared. The announcement was a product of aeronautics advocacy, the Strategic Defense Initiative's need for a cheaper launch vehicle, and President Reagan's love of political symbol.

The first of a series of reports advocating a continued NASA role in funding technology development and validation during the Reagan administration was the National Research Council's examination of the FY 1983 budget. The Committee on NASA Program Reviews had been asked to examine the projects that had been deleted from the budget proposal, estimate the impact of their not being completed, and assess the likelihood that industry would pick up the technologies on its own. They also chose to consider the larger question of whether government should help industry "bridge the gap" between NASA's internal research and technology activities and "early application" of NASA's results. This larger question was really the key issue in the conflict between OMB and the aeronautics lobby. The technology validation projects that served to "bridge the gap" were precisely what free marketeers perceived as corporate welfare. Unsurprisingly, the National Research Council's committee recommended no significant changes in the kinds of programs NASA supported. They recommended retention of several of the technology validation projects that OMB had tried to delete, including the controversial composite structures programs and the Energy Efficient Engine program.[29] Their rationale was that the extremely high cost of developing a new airliner required that the government mitigate the development risk through validation studies—otherwise, they implied, technological change in the industry would cease. Industry would not pick up a new technology until it had been rendered virtually risk free.

The NRC's report also attempted to justify continued government support for aeronautics research. Like the 1980 NRC study of NASA's role in aeronautics, the NASA Program Review committee relied heavily on the aircraft industry's importance to the U.S. economy and on the government's long support for it as primary rationales. Perhaps implicitly recognizing that this was an insufficient reason, the study's authors also sought to blur the OMB's defining line between military-related research and civil aircraft research. Claiming that "much aeronautical technology in transports, in the engine field, and in helicopters is equally applicable to defense as to civil aviation needs," the committee hoped to wrap support for the aircraft industry in the comfortable blanket of national security.[30] Their statement was certainly a true one. Technologies beneficial to commercial aviation had routinely come from military projects, although the direction of flow was beginning to reverse. But this was not an adequate argument for sustaining explicitly civilian research, either.

The White House Office of Science and Technology's study group also sought to answer the troublesome question of what the government's role in aeronautics research should be. The committee consisted of representatives from DOD, NASA, the Commerce and Transportation departments, OMB, and two observers from industry and the university. The NASA members of the committee's working group were Bill Aiken and Raymond Colladay, later associate administrator for aeronautics and space technology. OSTP had tasked the committee with two specific questions. First, was aeronautics a mature technology, or was continued investment justified by potential future benefits? And second, what were the proper government roles in aeronautics research?

The OSTP's report began by rejecting the earlier breakdown of the R&D process and reformulating it in three pieces: research and technology (R&T) development, technology demonstration, and system development. The committee's reconstruction of the R&D spectrum allowed it to further blur the line between what the NRC's earlier study called "technology validation" and "technology demonstration." The primary distinction between R&T development and technology demonstration in the new structure was *intent*. A program would belong to the R&T development category if its goal was basic research, design data, or validated design procedures, but would not if the program goal was demonstrated system performance. Reliance on subjective criteria like "intent" allowed technology validation and demonstration to continue as long as an appropriate intent could be formulated. Programs like the controversial composite structures effort could continue, for example, because while it was clearly demonstrating a specific technology's performance, one of the program's major purposes was data collection to support future certification by the FAA. Use of the new definitions allowed the committee to support continued government financing of civilian aeronautical R&T development, while restricting technology demonstration activities to military aircraft exclusively. In essence, it recommended no significant changes in the kinds of programs to be funded.

The report also denied the need for a restructuring of the nation's aeronautical R&T organizations. The group had examined other alternatives, including transferring all aeronautics research to the Defense Department, establishing a separate aeronautics research agency to conduct both civil and military research, and privatization of the aeronautics research establishment. Each of the alternatives had flaws. The first two, the committee believed, would result in Defense Department short-term needs dominating the research agenda, harming civilian aeronautical progress. The third would remove aeronautical expertise from the government, preventing it from being a "smart buyer" of aircraft and allowing it to be manipulated by unscrupulous contractors. Hence the report advocated maintaining the existing structure, with one minor exception. When the NACA had been transformed into NASA, the advisory committees that had allowed the Army, the Navy, and industry to influence the NACA's research agenda had been disbanded, and the group concluded that this had been a mistake. The committee thus advocated creation of a new high-level NASA-DOD-industry committee to review and coordinate all aeronautical R&T policies.[31]

Finally, the committee argued that aeronautics was still worthy of national investment by linking national security to research. Relying on the widely known fact that the Soviet Union maintained far larger numbers of vehicles, the committee argued that the United States had to meet the Soviet threat through superior technology. There were many opportunities available for ensuring the superiority of U.S. aircraft, including military transports: doubling the supersonic L/D, reducing drag 20 to 40 percent using laminar flow control, reducing specific fuel consumption 15 to 20 percent, advanced avionics, and hypersonic attack missiles, to name just a few. And, they implied, many of these advanced military technologies would find their way into commercial aircraft, preserving the nation's market advantage.[32]

Despite the committee's strong defense of the establishment, or perhaps because of it, its actual policy recommendations were weak. The United States should "maintain a superior military aeronautical capability," "provide for the safe and efficient use of the national airspace system," "maintain an environment in which civil aviation services and manufacturing can flourish," and, finally, "ensure that the US aeronautical industry has access to and is able to compete fairly in domestic and international markets."[33] And it should support aeronautical R&T to "ensure the timely provision of a proven technology base to support future development of superior US aircraft and for a safe, efficient, and environmentally-compatible air transportation system."[34] These policy recommendations were weak because they lacked specifics. And although these statements were characterized as goals, they provided nothing against which to measure progress. In short, the OSTP's study responded to David Stockman's threat by recommending return to the pre-Reagan status quo.

The one substantial institutional change that the committee had recommended, a new advisory committee, was the source of continued advocacy for change. In a letter to Keyworth, NASA administrator James Beggs and Richard DeLauer, the undersecretary of defense for research and engineering, recommended that a committee reporting to OSTP was the best way to ensure that the new policy was implemented.[35] Initially, the two recommended that the chair of the committee should be drawn from a university aerospace engineering department, but OSTP did not wish to restrict the chairmanship so narrowly. Instead, the initial nominees were from universities, the NRC, and from industry—although not, quite purposefully, from any of the major aerospace contractors.[36] Nonetheless, Keyworth appointed recently retired Boeing vice president Jack Steiner to chair the committee. During the next four years, Steiner's committee produced two more reports, each one advancing more specific criticisms and recommendations. As the recommendations from these reports were accepted, they significantly altered the way NASA performed its aeronautics research.

Steiner's first report, submitted in November 1983, argued that while the existing U.S. aeronautics research effort was "generally consistent" with the OSTP's policy statement, it was not consistent with the policy's "bold leadership spirit." NASA and DOD were not pursuing high-payoff areas of research aggressively enough to ensure U.S. preeminence into the twenty-first century. Especially important, in Steiner's view, was advanced research in systems integration, including an adventurous and stable program of flight research. Bereft of vision, the existing programs could cause the nation to fall short of its stated goals. Aeronautics was on the "threshold of a renaissance in research and technology accomplishment," he argued, and he laid out a series of visions for the year 2000. Stealth could revolutionize military aircraft design, composite structures could reduce aircraft structure weight by 40 percent and lower ownership cost, all-electric aircraft technologies would further reduce weight and increase efficiency, as would high-speed turboprops, ceramic engine components, and laminar flow control. Most important, however, was his proposal for manned hypersonic and transatmospheric vehicles. Hypersonic aircraft might be radically more effective interceptors and reconnaissance aircraft, while transatmospheric vehicles could allow entirely new mission concepts to evolve.[37]

Standing in the way of these leaps in technology was a set of roadblocks. These included lack of vision, an unbalanced R&T effort lacking in experimental aircraft and flight research, insufficient aggressiveness, inattention to the new discipline of systems integration, and the lack of long-term, high-risk R&T. NASA and DOD had become focused on incremental improvements in technology, not radical leaps, and this trend was exacerbated by DOD insistence on an "audit trail" that linked research to specific weapons systems. This impaired generic research that might be applicable to a wide range of future systems, but did not necessarily apply to anything current.

Steiner's report was thus an indictment of the status quo, and it recommended a series of changes. First, it called on NASA and DOD to construct a vision for the year 2000 and devise a development path to get there. "High payoff" areas should be chosen and pursued more vigorously. Systems integration studies should be emphasized. More research aircraft should be built and flown. And finally, the nation had to face that a steady real growth in resources would be necessary to maintain U.S. preeminence in aviation.[38]

Unsurprisingly, Steiner's report did not play well in either the Pentagon or in NASA headquarters. The acting undersecretary of defense for research and engineering, for example, responded that he could not agree that "the current DOD aeronautical research and technology program's content is so unimaginative, that its level is so low or its pace so slow that it places our country's future military aeronautical technical preeminence in jeopardy."[39] But unlike many other critical reports, this one was not buried. While protesting that what the committee seemed to want was far beyond their resources, the two agencies requested that the ASEB hold a workshop in which "innovative free thinkers" could project the state of knowledge in 2000.[40] The resulting reports validated the potential performance gains that the two previous policy review committees had asserted were possible with sufficient resources and focus. Focus was the key problem, however. The ASEB's workshop produced two volumes totaling more than three hundred pages, and they contained a huge variety of advanced aircraft concepts—supersonic V/STOL fighters, subsonic strike aircraft, hypersonic fighters and transports, supersonic transports, highly efficient subsonic spanloaders, turboprops, and several different rotorcraft.[41] Because resources were limited, the nation could not do all these things. Someone would have to choose the direction that U.S. aeronautical research would take.

Steiner's second report, completed in December 1984, provided a set of specific goals that the United States should work for. His committee had agreed upon "three broad flight regime-oriented" national goals. These were "transcentury renewal" in subsonic aircraft, long-distance efficiency in supersonic aircraft, and the preservation of U.S. options in transatmospheric vehicles. Attainment of each of these goals depended upon the achievement of "pacing technologies," such as laminar flow control for drag reduction, new "super bypass" or propfan engines, and metal-matrix composites. Achievement of the subsonics goal would halve the cost of air transport in the United States, the committee argued, while the supersonics technologies could produce the "starting basis for [a] Pacific Supersonic Transport." Finally, the United States should pursue the convergence of air and space technologies and operations. For this, the pacing technology was air-breathing hypersonic propulsion, which could eventually allow space operations with airplanelike flexibility and regularity.[42]

In his cover letter, Steiner noted that in constructing the three goals, the committee had engaged in a "spirited debate."[43] Indeed it had. While the prior year's report had focused on a lack of visionary spirit as the key impediment to aeronautical progress, this one homed in on "affordability" of military and civil aircraft as the major problem to be overcome. Technology validation was the key to achieving more affordable aircraft, he argued. This was the most expensive part of research, and it required "more time and money than many in program advocacy positions recognize or admit to." Technology validation was therefore being "short circuited" by national policies and procedures, including "short-sighted tax policies, acquisition procedures and the 'program' focus of special interests and congressional oversight."[44] The last was a thinly veiled criticism of NASA, and especially of the growing dominance of program managers, as opposed to researchers, within the agency, and NASA's representatives on the committee, Ray Colladay and John McCarthy, had not missed the implication.[45] Pressure from NASA caused the report to be rewritten, deleting the inference but maintaining the primary criticism of insufficient attention to validation.[46]

The final report justified the three goals on the grounds that "aeronautics [was] so woven into the mainstream of our national interest" that the nation could not allow its leadership position to continue to erode. The United States had to secure its preeminence in aeronautics by renewing its traditional strength: "the pioneering of new technology." Decisive technological and cost advantages would allow it to dominate over "foreign approaches distasteful to American culture and enterprise."[47] The committee was, in essence, arguing that the United States would achieve domination of the aircraft market through technological revolution, thus ensuring the success of the American "free enterprise" system over the statist Europeans.

The report was made public at a press conference on 1 April 1985, and OSTP initiated an advocacy plan including meetings with individual congressmen and senators, presentations to the relevant congressional committees, recruiting the support of industry leaders, and a meeting with the president.[48] Keyworth also asked NASA's Jim Beggs to assemble a set of "road maps" to guide the nation's efforts toward the three goals by 30 April 1985.[49] These were a key part of the overall effort, as they would define how the goals were implemented as programs.[50] The road maps that NASA delivered in mid-May presented a time-phased research agenda beginning with a "transcentury transport" whose product definition phase would begin almost immediately and go into production around 1993. For supersonic cruise technologies, the road map indicated beginning an "advanced technology" phase in 1985, proceeding to flight validation in 1995, with product definition to follow in 2000–2003. A supersonic transport could then follow around 2005. Finally, the hypersonic vehicle road map proposed that a research and technology program run through the early 1990s, followed by the construction of a research aircraft beginning in 1995, with potential production of operational hypersonic aircraft beginning around 2010.

In support of the supersonics road map, the study's authors argued that current market assessments showed the need for a transpacific-range SST. Year 2000 traffic projections on the North Pacific routes indicated that 32.4 million passengers would be flying by then, and current aircraft required nine to sixteen hours to make the flights. An SST would reduce these trip times to four to six hours. Achieving this goal would require an aggressive program to develop variable-cycle engines, supersonic laminar flow control, and high-temperature lightweight materials. The schedule the road map set out would require a "strengthened commitment to technology development," or the timing would prove extremely optimistic.[51] But the supersonic goal was not the one the White House chose to promote as a national program. The hypersonic transport received the White House's favor instead.

The idea of an air-breathing hypersonic vehicle was not a new one in 1985. During the 1960s, the Langley Research Center had designed and built a "hypersonic research engine" intended to verify some of the basic theory behind hypersonic propulsion. The major difference between a hypersonic jet engine and a subsonic one was the means by which air was compressed to reach ignition temperature. In a traditional subsonic and low-supersonic turbojet engine, a very high-speed rotary compressor powered by a turbine in the engine exhaust provided the compression. Above Mach 3, however, one no longer needed the compressor/turbine combination. By careful shaping of the inlet duct, one could achieve enough compression to sustain combustion simply from the vehicle's forward velocity. Engines operating on this "ram" principle between Mach 3 and Mach 6 were called ramjets. Above Mach 6, a variation on the ramjet idea called a *scramjet,* short for "supersonic combustion ramjet," would be the most appropriate concept. Langley's hypersonic research engine was a podded scramjet, meaning that like an airliner's engine, it was housed in a pod separate from the vehicle's primary airframe. Before the engine's testing was complete, however, many of the center's researchers had concluded that the podded scramjet would never produce enough thrust to overcome its own inherent drag—it would not produce "net positive thrust."[52] One alternative to the podded scramjet was an *airframe integrated* scramjet. This engine would not have its own compression chamber, as the HRE had. Instead, the vehicle's airframe became the "compressor," and fed the compressed air to the engine. To do this, the vehicle had to be very carefully shaped. Essentially, the vehicle's surfaces became part of the propulsion system. The airframe integrated scramjet erased the traditional distinction between airframe and propulsion, and taken to its logical conclusion produced radically different-looking aircraft. During the mid-1970s, the Langley Research Center had refurbished a disused facility to turn it into a testing area specifically designed to determine whether an airframe-integrated scramjet would produce positive thrust at Mach 7. During 1979 and 1980, a research group led by Robert Jones ran a series of tests that demonstrated an airframe integrated scramjet could produce positive thrust.[53]

Almost simultaneously, two other events occurred that set the United States on the hypersonics path. The first of these was an unsolicited proposal to the Defense Advanced Research Projects Agency (DARPA) by Tony DuPont, head of DuPont Aerospace. DuPont, a scion of the industrial chemicals giant's family, had been a program manager for AiResearch, the company that had manufactured the hypersonic research engine for NASA. DuPont had initially tried to design hypersonic vehicles using podded scramjets but had been convinced by Bob Jones that this would never work. So he had turned to designing a vehicle based on a proprietary integrated scramjet that could be converted to a rocket. Because scramjets produced no thrust below Mach 6, any scramjet-powered vehicle needed a propulsion system to boost it to that speed, and DuPont had chosen to try to integrate a rocket engine. This hybrid engine, he hoped, would permit a single-stage-to-orbit space vehicle to be built. DARPA manager Bob Williams liked the idea, and funded it as a "black" program code-named "Copper Canyon." The first phase of Copper Canyon supported DuPont's effort to produce a conceptual design of his vehicle. After consulting with several hypersonics specialists over Christmas 1983, Williams expanded Copper Canyon to analyze DuPont's design and compare it with others.

In England, roughly the same thing was happening. Alan Bond, a member of the British Atomic Energy Authority, had patented a new propulsion concept upon which British Aerospace and Rolls-Royce had based a new space vehicle design. This was to be a reusable, single-stage, horizontal take-off and landing aerospace craft. The concept was given the name "HOTOL," short for "Horizontal Take-Off and Landing." British Aerospace revealed HOTOL in August 1984 in a splashy marketing brochure, which focused on the vehicle's use as a communications satellite launcher. It did not, however, neglect the passenger possibilities—HOTOL could take a "people pod" to become a transatmospheric "skyliner" capable of flying from London to Sydney in forty-five minutes.[54]

When BAE unveiled the HOTOL concept, it had not yet received government sanction or funding. Its timing, however, indicated that if the United States were too slow in its pursuit of hypersonic technologies, HOTOL could become the first hypersonic transport, or at the very least, the first air-breathing transatmospheric vehicle. This raised some concern at OSTP, which was not briefed on Copper Canyon until early April 1985, when Langley's Bob Jones and DARPA's Bob Cooper briefed Keyworth and other members of the White House staff on the project. After hearing the briefing, Keyworth enthusiastically promoted a major hypersonic effort to Jim Beggs and to National Security Advisor Robert McFarlane.[55] He believed that Copper Canyon's aerospace plane could revolutionize the aviation industry and spur the commercialization of space. More importantly, he wrote to McFarlane, it would be a "beacon of U.S. technological strength and leadership."[56] The aerospace plane would be the kind of "bold national initiative" that President Reagan was famous for, and Keyworth advocated pushing it to the top of the nation's aeronautical research agenda.

At the same time that HOTOL and Copper Canyon were gaining advocates on opposite sides of the Atlantic, the White House was engaged in a set of national security studies devoted to the subject of space transportation.[57] The primary focus of these studies was on decreasing the cost of space access. This was a vital national security issue because President Reagan's Strategic Defense Initiative—Star Wars—called for placing a large number of very heavy ballistic missile defense satellites in Earth orbit. NASA's Space Shuttle had proven prohibitively expensive and unreliable, and thus something else was necessary to reduce SDI's cost to something remotely "affordable." These studies were not due for completion until May 1986, but well before then the idea of an aerospace plane as a shuttle replacement had already taken off.

The first public expression that the White House was considering a space plane came in hearings held in July 1985. After hearing the Copper Canyon briefers, Representative Dan Glickman (D-Kans.), chair of the House Subcommittee on Transportation, Aviation, and Materials, had called hearings designed as a forum to discuss the opportunities for a renaissance of "high speed commercial flight" that might derive from Copper Canyon's hypersonic space plane. At the hearings, Keyworth told Congress that the aerospace plane would skip over the supersonic transport and make a "true double jump" in technology. It would make travel to the Pacific Basin "routine and simple."[58] And its ticket prices would be competitive with the subsonic airliners.

At these hearings, Keyworth was supported by testimony from McDonnell-Douglas officials. At the conference, Douglas vice president Dale Warren promoted a Mach 5 airliner that derived from studies by the company's astronautics group. Retired Douglas vice president Paul Czysz explained later what the company had in mind.[59] Douglas had devised a hypersonic demonstration vehicle concept that they believed could, with minor changes to the propulsion system, serve as both a transatmospheric vehicle and as a Mach 5 airliner. (See figure 6.1.) Initially, the company intended to put the Mach 5 version of the demonstrator to use in the highly profitable small-package transport business, and Czysz recalled that Federal Express chairman Fred Smith had offered to invest a substantial sum of money in the idea. Understanding that the first-generation hypersonic transport would probably not be profitable, Czysz's group believed that these prototype-demonstrators were necessary to prove the technologies, and the knowledge gained in operating them would lead to a profitable second-generation hypersonic transport.

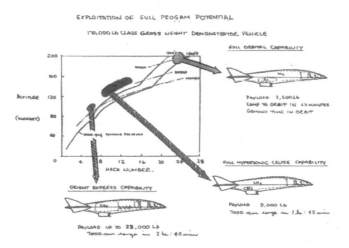

FIG. 6.1. The McDonnell-Douglas technology demonstrator concept. Three vehicles using the same airframe but slightly different propulsion systems could reach orbit with a small payload (upper right), achieve Mach 12 cruise with a somewhat larger payload (lower right), or using methane fuel, cruise at Mach 5 with a substantially larger payload (lower left). This last mission is what Paul Czysz had meant as the "Orient Express" mission. [Courtesy Paul Czysz]

At the hearings, Douglas's Warren had argued that the Mach 5 vehicle had a major advantage over a slower SST-like airplane. By flying above 100,000 feet, the aircraft's boom signature would be below 1 psf, which the company believed would be indistinguishable from background noise and thus publicly acceptable. This would allow the Mach 5 vehicle to fly supersonically over land, which a slower, lower-altitude SST could not. The vehicle's disadvantages were a substantially later availability date and a higher unit cost. Both of the disadvantages stemmed from the reality that there were no vehicles at all operating in the Mach 5 speed range, and hence the basic knowledge and technologies for the aircraft simply did not exist. While the SST was an extension of knowledge the aeronautical engineering community already possessed, the hypersonic transport was entirely new. The hypersonic demonstrator, however, would provide some of the technologies that a hypersonic transport would need, and thus Warren presented the concept as a post-2000 derivative.

In 1990, Keyworth aide Bruce Abell told a conference of the American Institute of Aeronautics and Astronautics that what had sold the White House on the National AeroSpace Plane (NASP) program was its potential for space access. "All our grand plans," he told the group, were "hostage to inadequate space access."[60] But the White House's strategy to sell the program to the public and to Congress was to emphasize its most visionary aspect, as a hypersonic airliner, not its potential to make Star Wars cheaper. Newspaper and magazine articles latched on to this hypersonic airliner idea, aided by Mc-Donnell-Douglas and Federal Express, which helpfully provided artists' conceptions of hypersonic airliners complete with airline livery. The result of the sales campaign, as one author put it, was "a remarkable amount of confusion . . . regarding the true purpose of the program."[61] But the confusion was deliberate. The administration had long been under attack for its massive defense expenditures, especially those stemming from SDI. It needed to wrap the National AeroSpace Plane in a suitably civilian cloak to win support for it.

President Reagan announced the national program with the following words: "We are going forward with research on a new Orient Express, that could, by the end of the next decade, take off from Dulles Airport and accelerate up to 25 times the speed of sound, attaining low-Earth orbit or flying to Tokyo within two hours." The "Orient Express" name came out of a nighttime meeting between Paul Czysz, former supersonic test pilot and then congressional aide A. Scott Crossfield, and Gus Weiss of the National Security Council.[62] It had not appeared in the draft speech insert that OSTP had provided the White House speechwriting office, but speechwriter Dana Rohrabacher had completely rewritten it, boiling a long-winded explanation down into a brief, visionary statement.[63] The "spin" the White House chose for the National AeroSpace Plane was as an airliner, and the choice of "Orient Express" gave the project an exotic, romantic sheen.

Finally, the Orient Express was a perfect expression of President Reagan's— and the American public's—preference for symbolism over details.[64] Keyworth himself had noted that Reagan's name was linked to "bold national initiatives," which at this point included two other space-related projects, the much-derided Star Wars and "Space Station Freedom." Reagan's presidential campaigns had been based on a sunny optimism about America's future, which he promoted as one without limits—technological, financial, or otherwise. His vision of an American resurgence after the pain of the Great Inflation of the 1970s was based upon gaining technological supremacy over the rest of the world, which, he believed, would lead inevitably to American military and economic dominance. The Orient Express, evocative of unlimited speed, power, and expansion, was the perfect symbol for Reagan's America.

THE ORIENT EXPRESS AND THE REVIVAL OF SUPERSONIC ENTHUSIASM

Despite the president's rhetoric, of course, the National AeroSpace Plane project was focused on achieving a launch vehicle, not an airliner, but the enthusiasm for speed that the Orient Express sales campaign had generated meant that the hypersonic airliner was not going to be ignored. Inside NASA, the technology road maps that Keyworth had requested after his aeronautics policy exercise had sparked the generation of a set of study contracts intended to define the "supersonic" goal in Steiner's policy, but due to the enthusiasm surrounding the Orient Express concept, they had evolved into studies of the entire supersonic/hypersonic speed range. Their purpose was to figure out what, if anything, about the NASP could be successfully commercialized into a new "High Speed Civil Transport." Almost simultaneously, a private organization began an effort to promote the idea of a HSCT, hoping to create enthusiasm and, ultimately, an international consortium to build a vehicle. By 1989, the hypersonic Orient Express had devolved into a supersonic transport and the private effort had foundered, but the HSCT had gained solid political support nonetheless.

Even before Reagan's Orient Express announcement, Boeing's John Steiner was angry about the hijacking of his carefully crafted, time-phased policy by officials who had "jumped on" the Orient Express. In an interview for *Science* magazine, he expressed his dismay about the sudden hypersonic focus. As a launch vehicle, the National AeroSpace Plane might be realistic, but as an airliner it could not be—it was, in his words, "baloney."[65] There was a solid and growing market in the Pacific Basin for an economical SST, he thought, and the Orient Express was distracting attention, and resources, from that substantial market opportunity. Others agreed with Steiner's position, and a bare two months after Reagan's speech, an organized effort to promote the idea of a "high speed civil transport" began.

The institutional home of the private effort to design and build an HSCT was the Battelle Memorial Institute in Columbus, Ohio. Battelle's management had established a "Center for High Speed Commercial Flight" in April. The center's mission was "to facilitate the identification, development, and marketing of a cost-effective, highly productive, environmentally acceptable high speed commercial transport under US leadership."[66] Their belief was that an HSCT would be so complex and expensive that it could be built only by an international consortium, and the center wished to play a role as the consortium's facilitator. The center's director, James P. Loomis, organized a conference in late October to begin defining what kind of HSCT airlines might want and whether or not there was a market.

The keynote address to the Battelle conference was given by Clarence J. Brown of the Commerce Department. He told the conference that the idea of an HSCT had the support of the administration due to the foreign threat to U.S. aeronautical leadership: "America must continue the technological progress that is represented by the aerospace industry. Both government and industry need to exert new efforts to maintain current aerospace markets and to open new markets for the benefit of our economy. We must respond to the challenge that has been posed by strong foreign competition from both Europe and Japan as well as from some of our southern neighbors."[67] The U.S. response to the foreign threat was to pursue technological progress in the form of higher speeds. These represented, he said, not only the "frontier of technology" but the "frontier of the human spirit."[68] More practically, an economical HSCT would increase American aircraft exports significantly, he thought, improving the nation's worsening balance of trade.

What form an economical HSCT would take was a central concern of the symposium. Northwest and Federal Express, convinced of the hypersonic vehicle's merit by McDonnell-Douglas, promoted a hypersonic transport that would travel at Mach 5 as the desirable next step in the aviation frontier. John Horn, Northwest's president, explained that he was interested Douglas's commercial, methane-fueled Mach 5 aircraft that would fly at around 100,000 feet. Northwest was the largest U.S. transpacific airline, and he believed that a hypersonic vehicle would revolutionize transpacific flying—if it were economically competitive with existing aircraft and if it met the standing environmental regulations.[69] Fred Smith, CEO of Federal Express, argued that a hypersonic vehicle would be more profitable carrying small cargo than carrying passengers. He believed that the Transportation Department should fund development of the commercial hypersonic transport, so that DOD and NASA could focus on the military and launch vehicle "versions" of NASP, and he thought that the first-generation HST should be designed as an express freighter. Federal Express would want three to five of these, and, he said, after some years of flying experience, the data generated would allow a second-generation passenger vehicle to be economical.[70]

Reflecting a strong schism in the symposium, Boeing's John Swihart argued for a nearer-term supersonic transport to appear in the late 1990s. He believed that a twin-engine Mach 3.2 aircraft could be profitable, if additional research into thermoplastic materials, supersonic laminar flow control, and variable-cycle engines was carried out expeditiously. With four hundred seats and a range of 6,000 n.m., the aircraft could have direct operating costs equal to a 747–400 and provide a superior investment return to its owners. Airports would not need new, cryogenic fuel installations to handle the SST he proposed (in contrast to the hypersonic transport), which would further reduce costs. Swihart, who had been a member of the original Boeing SST design team, also made clear that he did not want government financing beyond the "technical readiness" phase—he did not want all the strings that came attached to the government's dimes.[71] Swihart's proposal thus differed from the hypersonic transport both in the kind of aircraft favored and the means of financing its development. In essence, Swihart was arguing that because the SST was less technologically challenging, private financing would be sufficient to manufacture the product once a few remaining technical issues were resolved. This was clearly not the case for the hypersonic transport. The hypersonic would require government financing at least through extensive flight testing of a demonstrator vehicle and possibly beyond that, creating an uncomfortable similarity to the earlier U.S. SST program.

The various presentations regarding finance and economics given at the conference did not much help to resolve the conflict between supersonic supporters and hypersonic enthusiasts. A presentation by Japan Air Lines argued that, based upon travel time, there was no economic gain to be had from speeds beyond Mach 3.[72] Klaus Nittinger of Lufthansa argued that while the hypersonic transport might have slightly better economics than the SST, the market for the hypersonic transport was far too small to permit a manufacturer to make money.[73] Donald Schenk of Bankers Trust Company argued that the airline industry could not afford an Orient Express–like vehicle without a radical restructuring of the industry itself. In essence, a single global airline or airline consortium would have to be formed to permit the economies of scheduling that such a vehicle would require to have any hope of profitability.[74] These arguments suggested that the lower-speed SST would be a better choice, but did little to overcome the arguments against it—the sonic boom limitation and the ozone problem.

The Battelle conference ended with these and many other crucial questions unanswered, but it served as confirmation that the time for an (amorphously defined) HSCT had come. By the end of 1986, several nations had announced their own supersonic or hypersonic transport programs. HOTOL received funding from the British government following Reagan's Orient Express speech; in August West Germany proposed "Säenger," a slick two-stage-to-orbit launch vehicle whose first air-breathing stage could also be the basis of a hypersonic airliner; and the French company Aérospatialé proposed a "Concorde II." Japan had announced its own hypersonic research program shortly after Reagan's speech, too, making the enthusiasm for higher speeds truly global.

If the great rush to announce programs represented the arrival of a new era in commercial aviation, then the question still remained what that future would look like—supersonic, hypersonic, or transatmospheric. This was one of the central questions of the contract studies that NASA had awarded to Boeing and McDonnell-Douglas during the year. Beginning in October 1986, the contract studies lasted two years and encompassed three phases. In the first ten-month phase, the companies assessed the market potential of an HSCT, defined some vehicle concepts and assessed the technologies each concept would require, and studied the environmental issues relevant to each concept. In the second, six-month-long phase, the contractors selected specific concepts and performed more detailed examinations of them, while continuing the environmental and market assessments. Finally, during the third phase, the contractors focused on the environmental issues involved in each concept in order to verify that the technologies necessary to resolve the known environmental problems were feasible.

During the first phase of the studies, NASA research labs and the two contractors examined the entire speed spectrum, from Mach 2 to Mach 25, to identify the best candidate speeds for a future HSCT. But the highest speed ranges fell out of the studies very quickly. The researchers applied an acceleration limitation to the vehicle to ensure that passengers would not be subjected to undue discomfort during climb-out from an airport—an important consideration for a commercial vehicle. The number Boeing and NASA chose was .2g, 20 percent of the force of gravity, while McDonnell-Douglas chose .1g. At a .2g acceleration rate, an HSCT flying a 7,000 n.m. mission could just reach Mach 15 before having to decelerate to land, while at .1g acceleration the aircraft would barely reach Mach 10 before beginning deceleration. Strictly based upon these acceleration restrictions, designs beyond Mach 6 made little sense. Even a Mach 8 aircraft would spend more of its time accelerating and decelerating than cruising. Further, due to acceleration, deceleration, and ground time (passenger and baggage loading, fueling, etc.), productivity peaked at Mach 5. Speeds higher than Mach 5 would thus require a huge financial investment without producing a productivity increase over slower, cheaper aircraft. As Langley engineer Sam Dollyhigh put it jokingly, the world turned out to be too small for a Mach 25 airliner—or even a Mach 6 one.[75]

After phase one, all three study groups threw out speeds above Mach 5 and started to focus on specific lower speeds. McDonnell-Douglas's group examined vehicles designed for Mach 2.2, Mach 3.2, and Mach 5, while Boeing's team studied Mach 2.4, Mach 2.8, Mach 3.2, and Mach 4.5. In this phase, vehicle economics played a major role in weeding out certain concepts. Below Mach 2.4, the HSCT could use the same fuel that subsonic jets did. Between Mach 2.4 and 3.5, the aircraft would require a specially formulated jet fuel with greater thermal stability, to prevent the fuel from chemically decomposing in the tanks and fuel lines ("coking"). Above Mach 3.5, the extreme temperatures the vehicle would encounter required the abandonment of jet fuel entirely. Up to Mach 5, the aircraft could burn liquefied natural gas. Beyond Mach 5, it would have to use liquid hydrogen as both fuel and coolant for the aircraft structure. HSCTs designed for speeds greater than Mach 2.4, therefore, would require airports to install special fuel handling systems, imposing an additional cost that had to be factored in to the vehicle's overall value to the airlines.

Taking into account the acceleration and fuel constraints, the in-house Langley team quickly settled on a Mach 3.0 concept. Sam Dollyhigh and Warner Robbins, both veterans of the first Boeing SST program, explained that they had wanted to work on the most challenging technologies that might be commercially useful in the next fifteen years, and they did not think that a Mach 5 vehicle could appear until well after that. The Mach 3.0 aircraft was at the edge of what jet fuel and variable-cycle turbojet engines could achieve, and they could focus their efforts on the twin challenges of environmental and economic viability. Beyond that speed, the first challenge would be simply getting an entirely new engine concept to work, and then they would still face the economic and environmental problems. The configuration they derived was a six-engine, blended-wing, titanium aircraft. Without the additional weight penalties imposed by noise suppressors, the Langley group expected their HSCT to weigh about 620,000 pounds, including 250 passengers and fuel for a 6,500 n.m. mission.[76]

During phase three, while Langley settled on Mach 3.0, the two contractors diverged significantly in their choice of speed. McDonnell-Douglas remained interested in speeds above Mach 3, examining designs set at Mach 3.2 and Mach 5, while Boeing's team rapidly settled on the Mach 2 to Mach 2.4 range.[77] What Boeing found compelling at the lower speed range was the lack of additional fuel infrastructure cost and the simpler, less expensive technological task. Boeing manager Malcolm MacKinnon explained later that his structures engineers were extremely conservative and did not believe that sufficiently strong, light, and heat tolerant materials for higher speeds could be achieved. Hence his strategy had been to push for Mach 2.4, with the recognition that if the materials did not work out, a Mach 2 option would still exist.[78] Donald Graf's team at Douglas, on the other hand, wanted to take jet fuel to its maximum limit of Mach 3.2 to realize the productivity gain possible with existing airports. Their Mach 3.2 design was derived from the 1979 SCAR program reference aircraft and Pratt & Whitney's Variable-Stream Control duct-burning turbofan, and it included partial laminar-flow control. To get around the fuel "coking" problem, Douglas based the design on a high-temperature military fuel designed for the SR-71 aircraft, JP7. The additional infrastructure necessary to support the special fuel, they calculated, would raise the cost of the fuel one to five cents per pound.[79] Graf's group did not believe this would impair the aircraft's market potential.

Both Douglas and Boeing, in the end, then, rejected the "Orient Express" hypersonic transport. While Boeing had rejected it primarily due to its impact on airport infrastructure costs, Douglas's Bob Welge explained that liquefied natural gas did not have a high enough energy content to power its Mach 5 aircraft concept without making the aircraft too heavy for virtually all airports.[80] Liquid hydrogen did, but its high fuel and infrastructure costs and the vehicle's very high weight, over a million pounds, made the resulting product uncompetitive with advanced subsonic jets. Hence by 1988, the hypersonic airliner had already disappeared.

The Boeing and Douglas teams also agreed that a supersonic HSCT would face a profitable market if the well-known noise and emissions problems could be resolved. Both companies ruled out supersonic flight over land in their analyses of the potential market, which reduced the theoretical productivity gain by 10 to 20 percent. Douglas estimated that if no fare premium were required to make the aircraft economically viable, a market of up to fifteen hundred aircraft would exist between 2000 and 2025. They believed that their current design would not be viable at less than a 30 percent revenue increase over the subsonic competition, however, which would still generate a market of five hundred aircraft. Boeing's study found that their Mach 2.4 design would also require a 30 percent increase in yield over the subsonic competition to be viable.[81] This could be accomplished by raising first- and business-class fares 10 percent, reducing the number of economy seats slightly, and eliminating discount fares (which amounted to an economy fare increase of 18 percent). Under this scenario, the Mach 2.4 HSCT introduced in 2000 would capture half the long-range market, resulting in 650 to 750 unit sales. This was, Boeing pointed out, enough for one manufacturer to turn a profit on, but not for two.

One major inhibitor to the HSCT's chances, both companies also agreed, was the old noise problem. Post-1980 aircraft were required to meet stricter noise limits than the Stage 2 standards that the SCAR program had established as its goals. Under the Stage 3 rules, the sideline noise limit had been reduced from 108 EPNdB to 103 EPNdB, making resolution of the HSCT's noise problem even thornier than the 2707's had been. But tests on a series of scale models of noise-reducing nozzles had given some hope that a technological solution to the problem could be found. The companies also agreed on the NOx issue. The lean, premixed, prevaporized combustor concept from the SCAR/VCE program promised the lowest possible NOx emissions, a reduction to about one-sixth that of existing engines—if it worked. Because of the SCAR program's abrupt termination, no one knew whether it would achieve its promise on a real engine, and this was therefore a very high-risk piece of technology.[82] To further muddy the emissions issue, the companies could not know whether the LPP combustor produced low enough emissions *even if* it worked up to its theoretical potential, because the stratospheric ozone depletion models developed during the 1970s had been discredited by the British Antarctic Survey's discovery of the "ozone hole" over the Antarctic in 1982.[83] The hole was impossible, according to the models; hence the models were wrong. NASA's atmospheric sciences program, in company with the National Atmospheric and Oceanic Administration and other organizations were engaged in a research program to correct the models, but until new models were created and achieved scientific credibility, it was impossible for the companies to know whether an HSCT could be made ozone neutral. And credible ozone neutrality had to be the program's goal. Boeing and Douglas managers made clear to their NASA counterparts that they would not be branded enemies of the ozone layer.

By the end of the NASA contract studies in 1988, then, Boeing, McDonnell-Douglas, and the Langley Research Center had all concluded that an economically viable HSCT could be built beginning around 2000, if the noise and emission problems could be resolved satisfactorily, and if the vehicle's speed was restricted to between Mach 2 and Mach 3.2. The hypersonic "Orient Express" had not been economically competitive at any speed. Their finding was mirrored in the Battelle Institute's 1988 symposium on commercial high-speed flight. During the interceding two years, Battelle had continued evangelizing the HSCT idea with the aid of Federal Express and Japan Air Lines, and it had organized four international working groups to help refine the HSCT concept. In a substantial way, the Battelle effort paralleled NASA's, and its working groups achieved essentially the same results: an HSCT should be a Mach 2 to Mach 3 vehicle, with transpacific range and a passenger load of 250 to 300, and the noise and emissions problems were probably solvable with sufficient technological and scientific effort. If no fare premium was necessary, a market of at least seven hundred aircraft would exist by 2015.[84] Hence unlike the early 1960s, when a great many people in the airline business and in the aircraft industry did not believe in the supersonic future, by the late 1980s industry and government leaders had a consensus that a new supersonic transport was technically possible and would have a viable market.

All that remained to make the supersonic future real, then, was money. The HSCT would cost between fifteen and twenty billion dollars to bring to the point of delivery to its first customer—although the manufacturers would not publicly commit to a figure, given the industry's long history of inability to predict costs. And the manufacturer, or manufacturers in the case of a consortium, would not see a return on this huge investment for at least fifteen years. Because of this, governments would have to fund the research phase of the project, which would cost at least $1.5 to $2 billion. And government financing in the United States seemed to be possible. In early 1988, the Senate Subcommittee on Science, Technology, and Space had requested that NASA submit a plan to achieve technological readiness for an HSCT; simultaneously, Senator Ernest Hollings, chair of the full committee, had requested an analysis of the idea from the Congressional Research Service. The Congressional Research Service convened its own conference on the HSCT in 1988, and it found that many of its participants believed that government support of production would be necessary as well.[85] Nonetheless, the nascent HSCT program proved to have support.

On 4 April 1989, the Subcommittee on Science, Technology, and Space held hearings on the proposed HSCT new start in the FY 1990 budget. The subcommittee chairman, Senator Albert Gore of Tennessee, the Senate's leading environmentalist, was absent due to a family emergency, and Senator Hollings ran the hearings in his stead. In their testimony, NASA officials William Ballhous and Robert Rosen focused on the foreign threat to U.S. aeronautical leadership posed by "Concorde II" and found a sympathetic audience. In an exchange with the sole critic invited to testify, David Doniger of the Natural Resources Defense Council, Senator Richard Bryan made clear that if there were to be a technological revolution in aviation, he wanted it to "have an American flag on it."[86] Hollings expressed concern about the general deterioration of U.S. investment in high technology, including aviation, and declared the new initiative important. He too wanted to ensure that the future HSCT had an American flag on it.

General political conditions had also changed in ways favorable to a new

government-supported HSCT effort. The Democratic Party had begun advocating a form of "economic nationalism" during the late 1980s as a means of
regaining the support of unionists who had abandoned the party in the early
1980s for Ronald Reagan.[87] Under this label, the more liberal members of the
party pursued trade restrictions and aggressive use of "anti-dumping" laws
against trading partners, and particularly against Japan. Democratic centrists,
however, started advocated targeting specific high-technology industries
for national investment. Because the Democratic Party controlled Congress,
and because certain elements in his own party also supported the idea of
investing in high-technology industries, President George H. W. Bush accom-
modated the opposition under the more politically acceptable rubric of
enhancing "national competitiveness." One of the industries that the administration chose to favor with its support was the commercial aircraft business.
The new program, named "High Speed Research," thus had bipartisan support
and passed easily into the FY 1990 budget.

✦ ✦ ✦

The Orient Express had put all the political stars in alignment for a new SST
attempt. But it had had a great deal of help from Jack Steiner, George Keyworth, and Jim Beggs. Their coordinated counterattack against David Stockman's cuts in the NASA aeronautics budget had put in place a new policy of
explicit support for commercial aircraft manufacturers. But because U.S. cus-
tom forbade direct support to manufacturers in the form of low-interest loans
for new product development—the European "creeping socialist" way—that
support took the form of subsidies for new technology development. The
United States would foster continuous technological revolution in commercial aviation to ensure its dominance of the aircraft market.

This was a decisive rejection of free market ideology. The aircraft market
had never been free, and the OSTP/NASA counterattack had succeeded largely
by demonstrating that fact to other White House officials. Every state with an
aircraft industry subsidized it one way or another in pursuit of export revenue,
economic development, high-tech industry spin-offs, and skilled jobs. Hence
the aircraft industry actually operated under conditions more closely resembling the old mercantilist economic system that had prevailed until the
nineteenth century. Under mercantilism, each state sought to enrich itself by
fostering exports while limiting imports. During the 1970s and 1980s, many
nations had targeted specific industries for state-sponsored development,

including automobiles, electronics, semiconductors, and launch vehicles, causing the (second) Reagan administration and first Bush administration to promote federally aided "competitiveness" initiatives in the targeted industries while also beginning a diplomatic effort to restrict foreign subsidies.

By 1989, then, NASA's aeronautics program had been saved from dismantlement and launched on a mission to save the American aircraft industry from the Airbus threat. The mechanism for the industry's salvation was a new supersonic transport, reflecting the ease at which higher speed could be sold as a public good within the American political system. In the nearly twenty years since the defeat of the first SST program, the political conditions under which aerospace technology was financed had certainly changed. There had been other options for industry renewal, of course. The "transcentury transport" could have received government backing, and left entirely out of the policy discussions of the 1980s was the "very large transport," an airliner to replace Boeing's now twenty-year-old 747. But the vision of an Orient Express, regardless of its Mach number, had proven irresistible to industry engineers and, especially, to the political establishment. The National AeroSpace Plane quickly died, a victim of overpromising, underbudgeting, and the end of the cold war.[88] But the HSCT seemed to have an excellent future, if the thorny environmental problems could be solved. In his written questions, Senator Gore had made clear that they could not be ignored, but his implicit warnings were unnecessary. For the next four years, those questions dominated the NASA program.

7

TOWARD A GREEN SST

The new "High Speed Research" program that Congress had approved in 1990 was the product of three years' work by advocates in NASA and in the aircraft industry. The first step in the process of constructing the program had been the study contracts Boeing and Mc-Donald-Douglas had conducted. Put together by Neil Driver, Bill Aiken, Jack Suddreth, and Dominic Maglieri—all old NACA hands and veterans of the SCAR program—the contracts had served both to define the technological needs of a future supersonic transport program and to bring the two manufacturers into rough technical alignment. Getting the two remaining American manufacturers of commercial aircraft to agree on the need for a new supersonic transport research program also meant getting them to agree on what kind of SST should be researched. Just as it had been crucial to the Concorde program that Sud Aviation and Bristol Aircraft agreed on the basic goal of a Mach 2.2, aluminum-based, transatlantic SST, it was important that McDonnell-Douglas and Boeing agree on the basic technological goals of the program. The national aerospace policy crafted during the Reagan administration was aimed at the construction of a product aircraft that the two manufacturers would have to collaborate to build, and hence both companies had to agree on what the product would be. Further, without the two companies' support, the program had little chance of gaining the approval of either the White House Office of Management and Budget or of the Congress. The contract studies had been necessary to bring the two manufacturers into rough agreement on the SST technologies the United States should pursue, and they had the effect of making the companies "stakeholders"—to use the current management jargon—in whatever the future program turned out to be.

Cecil Rosen, NASA's director of aeronautics, had assigned Louis J. Williams the task of actually assembling the program in parallel with the contract studies in 1987. Working with Boeing's Michael Henderson, McDonnell-Douglas's Donald Graf, Sam Gilkey from GE Aircraft Engines, Richard Hines from Pratt & Whitney, Robert "Joe" Shaw from the Lewis Research Center, Jerry Hefner from the Langley Research Center, and Robert Watson, head of the Upper Atmosphere Research program at NASA headquarters, Williams had to balance the desire of the NASA centers to work on very advanced technologies against the corporate interest in working on technologies that might contribute to a year 2000 product launch. Initial ideas for the program's content had included metal-matrix composites for very high-temperature structures, air turboramjet and supersonic flow-through fan engines, and sonic-boom reduction techniques.[1] Advanced aerodynamic configurations were also proposed, including a strut-braced "gull wing" configuration and an "oblique wing" configuration, each offering substantially better aerodynamic efficiency than the arrow wing derived during the early 1960s.[2] These technologies could not be developed in time for the launch date that the program had been based upon, however. The schedule would not permit the pursuit of highly advanced concepts. Program planners ultimately adopted an arrow-wing configuration as their baseline, and by the time the High Speed Research program received its congressional blessing in 1989, their technology suite was aimed at a Mach 2.4, transpacific-range vehicle.

Williams recalls that, in contrast to the early 1980s, the Office of Management and Budget now strongly favored the program's focus on a product. OMB officials received extensive briefings from the four corporate members of Williams's team on the potential impact of an HSCT on the U.S. economy, particularly on the balance of trade. Mike Henderson and Don Graf had argued that if a European consortium produced a "Concorde II" and the United States did not, the U.S.—particularly Boeing—would lose about $200 billion in export sales between 2000 and 2020.[3] That "swing" in export sales would cost tens of thousands of skilled jobs in the aircraft industry and its thousands of subcontractors. This loss of sales would damage the American balance of trade and decrease tax revenue. The HSCT program was thus promoted as an investment toward future economic returns.

Political protection for the HSCT research program depended upon corporate support, demonstrated through industry leaders' testimony before Congress and through the four companies' program managers' presentations to the OMB. More specifically, the companies had to demonstrate their commitment to the program by spending their own money on parts of it—"cost sharing." NASA's intent was for the research program to be conducted under "no fee" contracts, which not only meant no profit margin for the companies but meant that they would actually lose money due to the reality that government procurement regulations disallowed certain kinds of charges. But the OMB required more than that loss. It wanted to see the companies contributing substantial sums of money, for the same reason Robert McNamara had demanded cost sharing during the national SST program of the 1960s. This was the only way to determine the true level of corporate support for the program's goals. The OMB therefore expected the companies to contribute a substantial portion of the technology development program's projected ten-year, $4.9 billion dollar costs.

That support, the four companies' representatives made clear to NASA and the OMB, was predicated upon resolution of the three great environmental questions that the SST development program of the 1960s had been challenged on: ozone depletion, airport noise, and the sonic boom. Everyone involved with planning the program had strong recollections of the Coalition against the SST's campaign, and the very high-profile current controversy over the "ozone hole" in 1985 had reinforced those memories.[4] SSTs had been widely derided as destroyers of the natural environment, in particular the ozone layer, and the corporate partners in the program's development were unwilling to risk hundreds of millions of their own dollars in an HSCT research program until they had reasonable certainty that the project could avoid a repeat of the SST debacle. That meant determining whether an HSCT could be made ozone-neutral, whether it could be made to meet the current airport noise requirements, and on whether its sonic boom could be made acceptable for overland flight. This environmental research was properly the government's role, the program planners agreed, and therefore the program was structured into two phases. In the first phase, a primarily NASA-funded research effort would tackle the environmental issues to determine whether or not they could be resolved. If this were successful, a second, much more extensive, and expensive, technology development phase would be financed jointly by the government and the four corporate partners.

The program Congress approved in 1989, therefore, was the first phase of what NASA named the "High Speed Research Program," or HSR, and it was focused tightly on environmental mitigation by the lingering memories of the SST debacle. HSR Phase I was scheduled to run from FY 1990 to 1995, with a cumulative budget of $284 million.[5] A collaborative effort between the Office of Aeronautics and the Upper Atmosphere Research Office named "Atmospheric Effects of Stratospheric Aviation," or AESA, was scheduled to receive $56.2 million over Phase I, while engine emission and engine noise research received the majority of the funding, $138.7 million. A community noise and sonic boom research element was budgeted for $75.6 million, with the remaining $13 million becoming the program reserve.[6] Lou Williams became the program manager for Phase I. At the Langley Research Center, Allen Whitehead was responsible for the airframe-related research, while Joe Shaw of the Lewis Research Center managed the propulsion elements. Finally, recognition that the future HSCT market would be profitable for one manufacturer, but not two, caused Pratt & Whitney and GE Aircraft Engines to "team up" on the project, creating a subunit of themselves in 1990 to work as a single entity on the propulsion research contract.[7]

HSR started with a concrete set of environmental goals that its researchers had to meet to justify the second phase of the program. First, and perhaps most importantly, the HSCT would have to be demonstrably ozone-neutral, and that status would have to be accepted by an internationally credible scientific body. Second, the vehicle would have to meet the existing Stage 3 airport noise standards. In practice, the program set for itself a slightly stricter goal of matching the existing 747-400's actual noise levels to make the HSCT "blend in" to the existing international airport noise environment.[8] And third, they investigated the sonic boom problem, trying to design low-boom aircraft configurations and identify an "acceptable" sonic boom level. By mid-1993, two of these goals appeared to be technologically attainable. Only the sonic boom could not be fixed, but the partners believed that a sufficient over-water-only market existed. Thus the major remaining question for the partners was whether the necessary technologies would result in an economical airplane.

THE OZONE PROBLEM

In 1982, researchers from the British Antarctic Survey had discovered a "hole" in the ozone layer above Antarctica. Occurring only in late spring, the ozone hole covered Antarctica and at its largest extent crept very close to inhabited areas in New Zealand and the southernmost tip of South America.

Within the hole's boundaries, the stratosphere contained up to 40 percent less ozone than the normal atmosphere. And further, the hole had not been predicted by any of the atmospheric chemistry models that had been constructed during the CIAP program of the 1970s or its successors. The survey's scientists did not immediately publish their results, hoping to be able to present an explanation of the hole along with their evidence of it, and it was not until 1985 that the hole became public knowledge.[9] The hole's existence was confirmed by NASA's Goddard Space Flight Center, based on data from the Total Ozone Mapping Spectrometer instrument on Nimbus 7.[10] The TOMS data had clearly shown the vast region of lower-than-normal ozone concentration, but because the instrument was experimental, and because the extant theories could not account for such low values, the computer software that ingested the instrument's data stream had flagged the Antarctic data as being probably erroneous. The TOMS project scientists had assumed that either the instrument or its inversion algorithms were producing spurious results and they had been studying the data in order to determine what was wrong with their technology, not with the ozone layer.

Revelation of the hole had caused a political frenzy much like what that had occurred during the mid-1970s when chlorofluorocarbons had been identified as potential ozone scavengers. The Antarctic Survey scientists had proposed that CFCs, or more precisely, chlorine derived from the photochemical breakdown of CFCs, were responsible for the hole as well. The United States had already banned the use of CFCs as propellants but most other nations had not; during the next two years, the United Nations Environment Program sponsored an effort to negotiate a worldwide ban on the manufacture and use of all CFCs. In 1987, most nations signed a "Protocol on Substances that Deplete the Ozone Layer"—the "Montreal Protocol"—that scheduled a phase-out of the chemicals over the ensuing twenty years.[11]

Political action against the ozone-depleting chemicals had evolved more quickly than had scientists' understanding of the hole itself. Scientists had proposed two general mechanisms to explain the ozone hole's occurrence. One was chemical, with different groups of scientists proposing different possible chemical reactions. Most chemical mechanisms focused on chlorine oxides, but a significant group fronted nitrogen oxides too.[12] The other potential mechanism proposed was a change in circulation that injected lower-altitude, ozone-free air upward into the stratosphere, displacing the usually ozone-rich stratospheric air. A ground-based experiment led by the National

Oceanic and Atmospheric Administration's Susan Solomon, the "National Ozone Experiment," in 1986 pointed strongly toward chlorine oxides derived from photochemical breakdown of CFCs, but it was not considered definitive within the scientific community. To figure out which mechanism was correct, scientists needed data on the precise composition of the Antarctic stratosphere throughout the year, and that meant sampling it. Hence during 1987, a larger group of scientists executed an experiment organized by NASA's Upper Atmosphere Research Office and NOAA's Aeronomy Laboratory called the "Airborne Antarctic Ozone Experiment" (AAOE), which used sounding balloons and two NASA aircraft, a modified Douglas DC-8 and an ER-2, to get better data.[13]

The concern over ozone depletion affected the structure of NASA's planned HSCT program, because everyone involved in the planning process remembered that the ozone issue had grown out of the old Boeing SST project. The HSCT's impact on the ozone layer had to be understood before manufacturers would commit billions of dollars to building the aircraft, and hence Cecil Rosen, the director of aeronautics at headquarters, had directed Lou Williams to set up an assessment program aimed at determining what level of HSCT emissions could be tolerated. Initially, Williams had assigned the program to Linwood Callis at the Langley Research Center. But Michael Prather, director of the upper atmosphere theory program, and his superior, Robert Watson, head of the Upper Atmosphere Research Program, had protested that decision, arguing that assigning the program to Langley would impair the assessment's scientific credibility.[14] NASA's atmospheric sciences program already had procedures in place for carrying out ozone assessments as required by its congressional mandate, and these procedures were designed to maximize scientific credibility by drawing on scientists from a variety of institutions inside and outside the government. Assigning the assessment to Langley, which was primarily an aeronautics laboratory, violated those procedures and created a perception of conflict of interest that opponents could use to attack the credibility of whatever results emerged. Instead, Prather & Watson argued, the assessment needed to be coordinated at the headquarters level to ensure that no one could accuse it of bias.

Williams accepted their argument and restructured the program. Cecil Rosen asked Howard Wesoky, a career propulsion engineer, to manage it from the headquarters Office of Aeronautics, and Michael Prather became chief scientist. In addition, Prather established a scientific steering committee composed of non-NASA scientists. One of the scientists Prather invited was Michael Oppenheimer, an atmospheric scientist working for the Environmental Defense Fund. He recalled that one of the steering committee's major tasks was evaluating research proposals and recommending which should be funded.[15] That way, the committee ensured that the major scientific questions were addressed. The new program, "Atmospheric Effects of Stratospheric Aircraft" (AESA), would run through 1995, when it would publish its final assessment.

One of the first things AESA did was ask Berkeley chemist Harold Johnston to write a review of the current state of knowledge regarding the SST-ozone issue and make recommendations about what the assessment program needed to investigate. Much of what was known, Johnston noted in his response, had come from the Climatic Impact Assessment Program that had followed the national SST program cancellation. This had predicted an average reduction of 12 percent for a large fleet of SSTs using 1970 state-of-the-art combustors. Although scientific interest in stratospheric ozone had turned away from nitrogen-oxide-induced depletion to the CFC problem after CIAP ended in 1975, the Lawrence Livermore National Laboratory had maintained a one-dimensional stratospheric chemistry model that was kept up-to-date. Using it, Johnston demonstrated that the aircraft-related ozone depletion problem had not changed much over the previous decade. Based on emissions and route data from Boeing and McDonnell-Douglas, Johnston found that an SST fleet would produce between 0.9 and 28 percent depletion, depending on the vehicle's cruise altitude.[16] This estimation closely agreed with one NASA had contracted from a private atmospheric research company.[17] But these two estimates were based upon models that did not include the chemistry behind the existence of the ozone hole, and that chemistry was bound to alter an SST's impact. In which direction—positive or negative—Johnston couldn't yet say. Hence adapting the atmospheric models to incorporate the new chemistry was an important early goal of the ASEA program.

By the time Johnston wrote his 1989 piece for the AESA program planners, early evaluations of the international Airborne Antarctic Ozone Experiment's data had indicated that the ozone hole's formation depended upon the presence of both chlorine from the breakdown of CFCs and of clouds of ice crystals containing nitric acid (HNO_3) forming at very high altitudes. These had been named "polar stratospheric clouds" by Langley Research Center scientist M. Patrick McCormick in a 1982 article.[18] The clouds had the effect of dramatically increasing the rate at which highly reactive chlorine oxide was produced. The increased amounts of chlorine oxide in turn led to greater rates of ozone scavenging, causing the hole. Johnston noted in his review that the HSCT's emissions could either increase the nitric acid cloud formation, causing even greater ozone loss, or the added NOx emissions could cause the nitric acid crystals to become large enough to precipitate out of the stratosphere, reducing the ozone loss. The AAOE campaign had not provided enough information to make a quantitative assessment of this effect, however.

Because the important reactions related to the ozone hole occurred on the surface of the crystals, this new chemical model was labeled "heterogeneous chemistry," to distinguish it from the previous gas-phase-only ("homogeneous") chemical model. In order to incorporate heterogeneous chemistry into the computer models, researchers had to determine the rates at which the chemical reactions occurred in the stratosphere, and they had to come to an understanding of how the nitric oxide clouds formed, moved, and dissolved. They also needed to know how the various chemicals involved would be distributed through the stratosphere. Finally, they needed to predict aircraft emissions far into the future in order to predict future damage. All of this information had to be built into new models, and the models then had to be verified against both the "real world" and each other.

Via the United Nations' Environment Program, the atmospheric science community was already organizing future sampling campaigns to provide the necessary raw data. AESA did not need to duplicate that research program; instead, Wesoky recalls, FAA associate administrator Tony Broderick, who had been involved in the earlier CIAP studies, suggested that Wesoky needed to put small amounts of money into existing lines of atmospheric research. AESA only needed to get scientists to focus on the question of how aircraft emissions would affect the chemistry of the stratosphere, and this was the most efficient way of getting the existing research community to attack the HSCT problem.[19]

AESA's top three goals at its beginning were the improvement of global chemical models for stratospheric ozone, the construction of a set of emissions scenarios for potential aircraft fleets, and refinement of stratospheric transport models.[20] But AESA did support several stratospheric sampling campaigns financially, and it funded instrumentation specific to HSCT emissions for a campaign in 1994. It also provided part of the funding for a powered, unpiloted ultralight aircraft to be used for high-altitude, long-duration sampling missions that was the precursor for NASA's Environmental Research Aircraft Sensor Technology program ("ERAST"). The AESA program produced a series of interim reports leading up to its mandated 1995 assessment, and the results were used to adjust the program's research priorities in future years, as were the results of reviews performed by the National Academy of Sciences.

The first major effort for AESA was the development of an aircraft emissions database known as "SEEDS"—the Stratospheric Emissions Effects Database.[21] Necessary early in the program's life because the modeling effort required quantitative data on future aircraft emissions, SEEDS was developed under contract by a team of Boeing researchers led by Steven Baughcum, an atmospheric scientist who became a valuable "bridge" between the scientific and engineering communities. The database included estimates of 1990 emissions based on existing Boeing data on current commercial aircraft and engine combinations and it projected scenarios for 2015 fleets, based upon a potential fleet of 500 HSCTs in addition to a projected mixed subsonic fleet of turbofan and turboprop aircraft. The HSCT emissions projections included Mach 1.6, 2.0, and 2.4 aircraft as well as HSCTs producing emission indexes of either five (best case) or fifteen (worst case) grams of NOx per kilogram of fuel burned. The database also structured the emissions in three dimensions, producing a global 3-D "map" of aircraft emissions. A separate group at McDonnell-Douglas produced a similar database representing nonscheduled, general aviation, and military aircraft emissions over the same period.[22]

In an interim report written for the National Research Council's Committee on Atmospheric Chemistry's review panel in mid-1993, the project's scientists were able to report that six photochemical models had produced a range of depletion values that were far less than the 1989 consensus had predicted. For the "worst case" emission index (EI) of 15 g/kg, the models estimated between 0.25 percent and 1.8 percent average depletion in the Northern Hemisphere, and if the Lewis Research Center's low-emission combustor research succeeded in lowering jet engine emissions to its goal of EI = 5 g/kg, the models suggested an essentially immeasurable ozone change of between a 0.25 percent increase and a 0.42 percent decrease.[23] In his introduction to the report, Goddard scientist Richard Stolarski reported that the large change in predicted impact from the 1989 report had come from the inclusion of reactions that occurred on the surface of sulfate aerosols.[24]

The sudden change in depletion estimates had come about after an examination of the effects of volcanic eruptions on ozone. David Hofman of the

University of Wyoming and Susan Solomon of the National Oceanographic and Atmospheric Administration's Aeronomy Laboratory had performed an extensive analysis of data from the eruption of El Chichon in 1982 that suggested the volcano had caused significant reductions in ozone. They argued that sulfate aerosols, like the polar stratospheric clouds, triggered the release of additional chlorine, resulting in increased ozone loss.[25] But the complex series of chemical reactions that the aerosols fostered also had the effect of neutralizing nitrogen oxides, substantially reducing the potential impact of HSCTs on the ozone layer. Incorporating this set of reactions into the stratospheric chemical models had produced the dramatic differences between Johnston's 1989 evaluation and the 1993 interim assessment.

The interim assessment was hardly the last word on the HSCT's ozone impact. The reaction rates involved in the new sulfate chemistry had been derived from idealized laboratory studies AESA had funded, and the rates needed to be verified in the "real world." Changes in the reaction rates could substantially alter the model results. The calculated effects also did not take into account the HSCT's own sulfate emissions or how those emissions would change the stratosphere's total inventory of sulfates, which could also substantially change the projected impact of an HSCT fleet. Other issues still outstanding included whether HSCT emissions would alter the formation rate of polar stratospheric clouds, which could change the size and depth of the Antarctic ozone hole. In addition, the role that water played in many of these reactions was not well understood, and the existing models did not yet take into account the reality that HSCT operations would be heavily concentrated in relatively small portions of the stratosphere, possibly causing substantially greater local changes in ozone density than suggested by the averages given in the report.[26] Finally, the program's scientists did not know whether HSCT emissions would be "lofted" from the lower stratosphere to the upper stratosphere, an effect that could make a large change in the damage assessment.

As is typical for large NASA scientific research programs, Wesoky and Richard Stolarski, who had replaced Michael Prather as chief scientist in 1992, submitted the interim assessment to a review panel formed by the National Research Council's Committee on Atmospheric Chemistry. In January 1993, they had asked the NRC to review the ASEA program and its assessment in order to answer a set of questions about the program's research priorities. In essence, Wesoky and Stolarski wanted the review committee's opinion on whether the program had identified the key uncertainties in aircraft emissions-related chemistry and was thus pursuing the right questions, and whether the program had adequate resources to meet the 1995 deadline for a "final" assessment. They also wanted the committee's opinion on whether enough uncertainty would still exist after the 1995 assessment to justify extending the program.

The NRC asked Thomas Graedel of Bell Labs to form a review panel to examine the program, report on AESA's achievements, and answer the program manager's questions. The panel he assembled consisted of ten scientists and engineers, four of whom were from Europe. They met five times at various locations to receive information about the program and drafted their report during the summer of 1993. The panel strongly supported the program and found that it "could rightly claim to have markedly enhanced the general understanding of atmospheric chemistry" in several areas.[27] They applauded the construction of the emissions database and related fleet operations scenarios; AESA's support for the Airborne Arctic Expedition, the 1993 Stratospheric Photochemistry, Aerosols, and Dynamics Expedition (SPADE), and the upcoming joint Antarctic Southern Hemisphere Ozone Experiment and Measurements for Assessing the Effects of Stratospheric Aircraft program (ASHOE/MAESA).[28] And they praised a 1992 modeling workshop for its efforts to reconcile the many different photochemical models of the stratosphere.

AESA came in for its share of criticism, too. Graedel's group believed that the program had focused too tightly on gas-phase chemistry, and had not been aggressive enough in pursuing the aerosol issue once it had been identified. AESA had also not yet incorporated a great deal of data that already existed from related programs, especially data from NASA's own Upper Atmosphere Research Satellite and from the European sampling campaign "SESAME"—the Second European Stratospheric Arctic and Mid-latitude Experiment. It had not dealt with the scientific possibilities inherent in the recent eruption of Mt. Pinatubo in the Philippines, which had injected billions of tons of sulfate aerosols into the atmosphere that could contribute to understanding the HSCT's aerosol impact. Finally, they believed that AESA had not placed sufficient emphasis on the possibility that HSCTs might cause climate change, and in particular the potential for changing the heat balance between troposphere and stratosphere.[29] But most of these problems, the report's authors believed, could be fixed within the program's remaining two years through changes in emphasis.

In answering the question whether significant uncertainty in the assessment would still remain after 1995, however, the panel suggested that it would. AESA could not, in their opinion, resolve five major causes of uncertainty in the results in the time remaining, including the transport problem.[30] This was largely due to the need for additional sampling campaigns to cover geographic areas and specific chemicals that the already planned campaigns would not. Without additional in situ measurements, the remaining known uncertainties in the models could not be resolved. While this report was being written, the question of whether AESA should continue past FY 1995 had become very significant, because HSR Phase II had not originally been planned to include it. AESA had to justify its continued survival. The NRC report's indication that the program could be usefully continued was one important voice in favor of continuation, and Graedel restated its conclusion much more strongly in congressional testimony in February 1994. Speaking before the House Subcommittee on Technology, Environment, and Aviation, Graedel contended that resolution of the remaining uncertainties required a minimum of two to three more sampling missions before a scientifically sound final assessment could be produced. Each sampling campaign took two years to organize, conduct, and evaluate, and therefore AESA needed to continue through at least 1998.[31]

At a meeting of the HSR program's Phase II planning team two weeks after Graedel's testimony, Stolarski and Wesoky made their pitch for continuing ASEA through FY 1998 with a $14.2 million budget. Their briefing relied heavily on the NRC review committee's report, quoting it verbatim in several places.[32] Citing the report, for example, they argued for a northern latitude, summertime measurement campaign aimed at determining the precise composition of aerosols in the region, and to make sure their mostly engineer audience did not miss the point, they turned the report's words into a diagram showing the "hole" in their sampling data. But while they relied heavily on the NRC report in their presentation, Stolarski recalls that what got them the program extension was support from the contractor representatives at the meeting, General Electric's Sam Gilkey and Boeing's Mike Henderson.[33] Gilkey and Henderson subjected Stolarski to a thorough grilling about the sources of the assessment's remaining uncertainty. They were particularly bothered by the fact that the "numbers kept changing," a reflection of the dramatic changes in the stratospheric ozone depletion projections over the program's short life. While those changes had been in the "right" direction for the program's hopes, one did not need much imagination to wonder if future research wouldn't cause the projections to veer off in the "wrong" direction and make HSCTs untenable.

And because any HSCT had to be approved by both the U.S. Federal Aviation Administration and the European Joint Aviation Authority to be a saleable product, the scientific case for an ozone-neutral HSCT had to be unimpeachable. Wesoky recalled having some "quite loud" conversations with German colleagues about whether an HSCT could ever be made ozone-neutral at meetings of the International Civil Aviation Organization's Committee on Aviation Environmental Protection in 1992 and 1994.[34] There was wide skepticism from the European members of the committee about the ozone issue. Part of this stemmed from protectiveness toward the European aircraft industry (and, of course, Airbus), which was not spending anywhere near the effort the United States was on the HSCT idea and was thus getting further and further behind.[35] But it also reflected a new political trend in Europe, where the Green Party was growing rapidly in strength, especially in Germany and at the supranational European Community level. The "Greens" were a party devoted to environmental protection and social justice that had exploded out of the widespread antinuclear campaigns of the 1980s, and they were making significant inroads in European parliaments and European bureaucracies. Airbus Industrie could easily use whatever uncertainties remained at the end of AESA to fight European certification of the aircraft, using the Greens as its allies. Hence the international politics of aircraft manufacture demanded that AESA continue until the major uncertainties were significantly reduced.

Despite this context, however, Stolarski and Wesoky both believe that they succeeded in winning funding in Phase II only because they were prepared to take an engineering approach to the ozone problem. AESA had funded a lot of different activities during Phase I, which helped determine what the biggest unknowns were but left many of the planning team members with the impression that AESA was unfocused and not making much progress. Hence they proposed that in Phase II, AESA concentrate its resources on the two biggest uncertainties, emissions transport in the stratosphere and the inability to make different models get the same results. This was a clear "path to progress" toward resolving the remaining uncertainty, and the planning committee agreed that the program should continue.

With the support of the contractors, AESA was funded through FY 2000, with a budget that was projected to peak at $11.5 million in FY 1997 and decline to $2.25 million in its final year. In 1995, it produced a second assessment that produced similar results to those of the 1993 interim assessment, but also found new sources of uncertainty. One of these had derived from a piece of luck during the previous year's ASHOE/MAESA sampling campaign. The Upper Atmosphere Research Program and the High Speed Research program had jointly funded this campaign to investigate the stratospheric transport questions. In essence, it was two campaigns, one in the tropics and one in the Antarctic, with one campaign essentially "piggybacked" onto the other. ASHOE had been designed to examine chemical transport in and around the Antarctic during all four seasons, while the AESA program needed to conduct the same experiments in the tropics. Hence MAESA was added to ASHOE by using the ER-2's transit flights to New Zealand from the United States as sampling missions instead of ferry flights.

ASHOE was conducted out of Christchurch, New Zealand, and an Air France Concorde had flown through on 8 October 1994 on a round-the-world charter tour. With some assistance from French scientific colleagues and New Zealand air traffic controllers, the sampling team had been able to arrange to fly NASA's ER-2 through Concorde's wake, getting the only in-flight measurements of the constituents of its exhaust. During the CIAP program in the 1970s, a Concorde engine had been tested in an altitude chamber to examine its emissions, but CIAP had not attempted to measure Concorde exhaust in flight. The serendipitous opportunity proved to be the source of significant new uncertainty, however, because the results were not entirely in accordance with the AESA program's expectations. Concorde's NOx emissions were in agreement with expectations based on the older test data, but its particulate emissions were not.[36] It had been expected to produce a large number of very small sulfate particles, but the ER-2's particulate counter had instead found a large number of large particles. Laboratory experiments on the particles indicated they were primarily sulfuric acid and water, and the program's scientists had not identified a mechanism that explained their presence. The lack of understanding of how these particulates formed opened up a source of new uncertainties for the modeling effort, because the size of the aerosol particles influenced their effect on ozone.

There were still other uncertainties in the assessment, of course, particularly the fact that the polar stratospheric clouds implicated in the Antarctic ozone hole were not yet incorporated in the models. Hence the HSCT's potential impact on the hole itself could not yet be estimated. But AESA's continu-

TABLE 7.1　Calculated steady-state total column ozone change in the Northern Hemisphere averaged over a year for several HSCT fleet scenarios

SCENARIO				2-D MODEL RESULTS
MACH NUMBER	Cl_y (ppbv)	EI_{NOx} (g NO_2/kg FUEL)	NO. AIRCRAFT	OZONE CHANGE (%)[a]
2.4	3	0	500	−0.3 to −0.1
2.4	3	5	500	−0.3 to +0.1
2.4	3	5	1,000	−0.7 to +0.03
2.4	3	10	500	−0.5 to 0.0
2.4	3	15	500	−1.0 to −0.02
2.4	3	15	1,000	−2.7 to −0.6
2.4 Cruise + 1 km	3	5	500	−0.5 to +0.02
2.4 Cruise − 2 km	3	5	500	−0.06 to +0.1
2.4	2^b	5	500	−0.4 to +0.02

Source: Stolarski et al., *Scientific Assessment of the Atmospheric Effects of Stratospheric Aircraft*, NASA RP-1381, 1995, p. ix.

Note: Second column is the expected value of stratospheric chlorine during the twenty-first century; third column is the engine emission data used as an input to the models; final column gives the depletion estimates. Note the difference between EI = 5 and EI = 15 cases.

[a]Range of average values obtained from the five 2-D models used in this assessment.
[b]Expected in a 2050 atmosphere.

ation into Phase II would ensure continued attention to these and other uncertainties. In the meantime, the dramatic reduction in depletion estimates was a highly positive development for HSR, suggesting that an ozone-neutral HSCT was possible. (See table 7.1.) HSR's managers in particular took the significant difference in depletion estimates between the EI = 15 g NOx/kg fuel case and the EI = 5 g/kg case as validation of their goal for the low-emission combustor research that was proceeding in parallel.[37] And that portion of the program was also demonstrating that the goal of ozone neutrality was within reach.

The second approach that the HSR program took toward the ozone problem was in low-emission combustor development. Existing engines produced considerably higher NOx emissions than appeared environmentally acceptable in 1990, with the most recent models producing emission indexes of over 40 g NOx/kg fuel. Richard Niedzwiecki, one of the advocates for HSR and the former head of the SCAR program's combustor effort, had chosen as the goal an emission index of 5 g/kg. At the beginning of the HSR program, Niedzwiecki recalled, he was given a one-year mandate to demonstrate that he could meet the 90 to 95 percent NOx reduction that NASA sought. To do it, Niedzwiecki had pulled out of storage the flame-tube experimental apparatus that had been built during the later years of the SCAR program but never used.[38] The rig was flexible enough to test several low-emission concepts and many variations on the basic concepts. The combustor research effort focused on three ideas. First, the lean premixed, pre-vaporized (LPP) concept from the SCAR program was favored by General Electric's research team. The second concept, primarily investigated in-house by the Lewis Research Center's own personnel, was a derivative of the LPP idea, "lean direct injection," or LDI. Finally, the concept favored by Pratt & Whitney was the "rich-burn, quick quench, lean burn" (RQL) combustor, and derived from a Department of Energy-sponsored research program in the early 1980s.

Each of the three concepts had strengths and weaknesses. Both the LPP and the LDI concepts worked on the same basic principle, that by burning the fuel in a very lean, fully mixed and vaporized fuel/air mixture, NOx formation could be substantially reduced. They differed in that the LPP concept required a set of premixing chambers that made the combustor section of the engine longer than that of traditional jet engines, resulting in a heavier engine. The LPP's premixing chambers also imposed new problems in combustor design, such as preventing "flashback," a movement of the flame from the combustion chamber upstream into the premixing chambers. Finally, depending on how large the premixing chambers could be and still provide thorough mixing, a product engine might have to have hundreds of "premixing tubes," making the engine very complex and harder to maintain. The LDI idea, in contrast, would inject the fuel into a high-velocity, very turbulent airflow, just ahead of the combustion chamber. Mechanically simpler than LPP, the LDI's primary challenge lay in achieving the same thoroughness of mixing as the LPP in a much smaller distance. The payoff for success, however, would be a lighter, simpler engine. (See figure 7.1.)

The third concept, RQL, relied upon burning the fuel in a very rich (high fuel/air ratio) mixture first, quickly quenching the combustion products, and then essentially burning it again in a very lean mixture. One difficulty both LPP and LDI had was that lean mixtures tended to suffer "blowout" at low power settings—they would stop burning, causing the engine to lose power. Blowout was much less likely in a rich mixture, and traditional jet engine combustors used "rich zones" of combustion to eliminate it. Hence the RQL represented an attempt to make use of traditional design ideas while still achieving low emissions. The idea behind RQL was to initially burn a mixture that was so rich that only half of the fuel would initially combust. Then this

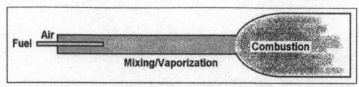

LPP combustion concept

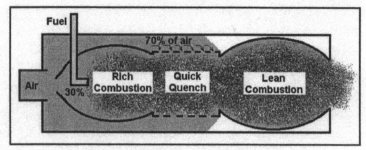

RQL combustion concept

FIG. 7.1. The two combustor concepts investigated under HSR. [Courtesy NASA]

partially burned mixture would be injected into an air-rich "quench zone," followed by the burning of the rest of the fuel under very lean conditions. The primary challenge in making this idea work was in moving large amounts of cooler air into the quench zone rapidly enough to prevent hot spots that would form NOx.

The initial flame-tube experiments with all three concepts in 1990 and 1991 were very successful, demonstrating that all three concepts might deliver the emissions reduction HSR sought.[39] During 1991 and 1992, then, the HSR combustor team faced the task of trying out various hardware implementations of each scheme to find the ones that produced the best balance of emission reduction and operability, since there were many possible variations on each idea. All of the lean burn concepts had demonstrated the ability to meet the NOx goal—in fact, beat it—by the 1992 annual program review. The LPP tests were much more successful than the LDI tests had been, although the LDI concept was not yet out of the running. The LPP rig had surpassed the program's goal across a wide range of temperatures, indicating that it was a "robust solution." It was likely to meet the emission criteria across a wide operating range. The results also indicated that LDI might not have that ability, although further efforts could improve its performance range. The Pratt & Whitney rich-burn research had also demonstrated that the RQL concept could meet the emissions requirement, although its performance was more marginal than LPP had been.

Successful demonstration that the emission target could be met led to the selection of several potential hardware configurations of each of the three primary concepts for a second round of tests. These test rigs were designed to represent one "sector" of an engine's combustor system. The flame-tube experiments, which tested only a single fuel injector and its associated components, did not provide all the data necessary to design a real engine's combustion apparatus. Primarily, this was because real engines had many injectors and their individual flow fields interacted. It was entirely possible that they would interact in ways that would invalidate the single-injector emissions results. More likely, though, was that operability problems with certain hardware configurations would become apparent in the sector tests and would lead to particular configurations being dropped from the program or subjected to redesign. Since some injector configurations might be more susceptible to blowout than others, or might not be stable at certain power levels, the sector tests were designed primarily to examine the operation of various configurations through a fuller range of conditions, while also making sure emissions did not suffer.

From 1993 to 1995, the combustor team investigated a wide variety of possible implementations of each concept. In the original plan for the HSR Phase I program, a series of "downselects" had been scheduled to successively weed out underperforming variations of the three primary concepts, ultimately leading to the adoption of one design in May 1997. By late 1995, however, the combustor program had to be rescheduled, placing the final selection between the two design concepts in 1998. Despite the setback, though, the Phase I combustor research effort had demonstrated that the emissions goal the program had set was technologically achievable, and the remaining technological challenges lay in devising a lining material that would enable a commercially useful life at the combustor's higher temperatures, as well as in replicating the operability of a traditional jet engine combustor.

From the standpoint of ozone protection, then, the HSR Phase I program had made substantial progress. The ongoing AESA research and assessment program had strongly suggested that an ozone-neutral HSCT was possible, if strict emissions targets were met. The combustor research program had similarly indicated that the emissions targets were well within reach. Both programs were to continue through the planned Phase II of the HSR program, with the combustor program intended to eventually result in a full-scale combustor rig and engine tests.

If the ozone problem affected both the structure of the Phase I research program and the specific design of engine combustors, the airport noise problem drove the development of the overall propulsion system. During the national SST program's last eighteen months, Bob Withington's attempts to resolve the 2707-300's noise problems had driven the aircraft's weight, and therefore its economics, far beyond that tolerable for a 1970 commercial aircraft. The Variable Cycle Engine program's propulsion research effort, therefore, had focused on noise reduction, pursuing a variety of engines and suppressors in the hope that one could be found that would permit a 1980s SST to meet the FAR Part 36 "Stage 2" noise standard. Cancellation of the SCAR/VCE effort in 1981 left supersonic engine noise research moribund until 1986, when the HSR precursor contract studies began. Those studies had focused on small-scale experimentation with the mixer-ejector type exhaust nozzles that McDonnell-Douglas had initiated in the late 1970s, and on cycle studies of new variants on the Variable-Cycle Engine concept from the same period.

As many participants in the Phase I effort recall, however, it was not at all clear as HSR began in 1990 that the Stage 3 standard could be met without making the aircraft highly uneconomic. Technologies in existence at the time would not produce an aircraft at all, according to some studies—the designs would not close, or they had infinite weight. Nor was there agreement within the propulsion research team what the best approach to resolving the noise problem might be. There were two approaches to solving the problem, a "novel nozzle" approach focused on a relatively conventional engine with a sophisticated, noise-reducing exhaust nozzle, and a "high-flow" engine approach that would result in a low-noise engine with a conventional supersonic nozzle.[40] And just as there had been many variations on the three basic low-emission combustor concepts, there were many possible combinations of engine and nozzle that might meet the noise criteria. The challenge was to identify concepts that would meet the noise standards with the least development risk.

The noise research effort had several facets. At the Lewis, Langley, and Ames research centers, and at nozzle test facilities at Boeing, Pratt & Whitney, and General Electric, investigators designed, built, and tested a series of nozzles that they hoped would produce the twenty-decibel or so reduction that a conventional low-bypass turbofan engine would need to meet the FAR 36 Stage 3 standard. A group of materials specialists at Lewis worked on high-temperature liner materials for the nozzle that could further absorb sound. At Langley, acoustics specialists led by Jack Preisser tried to develop a better understanding of mixing noise that would be generated within the nozzle. And at Lewis, a joint industry-NASA team supervised by Bill Strack explored a wide variety of engine concepts, seeking an engine cycle that would meet the noise goal at the lightest airplane weight.

During the pre-HSR contract studies, all four contractors had built and tested small-scale nozzles in search of concepts that would produce the necessary level of noise reduction, between five to eight decibels for the high-flow engines and twenty decibels for the "novel nozzle" engines. The early favorite had been the mixer-ejector nozzle. This nozzle was designed to reduce noise by supplementing the engine's exhaust airflow with cold outside air and mixing the outside air with the engine exhaust. Some of the engine airflow's energy would be used in entraining and accelerating the outside air, producing a higher-volume, lower-velocity exhaust. Since noise was a function of the exhaust velocity, this slower exhaust stream produced less noise. The nozzle would draw the outside air in through a set of doors that would be open for takeoff and climb-out but shut for cruise, and it also had two sets of mixers. The first mixer combined the engine's core airflow and its bypass (or fan) airflow into a single stream, while the second mixed the entrained outside air with the engine exhaust. The entire nozzle assembly would be roughly the size of a city bus, and preliminary estimates were that a complete nozzle would weigh around eight thousand pounds—each. As in the case of the combustor effort, of course, there were several possible ways to make a mixer-ejector nozzle, and Phase I focused on identifying which concept would be most effective at reducing noise levels.

The nozzle effort in Phase I examined two categories of mixer-ejector nozzles, rectangular (2-D) and cylindrical (3-D or axisymmetric), and the fluid-shield nozzle concept required by the high-flow engines. Both Pratt & Whitney and GE built 2-D mixer-ejector nozzles in the first two years of the program, while P&W also built a 3-D concept. GE was responsible for the fluid-shield nozzle. And Boeing brought to the program a "Near Fully Mixed" nozzle devised by Garry Klees that was a modified axisymmetric type. The nozzles were tested at a variety of facilities owned by NASA and by the contractors during 1992 and 1993. Of these "Generation 1" nozzles, only the GE 2-D mixer-ejector proved able to meet the noise requirement at a reasonable size. The fluid shield nozzle and Pratt's high-flow 3-D nozzle both missed their noise targets by four to five decibels. Boeing's nozzle could meet the target, but only at an unacceptably long length (and thus weight). The 2-D nozzle itself was successful, nozzle manager Bernie Blaha recalled, due less to superior noise suppression capability than to the way it radiated the noise that escaped it.[41] The axisymmetric nozzle radiated noise equally in all directions, while the 2-D nozzle radiated more of the noise out to the side instead of down. The resulting pattern meant less of the radiated noise reached the ground compared to the axisymmetric nozzle. But even GE's successful 2-D nozzle imposed an unacceptable thrust penalty at takeoff, more than twice the program's goal.

The success of GE's 2-D nozzle, though, indicated that it had the best chance of meeting both the noise and performance goals, and thus most of the nozzle research effort turned to ways of improving that nozzle's aerodynamic performance. The nozzle's liability in the Generation 1 (Gen 1) tests had been internal aerodynamic losses, and the nozzle teams scheduled an extensive series of "Gen 1.5" tests aimed at overcoming that problem. Three sets of tests were carried out during this phase. One of the difficulties the nozzle researchers had encountered was that Pratt and GE each designed their test nozzles independently, and there was no adequate way to compare results achieved with them. Hence one set of tests at the Boeing Nozzle Test Facility was intended to establish a database of nozzle performance by varying each of the key parameters of the mixer-ejector nozzle in a systematic way. This parametric test series encompassed sixteen designs, and it provided a common reference for future decisions. A second series of tests using General Electric's "Cell 41," an anechoic free-jet noise test facility, investigated the mixer's flow characteristics, through systematic variation of the mixer assembly. Finally, a series of tests by P&W at the NASA Langley Research Center's sixteen-foot wind tunnel was conducted to study the impact of various ejector configurations on nozzle drag and entrainment. A similar series of parameter variation tests was carried out on the fluid shield nozzle, using Cell 41 as well as GE's Acoustic Research Laboratory.[42]

Nozzle manager Bernie Blaha explained that while the four sets of tests were going on at the company noise facilities, the nozzle branch at the Lewis

Research Center made an intensive effort to develop better analytical tools. The supersonic nozzles that NASA and the engine companies had worked on in the past were designed for military aircraft and were far simpler internally than the mixer-ejector nozzles were, as they did not have to meet noise rules. Hence the existing nozzle design codes were not up to the task. Using data fed back from the Gen 1.5 testing program, the Lewis researchers improved the computational fluid dynamics codes so that they more accurately modeled the internal flow fields of the mixer-ejector nozzles.[43] They hoped that the improved CFD codes would permit the "Generation 2" nozzles scheduled for construction in Phase II to meet the noise goal without the high takeoff thrust loss demonstrated in the Gen 1 test series.

The 2-D nozzle's success at meeting the noise criteria strongly influenced the engine cycle selection process by indicating that the program would not need to take the "novel engine" approach to meet the program's noise goal. Because engines were far more complex than the nozzles were, a novel engine produced much greater development risk and cost than did the novel nozzle. The HSR program's goals included a measure of technology maturity that the chosen engine was to reach by 2000 in order to justify a product engine launch.[44] A novel engine concept could probably not be brought to that level within the program's schedule and budget, while even a highly sophisticated nozzle could be. The 2-D nozzle's success thus permitted the program to pursue a lower-risk, lower-cost path.

The cycle studies team led by Lewis's Bill Strack pursued their investigation of engine concepts through a downselect process like that used for the combustor research effort. They devised a wide variety of conventional, variable-cycle, "fan-on-blade" and inverting-flow valve engines, with and without afterburner, and subjected them to simulation studies to characterize their performance in a hypothetical aircraft. Their primary selection metric was takeoff gross weight. The engine/nozzle combination that provided the lowest gross weight would provide the best economic performance—assuming that they met the noise requirements, of course. The initial plan was to start with five engine/nozzle combinations in 1989, drop three in 1993, and then select the final combination in 1995 to take into Phase II. That's not what happened, however. Instead, some concepts were dropped earlier and new ones were added partway through the program.

The five initial concepts represented both the novel nozzle and high-flow approaches to the noise problem. The Boeing "Turbine Bypass Engine" concept from the later years of the SCAR program represented the most challenging nozzle-dependent approach, but was an early favorite because the engine itself was simple and light. To meet the Stage 3 noise criteria, it required a mixer-ejector nozzle that could more than double the engine's own airflow during takeoff. The Variable Cycle Engine concept represented an initial high-flow solution, as did the more recent "Fan-on-blade" (FLADE) concept. FLADE used extension tips on the engine's front fan to produce a third air-stream during takeoff. The extra bypass air was routed to a fluid shield nozzle that placed a mass of cold air under the engine core exhaust, reducing downward-radiated noise. The duct-burning turbofan, a high-flow concept, also reappeared from the SCAR/VCE program, as did a variant, the duct-burning turbofan with an inverter flow valve.

Two of the initial concepts were ruled out quickly and replaced. The most significant change was the replacement of the duct-burning turbofan with a common military engine type, the mixed-flow turbofan. The duct-burning turbofan, which relied upon an inverted velocity profile to achieve noise reduction, could meet Stage 2 but not Stage 3 without using a large mixer-ejector. The mixed-flow turbofan could achieve the same noise levels with a mixer-ejector nozzle with much lower development risk, so it replaced the duct-burner in the program. After a series of meetings in 1993, the mixed-flow turbofan engine combined with the mixer-ejector nozzle was chosen as the program's primary concept. Evaluated on the takeoff gross weight basis, it appeared to offer a significantly lighter airplane than the other concepts while offering a robust solution to the noise standard. It was also rated as the least risky of the five surviving concepts, with only the complex mixer-ejector nozzle carrying significant design risk. And the 2D mixer-ejector nozzle had already shown that it could meet the noise goal. The FLADE engine and its fluid-shield nozzle appeared to have somewhat worse performance, but it became the backup concept. FLADE, Bill Strack explained later, had had the advantage of being distinctly different from the mixed-flow turbofan, so that in the event it did not work out, FLADE still might.[45]

With the propulsion downselect finished, the propulsion research effort's first two major goals had been met. The possibility that a low-emission combustor could be built had been validated and a propulsion system concept capable of meeting the Stage 3 standard had been identified. There were, of course, still many outstanding issues by the end of 1993. The mixer-ejector nozzle's aerodynamic performance needed to be improved, a specific low-emission combustor design still had to be chosen, and the mixed-flow turbo-fan's cycle had to be specified in detail. New high-temperature, long-life materials had to be designed. And the entire propulsion system had to be held within strict weight limits. But from the propulsion system standpoint, the environmental criteria that were Phase I's major goals had been satisfied. The remaining propulsion issues belonged to Phase II.

SONIC BOOMS

The Phase I sonic boom research was initially focused on resolving two questions. The program's managers wanted to know whether an HSCT's sonic boom could be made acceptable enough for supersonic overland operation, and that meant determining what would constitute "acceptable" boom levels as well as whether one could design an economical aircraft that would achieve an "acceptable" boom level. Three different research efforts descended from these two questions, all of which ended in relatively pessimistic assessments. The one bright spot in the sonic boom program turned out to be its assessment of marine mammal response to the boom, a consideration that had not been part of the original program.

By the time HSR began in 1990, a great deal of effort had gone into sonic boom research over the preceding decades. In addition to the boom acceptability research that NASA and the Air Force had done during the 1960s at St. Louis, Oklahoma City, Edwards Air Force Base, and other sites, quite a bit of research had been done on boom mitigation. None of this research had been particularly conclusive, however. One way to reduce boom intensity, for example, was to project some sort of energy field in front of, or beneath, a supersonic aircraft, in effect tricking the air into behaving as if the aircraft was longer than it actually was.[46] This would produce a longer rise time, which was linked to the boom's perceived intensity. The heat field concept, though it was shown to work in 1974, required enormous energy consumption, making a vehicle using it uneconomical. A different method of reducing boom intensity had been described in the early 1970s, however, that involved distributing the aircraft's lift and volume in particular ways.[47] This was the method that drew the attention of the Langley Research Center's sonic boom reduction effort during HSR.

The sonic boom research plan had been to make an assessment of human response to the boom based on new research during HSR's first three years, to evaluate a series of low-boom configurations during 1994, and to measure the atmosphere's effects on boom propagation via a series of flight experiments ending in 1995.[48] The boom acceptability portion of the program was the responsibility of acoustician Kevin Shepherd of the Langley Research Center. His team designed a series of experiments to explore human subjective response to various kinds of sonic booms in order to establish a set of "boom criteria." The boom criteria, in turn, were supposed to allow aircraft designers to design supersonic aircraft with "acceptable" sonic booms for overland flight. Unrestricted overland flight would, of course, substantially increase the total market for the aircraft, and therefore was worth pursuing.[49] These experiments included use of a sonic boom simulator chamber to test individual responses to various booms, the installation of boom simulator systems in the homes of volunteers near the lab to test the "startle" reaction over long periods, and two large-scale community surveys to understand the reactions of people who had lived with military booms for more than thirty years.

The boom simulator chamber tests, Shepherd recalled, were designed to test the idea that one could reduce people's annoyance level towards the boom by reshaping the waveform.[50] A "natural" sonic boom took on the classic "N" wave shape during the shockwave's transit to the ground, and acousticians widely believed that people found the boom annoying due to the "N" wave's extremely rapid pressure buildup. The actual intensity of the "N" wave's pressure change was trivial compared to everyday things like the wind. What made the boom so startling, and annoying, was the suddenness of the pressure change, not its magnitude. Hence one obvious approach to mitigating the boom's annoyance level was to shape the wave a different way. The sonic boom simulator, which consisted of a boxlike chamber very similar to those used for hearing tests with a computer-driven set of loudspeakers to mimic sonic booms, could be programmed to produce a variety of potential reduced-annoyance booms.

During 1991 and 1992, the boom simulator team ran four studies on 212 volunteers in the search for more acceptable boom shapes. The first set of experiments was designed to validate a loudness model that would relate boom intensity to noise in decibels, so that annoyance levels could be predicted. Such a relationship was necessary to overcome the reality that neither rise time nor boom overpressure served as adequate measures of perceived intensity. The investigators needed to place subjective boom evaluations on the same foundation that subjective noise evaluations used. The scale they

settled upon, "perceived level" or PL, was chosen out of a field of five potential measures as the most consistent predictor of the test subjects' reactions.[51]

With the relative loudness relationship established, the investigators
examined human response to two different wave shapes, a "flat top" signature
and a "ramp wave." This set of experiments demonstrated that the ramp wave
produced a more positive response from the test subjects than did the flat-top
signature, and both received better responses than did the sonic boom's natural "N" wave.[52] Thus the boom-shaping experiments validated the idea that
shaped booms might achieve better public response.

Simultaneously, aerodynamicists at the Langley Research Center, Boeing,
and McDonnell-Douglas were pursuing configurations that could produce
the shaped waveforms being examined in the acceptability studies. This
effort, managed by Daniel Baize at Langley, involved designing various supersonic configurations with the aid of computational fluid dynamics codes,
building a few of the best-performing configurations in the form of wind-tunnel models, and testing them. A variety of low-boom configurations emerged
from this research effort, representing two general aerodynamic classes.
McDonnell-Douglas's low boom team pursued a Mach 1.8 configuration that
eliminated the conventional tail for a large, nose-mounted canard like the
XB-70's.[53] This approach produced a ramp wave–like signature, which was
validated in 1993 via testing in the Langley Unitary wind tunnel. Teams at
Boeing, Langley, and Ames all pursued more conventional tailed configurations, and achieved boom reduction through the use of very high wing sweep
angles.[54] The high sweep distributed the aircraft's lift farther aft, shifting
the pressure peak rearward and thus increasing the boom's rise time. (See figure 7.2.)

The boom simulator studies had suggested that a boom level of 92 to 100
PLdB might permit supersonic flight through "low population" corridors, and
some of the low-boom models approached this level. (On this scale, Concorde's boom measured 105 PLdB.) But the low-boom configurations presented new engineering challenges in addition to the program's existing
burdens. The very high wing sweeps resulted in poor low-speed and takeoff
performance, which in turn made the vehicle's airport noise signature worse.
The high sweep angle also made the wings structurally questionable, and
undoubtedly heavy. Hence boom mitigation imposed significant economic
penalties.[55]

To make the situation yet worse, the first set of data from the long-term

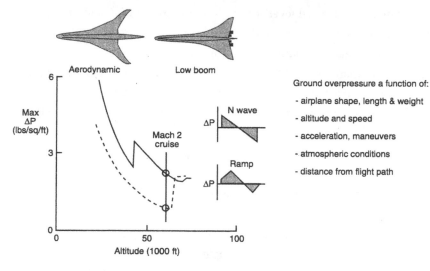

FIG. 7.2. Configuration shape effect on sonic booms. The "aerodynamic" configuration (left) produces the standard "N-wave" pressure signature (solid line). The "low boom" configuration (right) is optimized to produce a "ramp wave" signature (broken line), which lengthens the boom rise time and reduces annoyance levels at the cost of some combination of significant takeoff noise, weight, and/or stability challenges. [From Allen H. Whitehead, Jr., "Impact of Environmental Issues on the High Speed Civil Transport," paper presented at the Von Karman Institute for Fluid Mechanics, May 1998]

acceptability studies carried out in the communities surrounding Nellis Air Force Base came back "very pessimistic," in Malcolm MacKinnon's words.[56] Nellis Air Force Base, located outside Las Vegas, conducted the vast majority of the Air Force's supersonic training flights, and thus the population surrounding the facility was very familiar with sonic booms. As Langley's Kevin Shepherd relates, there had been some hope that familiarity with sonic booms would cause people to habituate to them over time, and the booms would become less annoying.[57] But the Nellis data, which was collected through extensive surveying in 1992 and early 1993, did not demonstrate such an effect. Instead, despite more than forty years of experience with the booms, people still found them annoying. Seventy-five percent of the persons interviewed still found the booms "a little" annoying, while 35 percent were still "very" annoyed by them.[58] This represented a more negative outcome than the Oklahoma City tests in 1964. (See figure 7.3.) Speaking before an audience at the Von Karman Institute for Fluid Mechanics in Belgium, Langley's Allen Whitehead commented that even the "low boom" configuration appeared to produce an unacceptable level of annoyance in the public.[59]

The combination of substantial performance penalties due to boom shaping and the high uncertainty regarding public acceptability convinced Boeing's program managers that overland supersonic flight would never be acceptable. Further, Henderson and MacKinnon believed that an economical HSCT would still generate sufficient sales to be highly profitable even if restricted to over-water-only flight. McDonnell-Douglas's market assessment was similar.[60] Using their own funds to protect the proprietary nature of the effort, the two companies had conducted marketing surveys on the long Pacific routes to understand passenger willingness to pay higher airfares for faster flights. Boeing's Henderson recalls that these surveys produced a surprising result: the passengers most willing to pay a higher fare for speed were the coach-class fliers.[61] Riding in coach for ten to fourteen hours was so uncomfortable that these passengers professed a strong willingness to pay more to halve their travel time. Hence HSR's partners believed that the

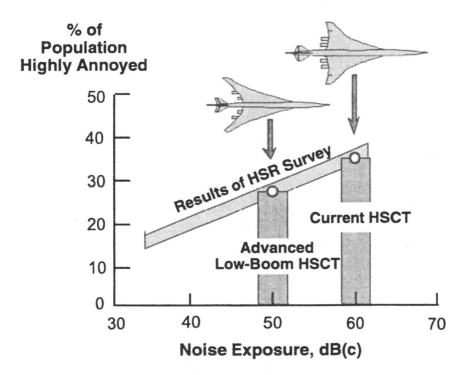

FIG. 7.3. Sonic boom annoyance, based on the HSR program's survey around Nellis Air Force Base. Even the "low boom" HSCT configuration was projected to annoy more than 20 percent of a boomed population. [From Allen H. Whitehead, Jr., "Impact of Environmental Issues on the High Speed Civil Transport," paper presented at the Von Karman Institute for Fluid Mechanics, May 1998]

potential market was more than sufficient to justify continuing the program, and during the planning for Phase II, the sonic boom program's focus shifted toward ensuring that over-water operation would be environmentally acceptable.[62]

Michael Oppenheimer of the Environmental Defense Fund had first raised the question of how sonic booms might affect sea life. Kevin Shepherd recalls that there were two widely circulated stories that suggested booms could produce unacceptable harm to animals. In one, a flight of military aircraft was believed to have caused sooty tern eggs in the Dry Tortugas to crack, leading to a massive hatching failure.[63] In the other story, low-flying supersonic military aircraft had appeared to cause a panic reaction in seals, which trampled their young in their rush to dive into the ocean. In both cases, sonic booms were the suspected culprits, although that was not at all clear—in the seal case, the aircraft were so low that the seals could easily see them, and the reaction could have been to visual, not auditory, stimulus.[64] Hence Oppenheimer had warned Lou Williams that NASA would have to show that sonic booms from high-altitude, and thus invisible, aircraft would not produce these kinds of damage.

The HSR program's management began considering the structure of a marine mammal research effort to be conducted during Phase II at a meeting in February 1994, held at McDonnell-Douglas's Long Beach facility. Shepherd had invited William Cummings of Oceanographic Consultants in San Diego and Ann Bowles of Hubbs–Sea World Research Institute to summarize the existing knowledge regarding bird and mammal response to sonic booms and to propose directions for a research program. They recommended that NASA experimentally validate theoretical predictions of sonic boom penetration into the ocean surface, determine the auditory thresholds for a select group of marine mammal species, and evaluate the behavioral response of seals exposed to routine sonic booms.[65]

By mid-1993, then, the sonic boom research program had suggested that supersonic overland flight was not going to be permissible, but a sufficient over-water market existed to justify continuing the HSR program into its second phase. The oceanic research effort was just beginning, but no one in the program really expected it would find evidence of substantial sonic boom damage to sea life. Concorde, after all, had been booming the North Atlantic for twenty years without a documented case of harm. The marine mammal research program was necessary to support the environmental impact assessment that the manufacturers might have to produce before the HSCT could be certificated, but because no one expected highly negative results, there appeared no reason to delay the beginning of Phase II.

PHASE II APPROVAL

By mid-1993, all of the three major environmental problems the HSCT posed appeared to have satisfactory answers. NASA and the four corporate partners had begun planning the second phase of HSR in 1990, and started advocating it in 1992 for inclusion in the FY 1994 budget. There was competition from other NASA programs for that money, one of which was President Reagan's "Orient Express." The National AeroSpace Plane project had been in political trouble for several years, in part due to the end of the cold war, termination of Star Wars, and the resulting contraction of the military space budget, and in part due to underdelivery on its grand promises. To afford Phase II, the money NASA was spending on NASP had to be reprogrammed to HSR, which meant the HSR advocates had to overcome a solid group of hypersonics supporters within the agency before headquarters could take the case to the OMB. NASA engineer Ming Tang, who had been part of the NASA advocacy team for NASP, believes that what ultimately convinced NASA leadership to dramatically scale back its support for NASP and hypersonic research in general was the "hockey stick chart."[66] (See figure 7.4.) The chart showed the relationship between the government's investment in HSCT technologies and the resulting corporate investment in the HSCT's production. The huge corporate investment in manufacturing the aircraft and the resulting sales would significantly improve the nation's balance of trade, fend off potential foreign competition, and, of course, create lots of skilled jobs. NASP could not make such a promise.

The HSR program's Phase II request also came at a politically propitious moment, just as Phase I had. NASA administrator Daniel Goldin had taken office in 1992 with an agenda to make the agency more responsive to its customers' needs, and a deep recession in the aerospace industry that had taken hold in 1990 had left the industry very needy indeed. NASP, whatever its merits, would not provide a product to sell in anything like the near term, while HSR promised a new product launch by 2000. Corporate support for HSR was much stronger than for NASP. Further, the new Democratic administration, like its Republican predecessor, supported national investment in advanced

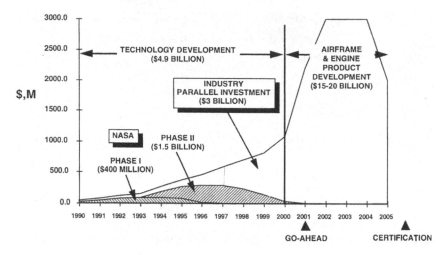

FIG. 7.4. The projected cost (in millions of dollars) of an HSCT launch. Note that in 1997, corporate investment would start to dominate the spending profile, and after 2000 government investment would essentially end. This chart and similar, more detailed versions were widely used in presentations to NASA senior management and to the Office of Management and Budget. [Courtesy NASA]

technology. It also promoted a strategy of export-led growth, which a successful HSCT would help foster.[67]

NASA's mission during the cold war had been to perform as a showcase for American technological prowess through its human (the new name for "manned") space flight program, but with the cold war's abrupt end, this mission was no longer valid. The core of NASA's program upon Goldin's appointment was Space Station Freedom, whose costs were soaring. But the agency's budget was shrinking. Beginning with fiscal year 1990, Congress had slashed it. President George H. W. Bush had requested $15.8 billion for fiscal year 1992 and Congress had provided only $14.3 billion. The president's FY 1993 budget request of $15 billion was also cut to $14.3 billion.[68] These cuts made clear to the administration, if not yet to NASA's professional staff, that the budgetary increases the agency had gotten during the late 1980s, and that its future planning was based upon, had come to an end.

The agency had long been criticized for the glacially slow pace of its scientific missions and their very high cost, and one of Goldin's intended reforms was to shrink the time scale and cost of these missions so the agency could afford more of them. Goldin also intended to make NASA more "businesslike," implementing a series of management reforms based upon corporate ideals. Most importantly for NASA's aeronautics program, however, Goldin wanted to make the agency more responsive to its "customers." In this case, those customers were the aircraft and aircraft engine manufacturers. To ensure that he understood what the companies expected from NASA, he and his staff arranged for a workshop on aeronautics that was held in Seattle on 23 October 1992. The centerpiece of this meeting proved to be the High Speed Research program, whose second phase was being planned but had not yet received Goldin's blessing.

To prepare for the meeting, Cecil Rosen, the director of aeronautics, had circulated a questionnaire asking the manufacturers to spell out their desired priorities for the NASA aeronautics program. There was a central conflict within the aeronautics profession about where NASA should concentrate its effort, with the manufacturers and the National Research Council's Aeronautics and Space Engineering Board (ASEB) holding divergent views, and it was this conflict Goldin wished to resolve. The aircraft manufacturers believed NASA could make its best contribution via the High Speed Research program, because the risk inherent in a High Speed Civil Transport aircraft was so great that the companies would not attempt an HSCT if NASA did not prepare the basic technologies for it. Boeing's Bert Welliver, the company's vice president for technology and the person whom the CEO, Frank Shrontz, depended upon for technical advice, explained that his company was "gun shy" about an HSCT.[69] If the company were not certain NASA would work the HSCT's problems, it would not want to invest its own money in that area. The ASEB, on the other hand, believed NASA could make major contributions to subsonic airliners, but the agency underfunded that research area.

At the Seattle meeting, the corporate representatives argued that NASA could not make significant contributions to subsonic jets, because the companies already spent so much on internal research in that area the small amounts of money NASA could make available would not make any difference. Furthermore, they contended, the United States' growing disadvantage in the subsonic airliner business was not a result of technology but of finance. One attendee from Pratt & Whitney pointed out that while Pratt made a significantly more efficient engine for Boeing's 757 than Rolls-Royce did, Rolls was able to offer much more attractive financial terms.[70] Advanced technology was thus not a solution to the industry's shrinking share of the subsonic market.

The companies argued that the HSR program was the United States' best chance of regaining the lead in commercial aircraft sales, and was therefore the best use of NASA's resources. Welliver believed that change happened most quickly in business when someone "upset the status quo."[71] An American HSCT would fundamentally alter the airliner market by establishing a new plateau of competition that the United States would dominate for many years, given its current lead in HSCT technologies. Once "Team America," as many attendees called the U.S. effort, introduced an HSCT, all the major airlines would have to buy it to remain competitive. The last time such a transformation in the business had occurred was the 1958 introduction of Boeing's 707—although, of course, Concorde and the Boeing 2707 were supposed to have had the same effect in the 1970s. They had not, due to their poor economics; hence success at creating the new high-speed market depended upon overcoming the severe economic penalties the supersonic future had foundered upon two decades ago. Researching the technologies necessary for economic viability, Welliver believed, was NASA's responsibility.

Fortunately for the assembled HSCT supporters, Goldin agreed. In addition to demonstrating NASA's ability to aid the aircraft industry, HSR was something he could sell to Congress and the public. The HSCT was something that the public could "touch and feel and see," he said, while subsonic technologies were "bits and pieces." They lacked "sex appeal."[72] The HSCT would be a visible symbol of aeronautical progress and U.S. technological supremacy in addition to its measurable economic impact, and that would make it an easy sell in Congress, despite the rapid shrinking of the agency's overall budget. Hence with NASA leadership and corporate managers in agreement, HSR became the agency's top aeronautical priority during the decade.

Once Goldin had been converted to HSR's cause, the next hurdle was convincing the White House Office of Management and Budget that HSR deserved to be in the president's FY 1994 budget submission to Congress. The OMB was not immediately enthusiastic. NASA had initially pitched HSR as a defensive technology program, insurance against the day the Europeans and/or Japanese decided to build a second-generation SST. But the OMB had not accepted that argument. It wanted, deputy assistant administrator Robert Whitehead reflected later, to know when the industry would actually build the HSCT.[73] The date had to be one that would interest the new Clinton administration and therefore could not be too far in the future. NASA and its corporate partners agreed to a 2006 initial service date, which implied a production decision around 2002. This was the date the partners gave OMB. Satisfied with this near-term goal, the agency put HSR's Phase II in the FY 1994 budget.

Finally, while NASA's budget was controversial within Congress, HSR proved not to be. At NASA's FY 1994 budget hearings in the Senate, Goldin was forced to defend NASA repeatedly against attacks based on the massive cost overruns and delays on what had started as "Space Station Freedom" and been redesigned into "Space Station Alpha," but HSR's dramatic expansion to a budget of $1.4 billion received only approbation. HSR was, Goldin told the Senate Subcommittee on Science, Technology, and Space, NASA's means to restoring the eroding aeronautics technology base of the nation. It was the space agency's "number 1 priority" for the rest of the decade.[74] And Senator Conrad Burns, a Republican from Montana, probably best summed up the committee's response. "The battlefield is no longer the Cold War," he said, "it is the global marketplace."[75] NASA needed to deploy its resources to that new front, to bolster national competitiveness in aeronautics and other advanced technologies. To the committee, HSR was a weapon in a trade war. Phase II passed without difficulty.

⊹ ⊹ ⊹

Three threads of activity converged during the late 1980s to make a third run at an American supersonic transport possible. The first was technological. New, and in the case of emissions, not all that new, technologies promised to eliminate the environmental questions that plagued the Boeing SST at the beginning of the 1970s. Further, it appeared possible to both fix the environmental problems and make the resulting vehicle economical. The second was political. Reagan's use of grand space visions to promote his politics had begun to lay the groundwork for a new run at the supersonic future, while both of his successors, George Bush and Bill Clinton, were faced with the problem of converting part of the vast national aerospace infrastructure to civilian use when the cold war suddenly vanished. Unlike the space station project, a successful HSCT manufacturing program would provide long-term, skilled jobs and contribute to the balance of trade, and both major political parties found this easy to accommodate within the their respective ideological frameworks. Finally, the aerospace corporations chose to support HSR over its competitors for funding.

Bringing about a successful HSCT meant keeping these three threads aligned for the next decade, so that the technologies could be brought to a level of maturity sufficient to justify a manufacturing program. That would not have been easy under any circumstances. There are always more advocates in search of funds than there are funds to dispense. But the 1990s were a tumultuous period in American politics, and HSR's technopolitical alliance did not survive it.

SIC TRANSIT HSCT

The HSR program's Phase II differed from Phase I substantially. The largest change was in focus. Whereas Phase I had concentrated on resolving environmental concerns, Phase II's mission was to address the problem of economic viability. Phase II's goal was to create the technologies necessary to permit a post-2000 HSCT to be competitive with year 2000 subsonic aircraft with a ticket surcharge of no more than 20 percent, and both Boeing and McDonnell-Douglas leaders agreed that this would limit a 300-passenger aircraft's maximum weight to 750,000 pounds.

That weight limitation directly affected the program's content. Phase II's planning had been the responsibility of Langley's deputy program manager Alan Wilhite, working with senior aerodynamicist Bob Welge from McDonnell-Douglas, deputy program manager Malcolm MacKinnon from Boeing, and deputy program manager Joe Shaw from the former Lewis Research Center, which had been renamed Glenn Research Center in honor of astronaut and senator John H. Glenn. They recognized that success in meeting this target required a comprehensive approach to weight reduction. They needed a 40 percent reduction in airframe weight compared to the Concorde and a factor of ten increase in propulsion durability. The largest weight reduction, they hoped, would come from the development of a new high-temperature composite material that would have to survive 60,000 flight hours at around 350°F. Developing, testing, and finding ways to manufacture parts from such a material became one of Phase II's major efforts.

Phase II also had to transform propulsion technologies to make an HSCT viable. In addition to continuing the development of the experimental mixer-ejector nozzles and low-emissions combustors started under Phase I, Phase II

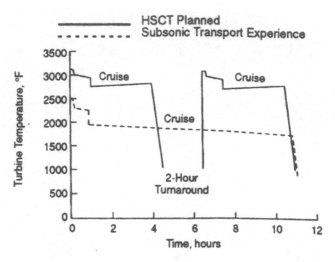

FIG. 8.1. Thermal profile comparison. In a subsonic engine, maximum temperatures occur only at takeoff. The supersonic transport mission requires maximum temperature operation for a much higher portion of the engine's life, making existing commercial and military engine materials unsuitable. [Courtesy NASA]

had to address the fundamental criticism that Robert McNamara had leveled at the SST program a generation ago: supersonic engines simply did not have the longevity required for a commercial airplane. As figure 8.1 shows, a supersonic airliner's engines would operate at maximum temperatures for a much larger portion of their lives than did military and subsonic commercial engines. HSR's propulsion goal was to generate new materials capable of tolerating the HSCT's flight conditions for 18,000 hours, with "hot parts" to be overhauled or replaced at 9,000 hours. The propulsion program thus had to achieve an order of magnitude improvement in engine life to make a future HSCT a viable product.

The Phase II detailed planning effort took place in a group of "temporary" office trailers that had been parked at Langley for many years. A significant challenge for the planners was keeping all four of the aeronautics centers (Langley, Glenn, Dryden, and Ames) involved in the program. Support from the centers was necessary in the annual budget battles, but the critical technologies were airframe materials, a Langley specialty, and several aeropropulsion technologies, all Glenn specialties. There was little for either Dryden or Ames to do. Ultimately, a supersonic laminar flow flight research program that had been part of the Aircraft Efficiency Program was transferred into HSR to keep Dryden interested; as Malcolm MacKinnon explained later, Boeing's engineers had wanted the chance to analyze the technology too.[1] Laminar flow promised a dramatic reduction in fuel consumption but came with unknown risks in integration and manufacturing, and modifying NASA's F-16XL for the flight research program would help define those challenges. But no one believed that laminar flow technology would be on the first-generation HSCT. This was one of the few pieces of long-term research in HSR. Ames became responsible for some wind tunnel testing.

A major change in HSR's structure under Phase II was its management scheme. Phase I was being run as a research program, with a generalized set of goals but without detailed performance measurements or a program control office to aggressively manage both goals and money; that management approach was no longer acceptable. The "space side" of NASA, which was primarily devoted to large-scale engineering, had long required rigorous management methods based upon explicit performance targets, or "metrics," for each portion of a program and extending to every level of a program—although extensive metrics in no way guaranteed adequate performance. The Langley Research Center's technologists had fought the extension of these engineering project management techniques into aeronautics research from the NACA's demise into the 1970s, when they began, grudgingly, to accept that they had value for specific endeavors. From the 1970s on, Langley lived a schizophrenic life, with an amorphous "wall" between its "research" and "project" areas. Phase II was too large and too hardware-focused to remain on the "research" side of that wall. For Wilhite, this meant putting a great deal of effort into finding a way to measure research progress using methods originally intended to assess the progress of large-scale engineering projects. Phase II therefore had much more detailed metrics and a significantly larger program control office to keep track of them than its predecessor. A decision by NASA administrator Goldin to remove project management from headquarters also resulted in the relocation of the program management task to Langley, which provided a program control office to handle the day-to-day oversight of the contracts. Initially, Ray Hook was chosen to replace Lou Williams as the program manager, with Alan Wilhite and Joe Shaw as his deputies at Langley and Glenn respectively.

Finally, a major structural change to Phase II came after what Wilhite recalled as a fiasco of a meeting in May 1994. This was to have been the major planning meeting for the Phase II airframe contract, which was being awarded to Boeing. Boeing was supposed to manage the contract for both itself and

McDonnell-Douglas, an arrangement that was itself a subject of some controversy. HSR's NASA managers had a program plan that specified which company would do each piece of the work, but the company representatives had come to the meeting thinking that all the funds were "up for grabs" and their job was to bring home as much money as possible.[2] When NASA's representatives made clear that Phase II would not be run that way, the contractor representatives just started to leave.

To recover, Wilhite accepted a suggestion from GE Aircraft Engine's Sam Gilkey, who suggested using the integrated product development team approach to manage the program.[3] The propulsion contract with the GEAE/ Pratt alliance was already being run under this system by the Glenn Research Center. Under this approach, the major decisions in the program would be made by a team composed of one member from each of the four prime contractors and two from NASA. This team, called the Integrated Planning Team (IPT), would essentially be replicated at the middle level of the program by "Technology Management Teams" (TMTs) and at the program's base level by "Integrated Technology Development Teams" (ITDs). (See figure 8.2.) Each TMT team would decide who did each bit of research within its area of cognizance and would monitor progress. Ideally, this would allow the teams to award each contract to the company best able to perform the work while keeping all of the program's partners in the decision-making process. This management approach was one that General Electric specialized in and it provided the expertise for implementation. Constructing functioning teams was difficult, because the team members were used to being competitors, not partners, but as Phase II evolved most of the teams developed a working level of trust.

HSR's second phase, then, was a partnership between NASA and its major contractors. Like any partnership, its success was predicated upon the partners continuing to pursue common goals. Unlike most partnerships, however, HSR's also had to satisfy the demands of an uninvolved third party, the Office of Management and Budget. During HSR's four-year life, the partners made great progress toward their goals of weight, noise, and emissions reduction and increased engine life. But it very rapidly became clear that they would not reach the level of technological maturity they required to make a production decision in 2002, the date they had promised to Goldin and to the OMB. Setting a new decision date for the HSCT launch triggered recognition that new

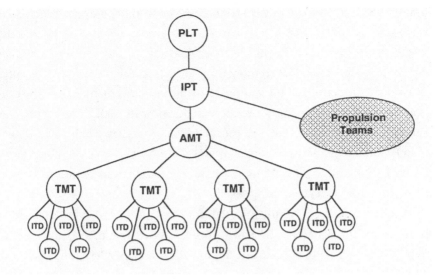

FIG. 8.2. HSR Phase II's team structure. The Integrated Planning Team (IPT) acted as the project manager via consensus-based decisions. The Airframe Management Team (AMT) oversaw the Technology Management Teams (TMTs) responsible for each technology area within its purview, while a Propulsion Management Team (not on the chart) had the same function for propulsion technologies. The TMTs were composed of researchers from the Integrated Technology Development (ITD) teams. [Courtesy NASA]

goals would be necessary along with the new schedule. But new goals meant new technologies, which meant another schedule slip. By 1998, the HSCT's launch date had moved so far into the future that HSR could not sustain the support of either NASA headquarters or the OMB.

COMMAND INNOVATION AND THE TROUBLE WITH SCHEDULES

The HSR program's mission was to "establish the technology foundation by 2002 to support the U.S. transport industry's decision for a 2006 production of an environmentally acceptable, economically viable, 300 passenger, 5000 nautical mile, Mach 2.4 aircraft." It was not to *develop* such an aircraft. In NASA jargon, HSR was a "research and technology" program whose purpose was to develop selected critical bits of technology for an aircraft meeting these specifications. Wally Sawyer, who became the Phase II program manager in June 1994, called these critical technologies his "nuggets." Within two years of Phase II's launch, however, it became clear to the program's managers that the original HSCT launch date was unachievable. The two that posed the biggest challenge to the schedule were the airframe and propulsion materials efforts. Despite "head start" money for materials research that had been provided in 1991, neither of these projects would result in mature enough technologies to justify making a production decision in 2002. The reality was simply that fully characterizing the material properties would take longer than HSR had. There was a fundamental disconnect between the research orientation HSR had at its working level and the production promises its partners had made at the leadership level.

HSR had chosen composite materials for its major materials research effort because the HSCT's Mach 2.4 cruise speed resulted in surface temperatures over 350°F, essentially eliminating aluminum from consideration. Titanium, which had been the subject of substantial research during the SCAR program, was a possibility if the program could develop coatings for it that would prevent corrosion over the aircraft's 60,000-hour life. It did not offer sufficient weight reduction to meet the program's goal, however, and titanium alloys were retained in HSR only because certain parts of the aircraft would need more strength than the program's preferred material, carbon-fiber composites, could offer. The primary drawback to composites was that when the HSR program had started in 1990 there were none in existence that offered both the temperature resistance the HSCT required and the desirable manufacturing qualities that the partner companies would need for successful, inexpensive high-volume production.

The class of composite materials that the airframe materials team investigated for the program had its origins in polymer chemistry of the 1950s. They were composed of two materials, a fabric (or "tape") made of carbon fiber, and a polymer-based resin "matrix." The carbon fiber provided very high tensile strength for its weight and also high temperature resistance, but like any other fabric had no significant stiffness. The matrix, once cured, provided the stiffness. Composite parts were made by soaking the carbon fiber "tape" in the liquid resin, building many layers of the wet tape up in a mold, and then baking the whole assembly at high temperatures and pressures. The baking process cured the resin by driving out the volatile solvent, producing a light, rigid part.[4] Because carbon fiber had high temperature tolerance by its nature, the HSR program's challenge was to develop a resin matrix that also had high temperature resistance.

Carbon-fiber composites had first been used extensively in military aircraft designed in the late 1960s and manufactured in the early 1970s. At the Langley Research Center, composite research had been part of the basic research program from the 1960s. After a joint planning study with the Air Force in 1972, composites research at Langley had been focused on polymer-matrix composites, and in 1975 the center had begun a project to develop polymer-matrix composites for high-temperature space vehicles. Their goal had been a material capable of withstanding 600°F temperatures for a few minutes at a time, and with a lifetime suited to the Space Shuttle, about 125 flights.[5] The following year, polymer-based composites for aircraft use received a major boost from the Aircraft Energy Efficiency (ACEE) project.[6] One portion of this effort was to make composite materials acceptable for commercial use in order to gain fuel economy from airframe weight reduction.

Commercial aircraft manufacturers had not accepted military-derived materials for two major reasons: they did not have sufficient proven durability, and they posed significant processing problems. The durability problem stemmed from the difference in expected life between military and commercial aircraft. Military aircraft were designed with lifetimes about one-tenth those of commercial aircraft, so military researchers had no reason to engineer very durable materials or even to test the materials they designed beyond 10,000 hours. The processing difficulties largely stemmed from the high viscosity of the resins in use. High viscosity meant that the resins did not flow or penetrate the fiber tape well, leading to uneven bonding between the layers of a built-up part, and during the heat curing process they tended to trap gas bubbles inside the material. Both flaws produced zones of weakness in the finished part. ACEE had focused on durability testing of selected composites beyond the 10,000 hours Air Force needs had required, on engineering lower-viscosity resins that would still provide the strength, stiffness, and resistance to environmental degradation that aircraft design required, on fabrication techniques for composite structures, and on developing testing procedures. Marvin Dow of the Langley center called ACEE the "golden age of composites research," but cancellation in 1986 left a great deal of unfinished business, especially in the area of large primary structures, such as wing boxes.[7]

The ACEE program had not attempted to investigate composites for high temperature use, however, as it had been focused on traditional subsonic aircraft. Neil Driver's SCAR program had funded an extensive study of extant composites by the Convair division of General Dynamics during its last few years, however, and this contributed some vital knowledge to HSR. Convair had tested representative samples of five general classes of composite materials, including the carbon-fiber, polymer-matrix type.[8] One of the major "lessons learned" on all five materials was that the 1970s-era predictions of high-temperature performance had been highly optimistic. Of the five systems, only the epoxy-matrix and polymer-matrix systems held up for significantly longer than 10,000 hours. The epoxy system, however, was limited to less than 250°F while the polymer system had satisfactory qualities at 450°F, although neither material had survived past 36,000 hours without severe degradation.

This was the state of the art when HSR began looking at high-temperature composites for the HSCT in 1991.[9] Langley's own substantial history of polymer-matrix composites research made this class of materials an obvious place to start. Using the head-start money, in 1991 a team of materials specialists managed by Paul Hergenrother began screening potential materials. The screening effort started out with about two hundred compositions made at Langley and by several other contractors, including Boeing, Lockheed, Grumman, and a number of smaller companies specializing in composites. Using the same sort of downselect process that the Phase I propulsion effort was, the researchers whittled the number of candidate materials down to thirteen. They subjected the survivors to a series of tests, including constant-temperature aging, environmental aging, fatigue life, and various strength and stiffness tests. In 1994 the thirteen candidates were again winnowed down to six; late in 1995, the team selected a single material known as PETI-5. Hergenrother recalls that one key factor that set PETI-5 apart from its closest competitor was that there were no toxic materials used in its manufacture, a quality that had been on Boeing's "wish list" for the material.[10]

PETI-5 had to be subjected to extensive durability testing over the next several years to demonstrate that it would have the lifespan a commercial aircraft required. It was here that the airframe materials program ran up against the schedule. The program's objective was an airframe with a 60,000-hour lifetime. One lifetime of data would take about ten years to accumulate. Because PETI-5 was a new material, its production process had to be scaled up enough to produce sufficient quantities for testing, and therefore the durability testing would not start until 1997.

Recognizing at the outset that new materials took a very long time to develop and characterize, HSR's planners had placed a great deal of faith in a related research program aimed at reducing the time needed to screen new composite materials. Using the head-start money, Langley's Tom Gates had established a research effort into accelerated testing, which uses increased temperature, oxygen, or mechanical loads to speed up degradation.[11] The accelerated testing research served two functions within HSR. The research team helped speed up the process of screening the hundreds of candidate materials by allowing relatively rapid collection of data regarding materials that failed relatively quickly. These could be "selected out" in favor of longer-lived candidates. The research program's longer-term goal was the construction of a detailed database that related the properties of many "accelerated" samples to those of many "real time" samples in order to understand how accurately accelerated testing results represented those of "real time" testing under many different environmental conditions. When HSR was cancelled, the testing team had subjected samples of two materials to more than 30,000 hours of "real time" testing and many lifetimes at accelerated conditions.

In parallel with the testing program, the materials team pursued further scale-up of the manufacturing process, to ensure that it could be manufactured in large amounts reliably. The structures teams used PETI-5 to make increasingly large parts, starting with small one-square-foot flat panels and working their way up to six- by ten-foot curved panels. They also built sections of fuselage barrel using traditional skin and stringer construction, to demonstrate that representative aircraft structures were possible. Ultimately, their goal was to build full-scale wing box and fuselage sections to subject to strength tests, and to demonstrate reparability of the material.[12] These last items, however, were not accomplished before HSR's cancellation.

The materials team also sought a suitable fuel tank sealant. Langley's Hergenrother had started his career working on a fuel tank sealant for the Boeing SST in 1967. That attempt had not succeeded before cancellation, and after the SST's demise no one had bothered to keep looking. The SR-71 did not use a sealant, and the CIA and the Air Force had simply tolerated leaky fuel tanks, while Concorde used a sealant that had to be reapplied every few hundred hours. When HSR began, Boeing had spelled out requirements for the HSCT, including a 60,000-hour life at the vehicle's cruise temperatures, but the researchers found nothing that came even close to this. Boeing cut the lifetime requirement in half and altered other required qualities too in order to make the task easier, but even at these lower requirements they were unable to find or make a suitable sealant during the program.[13]

The third major portion of the airframe technology program, the supersonic laminar flow control research, was cancelled early in HSR.[14] Laminar flow control research had a very long history at the Langley Research Center, where it had been investigated off and on since the 1930s. In the early 1960s, NASA, the Air Force, and the Northrop Corporation had modified a pair of WB-66D bombers to conduct flight research on a particular form of laminar flow control that used suction to keep the boundary layer attached to the wing.[15] Although the researchers generated a great deal of knowledge about laminar flow control in these experiments, the technology was never applied to a service aircraft and the program died out. Langley had revived laminar flow control research during the energy crises of the 1970s as a component of the Aircraft Energy Efficiency program. ACEE supported flight research on laminar flow control, including extensive experiments using a NASA-owned Lockheed JetStar, large-scale experiments using a Boeing-owned 757, and research involving various military aircraft.[16]

When ACEE was terminated in 1986, Langley continued supporting LFC research through a Laminar Flow Control Project Office. While most of the previous LFC research had involved subsonic aircraft, the beginning of the pre-HSR contract studies had caused the LFC Project Office to start examining the possibility of flight research on supersonic laminar flow control (SLFC).[17] There had been extensive theoretical and wind-tunnel investigations of SLFC, and Langley's researchers believed that development of SLFC technology would be crucial to an economically viable HSCT. In 1988, the LFC Project Office gave a contract to Boeing to study the impact of applying SLFC to a supersonic transport; using a descendant of the 2707-300 configuration, Boeing's investigators found SLFC could reduce fuel consumption 12 percent and maximum takeoff weight by 8.5 percent.[18] Given that the vehicle's design payload was only 7 percent of gross takeoff weight, this suggested a very significant performance improvement.

Lack of real-world flight experience with supersonic laminar flow systems was a barrier to adoption of SLFC for a future supersonic transport, and the Boeing contract study had recommended obtaining a suitable aircraft for flight research. Langley's Jerry Hefner recalled that the perfect aircraft for the job were a pair of F-16XLs that had an "arrow wing" planform like that of the baseline HSCT concepts, but they were about to be destroyed in live-fire exercises. With help from Wally Sawyer and from the North American Aviation division of Rockwell International, Hefner had been able to get Cecil Rosen at

NASA headquarters to intervene to preserve the aircraft and make them available for SLFC experiments.[19] The F-16XL experiments were managed under the LFC Project Office until 1994, when they were adopted by HSR.

The primary technical goal of the F-16XL flight experiments was to demonstrate the ability to maintain laminar flow over 50 percent of the XL's wing at supersonic speeds using a suction "glove" attached over a portion of the aircraft's original wing. Designing, building, and integrating the glove to the #2 F-16XL was a joint effort between Langley, the Dryden Flight Research Center, North American Aviation, McDonnell-Douglas, and Boeing, and took place during 1994 and 1995. The team conducted thirty-eight research flights with the modified aircraft, demonstrating that laminar flow could be maintained over most of the wing's surface.[20] There were unexpected results, however, that reinforced an existing belief that SLFC was still a high-risk technology. Despite extensive computational fluid-dynamics-based study of the aircraft and glove flow fields, shockwaves from the wing apex and canopy impinged on the glove region, causing transition to turbulent flow in areas where it was not expected.[21] The flight experiments therefore demonstrated that the CFD design tools were inadequate and would need substantial improvement before they could be used to design an SLFC-based HSCT.

Malcolm MacKinnon recalls that the primary difficulty with laminar flow for a commercial aircraft had not been with the aerodynamic results of the flight research, however. They could have been overcome with a reasonable effort. Manufacturing the SLFC technology was the "show stopper." The suction glove contractors had had to try four times to build a glove that met the stringent smoothness and stiffness standards required to maintain laminar flow. The contours of the glove had to be controlled to 1/1000th of an inch, which proved challenging to achieve even on the F-16XL's small, stiff wing. It was well beyond Boeing's manufacturing capabilities for a wing the size of a putative HSCT's.[22] There were other integration challenges, including the costs associated with ensuring that the tiny suction holes remained clear on an operational aircraft, but this one alone made the technology unattractive. It would cost far too much to develop manufacturing technology capable of reliably achieving this level of precision on a structure many times the size of an F-16.

Yet the promise SLFC had seemed to offer early in the program had led to an attempt to obtain laminar flow data using an aircraft larger than the F-16XL, and this effort outlived the F-16XL portion of the program. Late in

1992, Langley's chief scientist, Dennis Bushnell, had discussed refurbishing one of the surviving Russian TU-144 SSTs for use in supersonic flight research with the aircraft's manufacturer, Tupolev, and in March 1993 Joseph Chambers, head of Langley's Flight Applications Division, had formally proposed using a TU-144 for SLFC flight experiments.[23] The cost to restore the long-grounded TU-144 to flight status he estimated at $40 million, and he thought that once it was flying it could be brought to the U.S. for modification with suction gloves for another $50 million.

From this point, the TU-144 flight research project had taken on a life of its own. Lou Williams, the Phase I manager at NASA headquarters, had funded a study contract via the Air Force to North American Aviation to conduct a feasibility assessment of restoring a TU-144 for HSR's use, leading to a meeting in October 1993 at the Zhukovsky Flight Research Center.[24] In February 1994, Wesley Harris, NASA's associate administrator for aeronautics, had advocated use of the TU-144 as part of a joint NASA-Russia aeronautical research program established by U.S. Vice President Albert Gore and Russian Prime Minister Viktor Chernomyrdin in September 1993.[25] By this time, however, the original purpose of the TU-144 restoration had already been dropped. North American's analysis of the SLFC research potential of the TU-144 had been that a set of large-scale gloves would not provide the data regarding integration feasibility and scale-up that the aircraft manufacturers would need for an HSCT. Instead, they would ultimately have to replace an entire wing with a purpose-designed wing.[26] The TU-144's wing "rippled" in flight, a product of heating, and the ripples were large enough to destroy laminar flow regardless of the action of the suction system. Building a new wing would cost far more than HSR could afford. The political momentum in favor of resurrecting a TU-144 for joint experiments, however, was by now unstoppable. Instead, limited to a budget of $15 million, HSR used the TU-144 for a set of nine database-building experiments in cabin noise, thermal equilibrium, slender-delta ground effects, handling qualities, and CFD validation during 1996 and 1997.[27]

Without the large-scale TU-144-based SLFC experiment, and no realistic chance of putting SLFC on a future HSCT, there was little point in extending the F-16XL flight research program beyond its original goals. In September 1996, HSR's Airframe Management Team voted to end the F-16XL research program. The team members believed that the HSCT's basic economic mission could be accomplished without SLFC, based upon HSR's progress in the other technology areas. Years later Alan Wilhite recalled that SLFC had been in HSR as an enhancing technology that could be "pulled out of the bag" in case one of HSR's other "nuggets" fell short.[28] But it was not a near-term technology, and if it had a future at all, it was a future far outside HSR's time frame. It did not make sense in HSR's context.

PROPULSION

The "long pole" in HSR's "tent," to use a common metaphor in the program, had always been its propulsion technologies. While the airframe technologies were important, if PETI-5 proved unable to handle the full 350°F temperatures a Mach 2.4 HSCT would encounter, Boeing could easily reduce the speed to Mach 2.2 or even 2.0 to bring the vehicle within the material's capabilities. An economical Mach 2.0 vehicle would still be worth building. But without the propulsion technologies in HSR, an economical HSCT at any speed was unachievable. Missing the targets for the long-life turbine blade or combustor liner materials would make engine maintenance prohibitively expensive. Missing the noise targets would make the vehicle uncertifiable. Missing the nozzle weight targets would make the vehicle uneconomical and possibly impossible to actually build. Hence these were HSR's major challenge.

Phase II's propulsion research effort was divided into two major subcomponents. The "Critical Propulsion Components" program contained the low-emission combustor research, the exhaust nozzle aeroperformance and acoustics studies, inlet and low-noise fan research, and the propulsion cycle studies continuing from Phase I. The second major portion of the propulsion effort, the "Enabling Propulsion Materials" (EPM) program, encompassed the new Phase II research into combustor liner materials compatible with the low-emission combustor concepts, materials for the large mixer-ejector nozzles, and long-life materials for the engine's turbomachinery. Each of these applications had unique requirements. The combustor liner, for example, had to function without air cooling of its front surface to prevent NOx formation, unlike the combustor liners in a conventional turbofan engine. It therefore had to withstand much higher temperatures than existing combustor linings, but last at least until the 4,500-hour overhaul that normal airline engines underwent.

Like the airframe materials research at Langley, the propulsion materials research had started several years before Phase II. The propulsion contract for Phase I had contained money for systems studies, most of which was for conducting the engine cycle studies discussed in chapter 7. Some of the systems study money, however, went to small-scale materials screening efforts so that when Phase II started, materials suitable for testing at larger scales would already have been identified. The "head start" money provided in 1991 enhanced this effort. When this early research phase started in 1991, the materials research was focused on two areas, materials applicable to the mixer-ejector nozzle and those suitable for the low-emissions combustors.

GE materials program manager Andrew Johnson recalled that receipt of the NASA request for proposals for Phase I late in 1990 had sent him and his counterpart at Pratt & Whitney, Ralph Hecht, into a flurry of planning.[29] They had started early, defining the nozzle and combustor technologies they wanted to work on based upon what they had learned from their contacts with the aircraft manufacturers and with Glenn Research Center personnel before the mandatory "black out" period began with the issuance of the formal request for proposal. Because in 1990 the turbine bypass engine concept was the expected favorite, Johnson and Hecht identified a novel high-temperature composite system, intermetallic-matrix or IMC, for the mixer-ejector nozzle because the TBE concept produced a very high temperature exhaust. Existing materials would not suffice, while what little was then known about IMCs suggested that they would have the necessary heat resistance and light weight. For the combustor liner, the two agreed to propose a ceramic-matrix composite (CMC) based on silicon carbide. Both companies had been working on silicon carbide–based CMCs for many years, using internal R&D funding and under other Defense Department and NASA contracts, although they were not yet well enough understood to integrate into products.[30]

The pre-Phase II materials effort was substantially redirected in 1993, when it became clearer what sort of engine was likely to be the program's major focus. Only one of the several engine types under investigation at Lewis needed intermetallic materials, and once it was dropped these challenging materials were unnecessary. Conventional "superalloys" already in use for other applications would suffice for the nozzle, but they had never been used to manufacture very large parts.[31] The nozzle's internal mixer, for example, which would be made of a superalloy like that used in the engine's low-pressure turbine, was far larger than any previous application of the metal. Hence the propulsion materials team redirected the nozzle materials research toward developing large-scale casting techniques for existing materials, and toward devising repair processes for fixing manufacturing defects that inevitably arose in large castings.

The Phase I cycle studies tasking had also led the propulsion systems team to add three new components to Phase II's EPM program. One new task was to develop turbine blade materials and thermal coatings capable of withstanding the engine's maximum temperature for 4,500 hours, an order of magnitude improvement over military or commercial subsonic engine requirements.[32] A second addition was long-life turbine disk research. The turbine disk, like the blades attached to it, had to withstand much longer exposures to maximum temperatures than any previous design.[33] Finally, the EPM team added a fan containment system research program. All turbofan engines had fan containment systems to reduce the potential for damage to an aircraft after a fan blade failure, but the containments were heavy. The researchers believed that lightweight materials could be used to replace the traditional steel, providing significant weight reduction to the aircraft overall. Initiated in 1995, this portion of the EPM program was terminated later when higher priority research needed the funds.

During Phase I, HSR had put a great deal of effort into designing low-emissions combustors to mitigate the ozone depletion problem, and a crucial enabling technology for them was a combustor liner that could withstand combustion temperatures without requiring cooling of their interior faces. GE's Johnson recalls that the two companies had already done a good deal of research on ceramic composites based on silicon carbide (SiC), and this was the technology that they wanted to pursue in HSR.[34] There was a great deal of skepticism within the program that the new material would work out, however, and as a risk reduction strategy HSR also pursued a backup, a segmented metal liner protected by a ceramic coating. By 1998, though, the main ceramic matrix composite liner effort was proceeding well enough that the propulsion management team terminated the metal liner project.

Silicon carbide had been chosen because it clearly possessed the temperature capabilities that HSR required. Particularly important was its high thermal conductivity, which would allow it to transfer heat from the liner's uncooled front side to its cooled backside relatively efficiently. In its existing, commercial forms, however, the material did not have sufficient durability. Existing fibers were not stable enough at high temperatures, or strong enough, for the program's needs. And current production processes resulted in a low-density form of the material that did not make maximum use of the basic material's high conductivity. So the ceramic matrix liner team sought an improved fiber, which Dow Corning eventually developed for them, and a new manufacturing method that produced a denser material. Short-term testing of HSR's new version of the material in 1996 showed a doubling of thermal conductivity and fatigue life, two important measures of progress.[35]

Glenn's Ajay Mishra recalls that the tests also provided the ceramic liner team with its major surprise during the program. The silicon carbide material proved to be very sensitive to moisture in the combustion gases. Because the basic jet fuel combustion process produced large amounts of water vapor, during the 1996 tests the material had suffered rapid erosion of its surface. To prevent this from happening, the team needed a barrier coating that would prevent moisture from reaching the silicon carbide material without impairing the material's other qualities. This was an entirely new research area for the partners that sent them back to first principles. They tested various potential materials on small panels for up to 2,000 hours in simulated combustion environments, settling on a coating in 1997 that provided nearly complete protection of the underlying material.[36] When HSR ended in 1999, the partners had achieved about two hundred hours of testing on candidate liners in the combustor program's RQL sector test rig, not enough to achieve real confidence in the material's ultimate viability. But they had also provided an early version of the material system to the Department of Energy for testing in an experimental large-scale power turbine, and there the silicon carbide liners survived 4,000 hours of testing, quite enough to demonstrate that the composite liner program had met its goals.

The long-life turbine materials research, like the ceramic matrix composite liner project, began in the "head start" program that preceded Phase II. The two engine manufacturers brought into the program their existing best materials, and in addition the researchers initially looked at some new materials systems. By mid-1993, however, they had realized that modifications to the existing systems were all they would have time for within HSR's horizon. Existing commercial turbine materials systems were composed of three elements: a metal "superalloy" blade, a ceramic coating that reduced the temperature at the metal's surface (called a "thermal barrier coat" or TBC), and a "bond coat" between the blade and the ceramic material that served as the system's glue. The thermal barrier coat, a product of research at Glenn in the

1970s, had the effect of lengthening the metal blade's lifespan at a given gas temperature.[37] Their introduction into commercial engines had allowed engine designers to increase the engine's internal temperatures without reducing blade life. Higher temperatures produced better fuel economy, a major focus of propulsion research in the 1970s and early 1980s.

The existing blade material systems were not sufficient for the HSCT's engine cycle, however. They had been optimized for a subsonic airliner engine's operating cycle, which included a short period at maximum temperature during takeoff, while cruising at a significantly lower temperature.[38] Beginning in 1993, the research teams sought to reoptimize each of the material system's three components to achieve HSR's hot life goals. Glenn Research Center's Rebecca McKay managed an effort to produce a new superalloy with superior temperature capabilities, while Bob Miller managed the coatings research. McKay recalls that designing the new blade material proved unexpectedly challenging.[39] Ultimately 102 alloys were developed and tested through numerous testing "rounds," with the selected material being a hybrid of the chemistries of two other promising experimental alloys. Beginning with the fifth round, McKay's team began assessing castability of the alloys by having full-size blades of various existing designs cast. Castability of the materials turned out not to be a significant challenge, but the materials did not meet all of the goals HSR had set for them. The key measure of improved blade material life the team had chosen was a 75°F increase in temperature resistance over the engine companies' current alloys, and the team did not quite achieve it. But the new material did provide substantial improvement over the alloys the program had started with.[40]

The bond coat and TBC research, while initially planned as separate activities, were conducted together. During 1993, the research group's first task had been to standardize test methods and subject a baseline TBC/bond coat system to those tests. The baseline, a ceramic thermal coating deposited on a platinum aluminide bond coat, was a well-known system. During 1994 and 1995, the researchers conducted an extensive survey of potential new materials for the thermal barrier and bond coats, only to resolve that modifications to the chosen baseline system promised better performance than any of the other materials they had tested. The tests that they had subjected the baseline to had shown that the limiting factor in the TBC's life was failure of the bond coat, not of the thermal coat itself, and in 1997 they stopped looking for a new thermal coat to concentrate on improving the bond coat. By the time HSR ended in 1999, Miller's coating team had essentially tripled the life of the bond coat over the baseline material.

Even as late as 1999, however, the airfoil materials program team had not demonstrated that their individual improvements to the bond coat and to the superalloy would result in the order of magnitude improvement in component life that was their overall goal. The new alloy, bond coat, and TBC had not been combined and tested as a single system, and there was a possibility that they would interact in a way that impaired their individual performance instead of the hoped-for collective improvement.[41] Furthermore, each of these projected improvements was based upon accelerated testing, and the team knew that the ultimate determinant of success had to be a full-scale, real-time test in an engine. That could not happen within Phase II's time frame, and it had not been budgeted for since it could not be shoehorned in before the 2002 production decision date.

One of HSR's top-level goals had been to bring its technologies to a level of maturity that would allow the manufacturers to be able to make a production decision in 2002, and by 1996 it was already clear to the program's managers that they would not meet this challenge. The program's time horizon was too short to complete the airframe and materials durability testing and validation, and it lacked both the time and money to do engine testing. HSR needed a Phase III to meet its maturity goal, which would push the decision date out to 2007. That reality introduced a new challenge, however. A date that far in the future invalidated their assumption that the international noise standards would not change before the HSCT's arrival. Instead, the opposite was true: the noise standard almost certainly *would* change before 2007. Hence during 1996 and 1997, HSR's environmental assessment and technology integration teams struggled to understand the impact a tighter noise restriction would have on their HSCT.

MOVING GOALPOSTS

During the national SST program of the 1960s, the growing concern over airport noise had led to the imposition of noise standards on aircraft. While these regulations dramatically improved the noise situation around airports over time, they devastated the economic potential of the SSTs designed during the decade. Lingering memories of this had caused the HSR program's planners to adopt noise reduction technologies as a centerpiece of their effort, leading to the successes discussed previously. But as the 1990s wore on, the noise goal Lou Williams, Joe Shaw, Alan Wilhite and their corporate partners had chosen for HSR became increasingly untenable. A 1997 assessment of the future HSCT's noise indicated that despite their technological successes in noise reduction, the HSCT would still be the loudest airplane in the skies because the noise levels of other aircraft were decreasing dramatically. At the same time, European nations and some major American cities were pushing to further tighten aircraft noise standards. As the HSCT's launch date receded further into the future, the program's internal environmental assessors began to argue that the program's noise goals were too lenient to suit that future.

In 1977, a new set of noise standards, known as "Part 36 Stage 3" or simply "Stage 3," had gone into effect for new aircraft designs, but a phaseout plan for older aircraft was not part of the new rule. The problem of what to do with the older aircraft remained unsettled for more than a decade. President Carter signed a law in 1979 that extended the deadline for modifying the oldest aircraft to meet the 1969 "Stage 2" standard to 1988, extending the lives of the noisiest airliners by several years.[42] The anti-regulation bias of the Reagan administration, combined with an airline-led counterattack against the Stage 3 rules, ensured no regulation forcing modification of older aircraft to Stage 3 standards was issued during Reagan's watch. His successor, George H. W. Bush, finally signed a law in November 1990 that required all commercial aircraft to be either retired or modified to meet Stage 3 by 1999.[43]

By this time, the nations of the European Community had begun to implement restrictions on Stage 2 aircraft, reflecting a shift in leadership in the aircraft noise issue. During the 1970s, the United States had led the aircraft noise regulation effort, but by 1988 the European Community had taken the lead in promoting greater stringency.[44] At the International Civil Aviation Organization Assembly in 1986, the EC had tried to force adoption of operating restrictions for Stage 2 (referred to as "Chapter 2" in international parlance) aircraft leading to a complete ban by 2000, but had been blocked. The EC was again blocked in 1989, but having been foiled at the ICAO level, its governing body, the European Council, passed a "no addition" rule preventing the registry of any more Chapter 2 aircraft by its member states.[45] Finally, at virtually the same time as the United States passed its law, the EC succeeded at getting ICAO approval of a set of rules governing the phaseout of Chapter 2 aircraft internationally.[46] All Chapter 2 aircraft had to be either "hushkitted" or retired by 2000 under the rules.

The intensive noise research that NASA had funded during the 1970s undermined the Stage 3 standard even before it was fully implemented, however, because new aircraft and engine designs that came from the manufacturers beginning in 1991 were dramatically quieter than Stage 3 required. The first of these aircraft, the long-range widebody Airbus A330/A340 series, beat the standard by more than twenty decibels using an essentially new version of General Electric's CF6 engine. Boeing was also committed to a new, large widebody airliner using new engine designs from Pratt & Whitney and General Electric that would produce noise levels about twenty decibels below the standard. Reflecting what was now obviously technologically possible, in 1993 the EC issued draft proposals calling for another reduction for aircraft purchased by European airlines.[47] These regulations were never implemented, but they made clear within the industry that the EC intended to continue pushing for greater stringency despite the newly minted Stage 3 implementation scheme.

When HSR's leaders had planned Phase II, they had agreed to a program goal specifying that their technologies should permit achieving 1 EPNdB less than Stage 3 at the sideline noise measurement point, 3 EPNdB less than Stage 3 required at the cutback, or "community," noise measurement point, and 1 EPNdB less than required for approach noise (abbreviated -1, -3, -1). (See figure 8.3.) This was approximately the noise level produced by the aircraft the future HSCT was likely to replace, the 747-400. Although this goal had appeared very challenging when set in 1989, by 1995 the HSR partners had clearly met it and the Phase II nozzle engineering effort had turned toward improving the nozzle's aerodynamic performance and reducing its weight. Phase II's increased emphasis on metrics had led the Integrated Planning Team to form a team assigned the task of continuously reviewing the HSCT's overall environmental impact, and this "Environmental Impact" team mounted an internal challenge to the project's noise goal.

The EI team's members were Langley's Allen Whitehead, Boeing's Gene Nihart, and McDonnell-Douglas's Alan Mortlock. One of the team's first efforts after it formed in 1994 had been to investigate and analyze the regulations governing noise and evaluate the HSCT's "fit" within them. The noise regulations the team had to address included not just the Part 36 standards

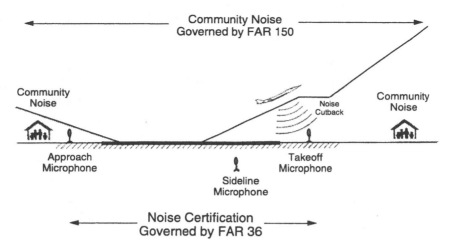

FIG. 8.3. Noise measurement points for aircraft certification. The certification standards, regulated under Part 36 of the Federal Aviation Regulations (FAR), apply to individual aircraft and must be demonstrated to the government's satisfaction before the FAA will issue a *type certificate* to a manufacturer. Typically the first production aircraft of a new type is used for the required testing. The community noise regulations contained in Part 150 apply to airports and airline operations, not directly to individual airplanes. However, if a new aircraft type will significantly increase the average noise exposure in the airport community *even if* it meets the Part 36 rules, the FAA can limit the operation of the new type. Operating limitations are undesirable to airlines and have the result of reducing the potential market for a new type. [From Allen H. Whitehead, Jr., "Impact of Environmental Issues on the High Speed Civil Transport," paper presented at the Von Karman Institute for Fluid Mechanics, May 1998]

that governed noise certification for a new design but the FAR Part 150 regulations governing total noise exposure to the communities surrounding airports. To understand the HSCT's impact under these regulations, the team had to "fly" an HSCT fleet on a simulated commercial schedule through some of the airports the aircraft was expected to serve and estimate the average noise levels the aircraft would produce in the airport community. This technique resulted in a set of noise contour lines, referred to as LdN curves, very similar to elevation lines on a topographic map. These could be overlaid on a scale map of the airport's surroundings so that researchers could estimate the geographic area in square miles subjected to a given average level of noise.

Transportation Department regulations specified that the operation of a fleet of new aircraft could not increase the geographic area encompassed by the 65 LdN contour (roughly an average noise exposure of 65 decibels) by more than 17 percent without triggering a full environmental impact statement, a very expensive and potentially politically damaging process.[48] Writing to the EI team in April 1995, Boeing noise specialist Bob Cuthbertson had argued that the program needed to adopt a tougher standard for itself because it was likely to violate this limit. He compared the HSCT's potential problem with what Concorde had faced when introduced. It had been designed when there were no noise standards, but by the time it was certificated, the Part 36 standards resulted in substantial operating restrictions on the Concorde, in that it was by far the loudest aircraft in the air. The HSCT, he believed, had to be able to operate unrestricted in a post-2006 aviation environment. The loudest aircraft still in production at that point would be the Boeing 747-400, and all other aircraft would be 5 to 8 EPNdB quieter at *each* of the measurement points.[49] (See figure 8.4.) Without a stricter goal, the HSCT would probably face operating restrictions at noise-sensitive airports like London's

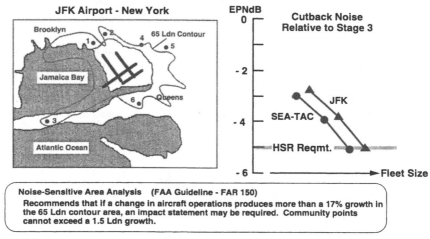

Noise-Sensitive Area Analysis (FAA Guideline - FAR 150)
Recommends that if a change in aircraft operations produces more than a 17% growth in the 65 Ldn contour area, an impact statement may be required. Community points cannot exceed a 1.5 Ldn growth.

HSR airport noise studies have projected the growth of 65 Ldn contour area with assumed HSCT operations. The FAA guideline could establish potential limit on number of HSCT operations allowed at a given airport.

FIG. 8.4. Potential effect of FAR 150 on HSCT fleet size. To the left is the 65 Ldn contour overlaid on a sketch of John F. Kennedy International Airport and surroundings. The numbers on the map represent the locations of permanent microphones used by the airport authority to monitor airplane noise. To the right is a chart suggesting the impact of HSCT noise levels on the potential market. HSR's original "cutback" goal of −1 EPNdB relative to Stage 3 rules results in a far smaller market than achieving more stringent goals in both Boeing (SEA-TAC) and McDonnell-Douglas (JFK) analyses. By the time this chart was made in 1998, −5 EPNdB relative to Stage 3 at cutback had become the program goal. [From Allen H. Whitehead, Jr., "Impact of Environmental Issues on the High Speed Civil Transport," paper presented at the Von Karman Institute for Fluid Mechanics, May 1998]

Heathrow and New York's Kennedy because it would be noticeably louder than the other traffic.

Changing the noise goal was not a simple matter, though. As Langley technology integration team leader Bill Gilbert reflected later, virtually everything about the HSCT was influenced by noise considerations. In addition to the obvious impact on the nozzle research program, noise affected the chosen engine concept's thermodynamic details, and beyond a certain level of stringency, a new goal could force the selection of a new engine concept. The noise goal chosen also essentially determined the size and shape of the HSCT's wing, which in turn impacted the aircraft's cruise performance, weight, and thus its overall economics.[50] Again, at some level of stringency, the HSCT would face the same problem Boeing's 2707 had faced thirty years before—a design that would not "close" under the new noise goal. And while the HSR program's goal was not a finished aircraft design, the program's existence was predicated upon the corporate partners' intention of ultimately building an HSCT using HSR's technologies. For the program to survive, Boeing and McDonnell-Douglas had to remain satisfied that the program's technologies would lead to a successful design. Hence the Integrated Planning Team had Whitehead's group analyze the impact of a slightly stricter noise goal on the HSCT during 1996.

A configuration known as the "Technology Concept Aircraft," or TCA, was the concept the EI team subjected to evaluation under the proposed new goal. The Technology Integration team had laid out the TCA concept in a series of meetings in Long Beach. Earlier in the program, the companies had used a Boeing configuration in their analyses, but this had not worked out. The Douglas engineers could not have the same level of knowledge about the configuration that Boeing's did, and thus they had been at a disadvantage. The TCA's purpose was to serve as a baseline configuration for the program that all three partners could address as equals. The EI team imposed a ground rule that they could not violate the 17 percent increase in 65 LdN contour size and, working on the proposition that the TCA would meet the original (Stage 3 minus 3) program goal, they studied the impact on two major international airports expected to have intensive HSCT operations, Seattle-Tacoma International Airport and New York's John F. Kennedy International Airport. These studies were done under contract, with Boeing conducting the Sea-Tac study and McDonnell-Douglas the JFK study. Both studies suggested that meeting only the Stage 3 minus 3 goal would result in a total HSCT market of fewer than six hundred aircraft, due to operating restrictions necessary to meet the "no significant impact" goal—too small a number given the required investment.

These studies also revealed a new noise challenge for the program in that the TCA's noise signature remained far beyond the traditional noise monitoring points. The vehicle climbed much more slowly than a subsonic jet, so its noise footprint was much longer. The NASA computer program used to predict noise propagation indicated that it would be audible on the ground more than twenty miles into its climb.[51] While there were no standards governing areas so far outside airport boundaries, the program's partners believed this would be unacceptable to the affected populations. Existing subsonic aircraft at that distance were generally inaudible on the ground, but the HSCT would be a substantial noise burden to these areas. The EI committee concluded that, at the very least, they had to achieve a stricter Stage 3-5 EPNdB cutback goal.

By the time the EI team's assessment had appeared in December 1996, however, there was already strong opinion within the program that this stricter goal was still too weak. The International Civil Aviation Organization's "Committee on Aviation Environmental Protection," or CAEP, which was the international rule-making organization with authority over noise rules, had met in Montreal during December 1995, and the majority of its members had favored increased stringency. They had not, however, reached an agreement, owing to disputes between the European members who favored greater stringency and non-European members who did not. Due to that failure, various European regulatory bodies were considering regulations that would tighten standards within Europe. These might include "non-addition" rules that would bar registry of noisier aircraft by European airlines or a ban on aircraft that had been "hushkitted" to meet Stage 3. While these rules would not affect the HSCT, and in fact were never implemented, they indicated that Europe was unlikely to accept a noisy HSCT.

During 1996, the Technology Integration team produced an updated concept vehicle, the "Preliminary Technology Concept," or PTC. The PTC differed from the TCA quite significantly. (See figure 8.5.) During the preceding year, refinements to the program's acoustic projections had shown that the TCA's noise level would not be within the program's goals. Gilbert's TI team had responded by altering the TCA's planform to improve the configuration's climb performance. A study of various alternate planforms had indicated that a higher aspect ratio wing with a substantially reduced outboard sweep would

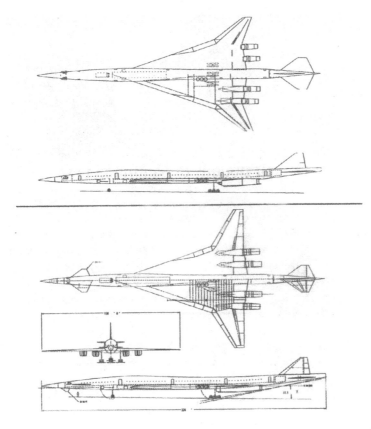

FIG. 8.5. Comparison of the Technology Concept Aircraft (TCA, top) and the Preliminary Technology Concept (PTC, bottom). Note the addition of the canard and the reduction of sweep angle of the outboard wing. While the aircraft are the same length, the PTC has greater wingspan and larger overall wing area, changes necessary to improve noise performance. (The drawings are not the same scale.) [Courtesy NASA]

restore the vehicle's noise performance.[52] Similarly, a program review by the National Research Council in 1996 had led HSR's leaders to recognize that the long, thin, highly flexible fuselage imposed control problems that had to be overcome, and a study of alternate control concepts caused the TI team to adopt a small canard.[53] By distributing control forces along the entire aircraft, instead of concentrating them at the tail, the canard reduced bending forces in the fuselage. This was the configuration the EI team used for their 1997 environmental assessment.

After the CAEP meeting in December 1995, Whitehead's EI team understood that the Stage 3 minus 5 goal would almost certainly not be acceptable to European regulators. They therefore ran the new configuration through

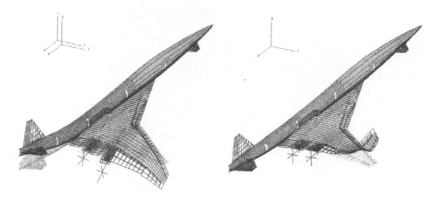

FIG. 8.6. The PTC's flutter modes. Left, a torsional mode produced by an interaction between the aft placement of the engines and nozzles and the wing's flexibility. Moving the engines inboard, forward, or off the wing entirely (all violations of HSR's design ground rules) were the only viable solutions. Right, a "classical" bending mode. A product of the wing's flexibility and the PTC's reduced outboard sweep angle, this mode was curable via substantial stiffening of the outboard wing structure (and a considerable weight penalty) or by increasing the sweep angle, which impaired noise and takeoff performance. [Courtesy NASA]

noise and sizing routines for several potential stringency increases. The strictest of these, a 13 EPNdB reduction from the program's official goal, raised the PTC far above the weight goal that the corporate partners had defined as the program's measure of economic success. Yet this level of noise reduction would still have left the PTC the loudest airliner in service.

To make things worse, a structural analysis of the TCA and PTC indicated that neither configuration was actually viable.[54] The position of the engine/mixer-ejector combinations on the trailing edge of the wing, combined with the thin, flexible wing and fuselage, triggered an aeroelastic condition called flutter when the configurations were "flown" in computer analysis. (See figure 8.6.) Flutter, a self-reinforcing vibration in the structure, could destroy a real aircraft in a few seconds and could not be allowed to occur. But fixing it, Rodney Ricketts, HSR's airframe materials and structures manager recalled, meant adding large amounts of additional stiffening to the wing structure.[55] The PTC was actually worse in this regard than the earlier TCA configuration had been, because the lower sweep angle of the outboard wing resulted in a more severe flutter condition. The structure needed to resolve the PTC's flutter problem simply could not be built. There was not enough space inside the wing to put the necessary material. Even if the structural engineers had been able to design a manufacturable structure, the resulting vehicle's weight would be so high that its economics would be terrible. It would be a million-plus-pound aircraft. Hence any increase in noise stringency that required heavier nozzles, engines, or further unsweeping of the wing could not be achieved using the program's baseline concepts. Something entirely new would be necessary. These new technologies and configurations would also have to be matured, postponing the HSCT's arrival by many more years. A substantial stringency increase meant, in essence, starting over with a new schedule, new technologies, and new investments.

A major increase is what Boeing's own team members increasingly believed was necessary to justify the huge HSCT investment. The company maintained an internal noise analysis team, and this group argued for a much stricter standard based on the fact that many cities were imposing local regulations that discriminated against noisier aircraft. London, for example, would count an aircraft having the PTC's projected noise signature as four "regular" airliners under its quota system, effectively reducing the number of flights an HSCT-owning airline could operate out of the city's airports and discouraging airlines from buying the HSCT. Late in 1997 they succeeded in convincing Mike Henderson and Boeing's senior management council that the HSCT needed to be no noisier than existing subsonic aircraft to avoid being singled out for discriminatory regulation.[56] Because new subsonic airliners were a cumulative twenty decibels below the Stage 3 standard already, HSR needed to pursue a whopping fifteen-decibel reduction below its existing target to meet Boeing's new demand.[57]

Boeing's decision to push for a new target of "blending in" to the post-2006 traffic meant an enormously difficult task for its engineers. Boeing's MacKinnon recalled that NASA then inadvertently made things worse. Administrator Goldin announced a technology program aimed at reducing the noise of new subsonic jets by another ten decibels as part of his "3 Pillars, 10 goals" initiative. Boeing's president, Harry Stonecipher, would not accept an HSCT that was louder than its subsonic competitors, so the HSCT had to meet this new goal too. Thus the total reduction necessary to meet Stonecipher's requirement was around twenty-five decibels. MacKinnon reflected later that this was a show stopper. The program's technologies could not be stretched that far. "We needed a breakthrough that we couldn't even envision. No combination of technologies we could see would do it."[58]

Boeing management's decision to require a future HSCT to be no louder than subsonic jets regardless of the legal standard was ultimately fatal to the

HSR program. Henderson and MacKinnon reflected later that what Boeing had really wanted after 1997 was for HSR to redirect its focus to research on advanced concepts that might eventually meet this new goal with a viable aircraft sometime after 2010. But this was not what NASA or the engine companies wanted. NASA headquarters and the OMB had supported HSR because of its focus on a near-term product, while Pratt & Whitney and GE wanted HSR to continue developing its propulsion technologies. Unlike HSR's airframe technologies, the propulsion materials were very valuable to the companies' other subsonic product lines because they would substantially improve the operating economies of these engines as well, and hence they wanted HSR to run out its course—and preferably be extended through a demonstration phase. Without agreement on the program's basic goals, however, the HSR partnership could not survive.

THE ROAD TO CANCELLATION

Engineering a complex technological system such as the HSCT required both a web of successful technologies and a robust network of institutional commitments. HSR had been funded because it had the solid support of The Boeing Company, General Electric Aircraft Engines, Pratt & Whitney, and the McDonnell-Douglas Corporation. Together, they had been able to convince administrators Dick Truly and Dan Goldin and the Office of Management and Budget that HSR was worth the nation's investment. It would, they had argued, result in a product with enormous export potential and they had supported that argument by adding their own money to the funding "pot" for Phase II. And they had found strong support from a Congress concerned about insufficient national investment in advanced technology. Sustaining HSR meant maintaining that network of support. But the eight years HSR had lived had witnessed profound changes in the American political and commercial landscapes. By 1998 the corporate alliance that had sustained HSR was collapsing, and the program could not survive its loss.

The NASA/industry High Speed Research program had been founded at the end of the cold war and the beginning of a brief political era in which politicians were concerned with "national competitiveness." By 1998, however, that short-lived era was already long over. Instead, the major political drive had become deficit reduction. President Clinton's chosen approach to deficit reduction had been a combination of modest tax increases and budget cuts. A disastrous rout of his Democratic Party in the 1994 congressional elections, however, put both chambers of Congress in the hands of conservative Republicans. A "Republican Revolution" in the House of Representatives led by the new speaker, Newt Gingrich, launched an attack on the civil government via a "Contract with America." One of the Contract's centerpieces was a demand for a balanced budget amendment, and this was supplemented by a goal to achieve a balanced federal budget in seven years while also delivering large tax cuts and military spending increases.[59] Because more than two-thirds of the federal budget went to entitlements, the Defense Department, and interest payments on the federal debt, balancing the budget while cutting taxes and raising defense spending meant very deep cuts in all other domestic discretionary spending—including NASA.

The space agency was in a particularly poor position to fend off budget cuts as the struggle between Clinton's White House and the Congress played out during 1995. The Clinton administration was not interested in NASA's primary mission of human spaceflight except as a diplomatic symbol of improving relations with the former Soviet Union.[60] They had no interest in the agency's long-cherished goal of a crewed mission to Mars, a program that the Bush administration had started planning. And they had left in place President Bush's NASA administrator, Daniel Goldin, who had promised both Bush and Clinton that he intended extensive reforms at the agency to make it more efficient and less expensive.[61] Hence when Clinton's OMB fashioned his presidency's first full budget, FY 1994, the agency took a large cut below its 1993 budget and saw its projected increase slashed without complaint from Goldin. While the FY 1993 budget had projected the NASA budget rising to $21 billion in FY 1999, in November 1994, the Office of Management and Budget told NASA HQ to expect a budget flatlined from FY 1995 onward at about $14 billion.[62]

If NASA had few friends in the administration, its position in Congress was no better. Speaker Gingrich and some of his key allies were actually space enthusiasts—Gingrich had written a book that extolled space settlement—but they did not like NASA.[63] They had a laundry list of grievances. NASA, in their view, suppressed innovation by not supporting small businesses. It distorted the space industry by picking winners and losers. It had deliberately blocked privatization of launch services to protect its Space Shuttle. It built a few expensive, gold-plated "Cadillac" space probes instead of many simple, cheap ones. It was not funding technologies necessary to reduce the cost of space access, a prerequisite to private development of space.[64] By all these means, NASA was preventing rapid settlement of the "space frontier." Gingrich and others wanted dramatic reforms at the agency to break it of its bad habits in pursuit of what historian Andrew Butrica has called a "conservative space agenda."[65] They intended to use their power of the purse to force the agency to reform.

If NASA's fiscal circumstances changed dramatically during the HSR program's years, so too did the commercial circumstances in which its corporate partners operated. The collapse of the Soviet Union in 1991 and its replacement by a set of small, weak, and far less hostile successor states had abruptly ended the cold war and led to a declining spiral in defense procurement. The overall defense budget fell 35 percent between 1985 and 1997. The procurement budget dropped from $125 billion to $44 billion, leaving the aerospace manufacturers with massive overcapacity.[66] The need to eliminate excess capacity produced a wave of mergers in the industry, leading Grumman and Northrop to merge in 1994, Lockheed to acquire Martin in 1995, and Boeing to purchase the North American Aviation component of Rockwell International and to merge with McDonnell-Douglas in 1996. The shrinking defense budget also produced a decline in aerospace research and development spending. As a percentage of net sales, R&D spending dropped from 16 percent to 11 percent between 1988 and 1997; put another way, the industry had spent nearly twice as much on R&D in 1988 as in 1997. A sharp rise in NASA aeronautics spending beginning in 1992 and peaking with the FY 1994 beginning of HSR Phase II had only partly offset declining Defense Department spending.[67]

The commercial aircraft business had also not been kind. But whereas military spending had trended uniformly downward, Boeing suffered from a trauma induced by a long airline recession followed immediately by an unexpectedly large boom. Between 1990 and 1994, an airline recession had led Boeing to shed workers and suppliers to cut its costs and streamline its production processes. During 1995 and 1996, however, the company's sales force was enormously successful at winning new orders, particularly for three improved versions of its smallest aircraft, the 737. They promised far more aircraft than the company could actually deliver. Instead of acting like the monopolist it was often accused of being and raising prices to restrict demand, the company resolved to keep selling and to dramatically expand production rates to keep up.[68] The result was a financial catastrophe. Manufacturing costs soared as newly hired, less efficient workers entered the assembly lines and as parts suppliers could not keep up with demand. Instead of earning the record profits analysts expected from the sales boom, Boeing took more than $2 billion in charges during 1997 and 1998 and reported its first annual loss since 1946.[69] Fixing the manufacturing disaster consumed management's attention during these two years, and repairing the balance sheets caused them to substantially reduce R&D spending. According to one source, the company reduced its internal R&D spending by 25 percent, cutting it from $2 billion in 1998 to $1.5 billion in 2000.[70]

The threat of a "Concorde 2" also receded during HSR's lifetime. In 1990, two international "study groups" had formed to examine the technologies and potential markets for two new classes of aircraft, "High Speed Transports" and "Very Large Aircraft," or "superjumbos."[71] These two groups, composed of senior executives from Boeing, McDonnell- Douglas, and the Airbus partners, had served several purposes. Both aircraft types would face a relatively restricted market that would not be profitable for two manufacturers, and these study groups would help smooth the way to cooperative manufacturing ventures later. They kept all the participants up-to-date on what each company was planning, and they served as a forum for discussing issues common to all manufacturers, such as the challenge of integrating either aircraft concept into existing infrastructure.

They also enabled the participants to gauge each other's level of effort. As the two groups met during the first half of the decade, it became clear to the American participants that the European "Concorde 2" partnership between British Aerospace and Aerospatiale was much less active than the VLA (European Very Large Aircraft—later A3XX and A380) studies. The Concorde 2 studies had not gained significant government backing and were falling behind the better-funded U.S. effort. Too, the Concorde 2 group were pursuing a significantly different aircraft, focusing on a Mach 2 aircraft made mainly of advanced aluminum alloys, not a Mach 2.4 composite vehicle. The HST study group disbanded after 1994, as it became clear that Concorde 2 was not a real threat and that the Europeans were pursuing technologies that would not make them useful partners in any American HSCT effort.

The VLA study group also disbanded, but not over technical disagreement. The European and American representatives disagreed on the size of the future market. The American companies believed that there would not be a large enough market to make a 600-seat-plus "superjumbo" profitable for even a single manufacturer. Boeing in particular forecast continued "route fragmentation" brought on by passengers' increasing desire for nonstop flights. Airlines had responded to increasing traffic over the preceding two decades by buying more smaller aircraft, spreading passenger traffic out by scheduling more flights, and adding service at new airports. Hence while Americans flying to Europe during the 1960s had to embark at one of six international "gateway" airports, by 1990 more than thirty U.S. airports offered direct service to Europe using aircraft in the 200-to-250-seat class. Fragmentation of traffic substantially undermined the case for an aircraft larger than the 400-seat 747 in their analysis. (It also undermined the case for a 300-seat HSCT).

The Europeans, on the other hand, generally believed that the fragmentation trend would not, and could not, continue. The proliferation of smaller aircraft would overwhelm the air navigation and airport infrastructure, whose limiting resource was runways. A 100-seat aircraft consumed as much runway time as a 600-seat plane, so unless one could guarantee endless access to new runways, fragmentation was a temporary phenomenon. Airlines would have little choice but respond to increasing congestion by buying large aircraft again.[72] Excess congestion would start in the Pacific, where a number of important routes already utilized specially modified 747s with reduced range and up to 550 seats, but would eventually spread worldwide.

After 1994, therefore, the Airbus consortium members pursued the Very Large Aircraft concept instead of Concorde 2. The basic criteria for this aircraft design, sketched out by British Airways in 1994, were 600 to 900 passengers, a 20 percent reduction in direct operating cost over that of the 747-400, and compatibility with a future "Chapter 4" noise standard set three to four decibels below the 747-400.[73] They began promoting specific designs to potential launch customers during 1997 in hope of obtaining enough customers to formally start the development project during 2000. This decision eliminated the "supersonic threat" that had been so useful in selling the High Speed Research program to Congress in 1989; instead, the new Airbus threat was in a vehicle that Boeing had had no interest in building. Boeing attempted to save face by promoting another "stretch" of its venerable 747, but without success. While the 747 stretch was arguably as economical as the new Airbus "superjumbo," airlines had never liked having to pay monopoly prices on the 747 line to Boeing and wanted Airbus to introduce a competing aircraft as a means of gaining negotiating power.

The problem NASA's HSR managers faced, then, was in getting Boeing to continue investing part of its shrinking R&D budgets in the HSR program, given the disappearance of a "Concorde 2" and the Airbus decision to pursue an aircraft that threatened Boeing's most profitable aircraft. In years of declining resources, companies sought to invest in research that would produce near-term returns. But the HSCT's launch date seemed to be receding into the remote future as the potential impact of greater noise stringency became understood within the program. This led to waning support from Boeing for continuing the program into a third phase; in turn, NASA headquarters, faced with large cost overruns in its space station program and a hostile OMB, cancelled HSR and redirected most of its funding into space.

In early 1996, NASA had asked the National Research Council's Aeronautics and Space Engineering Board to perform an independent evaluation of the High Speed Research Program. A panel chaired by retired USAF general Ronald Yates met several times with the program's leaders between June 1996 and January 1997, and at a meeting later in 1997 it presented its findings to the agency. HSR, the panel believed, could not achieve the goals it had set for itself. "In order to achieve the vision of the HSR program," the committee wrote, "we believe it is essential that ongoing technology development be supplemented by corresponding technology maturation and advanced technology demonstration."[74] These new efforts could not be completed within HSR's time frame; indeed, they would extend the federally funded portion of the program through 2010, and a first flight of the production aircraft would not occur until 2014.

The new post-2002 phase of the program that the panel was proposing would be focused on major issues that Phase II had not addressed. First, HSR would not result in an engine demonstration. The HSCT's propulsion system was the most challenging piece of the aircraft, and given the enormous amount of money that the corporate partners would have to spend to bring about a marketable HSCT, few believed that the companies, or Wall Street, would invest until the engine technology had been validated in full-scale tests. Second, the HSCT's composite airframe required new manufacturing technologies, and these the companies also needed to build and validate before they could make reasonably accurate assessments of aircraft construction costs. Without credible manufacturing costs to present to potential financiers, securing the estimated $15 billion in development costs would be impossible—especially with Boeing's demonstrated inability to assess and control costs on even its very well-known product line fresh on analysts' minds. Third, the highly flexible airframe technology needed to be examined in a full-scale flight research program. The long, thin structure promised to induce control problems that could only be understood and resolved in flight. Hence HSR needed to build a full-scale technology demonstration aircraft. But the new program would be far more expensive than Phase II had been. While the panel did not give an estimated cost, they suggested that this follow-on program would cost several billion dollars.

The possibility of a follow-on "Phase III" had come up during preparation of the FY 1997 budget during 1995. The Office of Management and Budget had allowed administrator Goldin a "wedge" of funding for new initiatives in FY 1999 and FY 2000, and one potential candidate for that funding was a supersonic cruise engine demonstration.[75] To win the money, HSR had to compete against other candidates, and thus it had to construct a credible program to propose. Starting the following May, the Integrated Planning Team began working on a Phase III proposal in earnest. Guidance from the headquarters Office of Aeronautics had been that the planners should concentrate on the technologies necessary to get "senior management to continue investment through product launch"; a series of ground rules limited the scope of the future plan.[76]

Even this early in the planning stage, though, money was already a problem. The general plan for a propulsion demonstrator laid out by Shaw's group at the Glenn Research Center exceeded the available funding wedge by more than half. The airframe portion of the program, while it would be less expensive than the propulsion element, would still result in costs above what was going to be available.[77] All in all, despite some descoping of the propulsion plan, by June the complete HSR III plan looked like it would cost a bit more than twice what the funding wedge contained.

The difference between the resources available and HSR III's needs caused NASA HQ to direct the Integrated Planning Team to redo the plan with a significant reduction in breadth. This new "Phase IIA" plan eliminated the propulsion materials, inlet, and aerodynamics research that had been part of HSR III, while retaining the engine demonstrator and some of the production technology validation efforts. The planning team projected Phase IIA's cost as $774 million through 2007, with the costs broken down roughly 60/40 in favor of propulsion. In addition, the two engine manufacturers had pledged

$120 million in independent research and development funds (IRAD) to support Phase IIA technologies.[78]

In 1997 Phase IIA was approved to begin in FY 1999, but in approving it the OMB reduced its funding even further. The balanced budget agreement worked out between the administration and Congress during 1995 and early 1996 had shrunk NASA's funding below the levels projected for the agency in the FY 1995 budget. The agency was to receive only $13.4 billion in FY 1999, about $900 million less than expected, and $13.2 billion each year thereafter through 2003.[79] Simultaneously, the agency's major project, the International Space Station (for which Boeing was also the lead contractor), was experiencing substantial cost overruns that had to be funded from the shrinking budget.[80] The administration's policy had shifted in 1997 to allow aeronautics programs to be cut to make up other shortfalls, and $66 million was deleted from Phase IIA. Due to the importance of the propulsion demonstrator, NASA's Office of Aeronautics allocated most of the cuts to the airframe manufacturing technology portion of the program.[81]

The reduction in Phase IIA's funds cost some support at Boeing, but Mike Henderson recalls that the event that finally cost him HSR's remaining support within Boeing's senior management council was an engine performance update that arrived in May 1998. Due to the engine demonstration portion of the proposed Phase IIA, Pratt & Whitney and GE had frozen the engine design and done a high-fidelity detailed design. But the design's performance was much poorer than projected and its weight was higher. Because each pound of extra engine weight translated into several pounds of aircraft weight, this was devastating. The engine put HSR's PTC configuration well over a million pounds, and the company's "internal airplanes" suffered equally dramatic gains. When Henderson briefed his management council on the new update, he received very rough handling. Neither Harry Stonecipher nor Boeing's CEO, Phil Condit, had been strong supporters of the HSCT concept in any case, and this result confirmed for them that the HSCT was nonviable. Given the HSCT's perceived inability to meet acceptable noise targets and the company's financial woes at the time, this far-out-in-the-future aircraft project was voted down. HSR would be one of the programs cut to reduce R&D spending.

The company dispatched vice president Bob Spitzer to tell NASA that the company wanted HSR significantly reduced in scope, and that it would no longer provide funding for it. In a meeting with NASA associate administrator

Sam Armstrong, GE Aircraft Engine's vice president Mike Benzakein, and Lee Koons from Pratt & Whitney on 9 July 1998, Spitzer argued that HSR had to get away from its product focus and look for new technologies to meet the far stricter noise targets his bosses wanted. Unsurprisingly, Spitzer's words set off an argument. The two engine manufacturers wanted to pursue the propulsion portion of Phase IIA as planned, and Boeing's MacKinnon recalls that Boeing wanted that too. Harry Stonecipher, the company's CEO, had told him and Henderson not to get in the way of the propulsion funding, since ultimately the materials technologies would benefit Boeing as well. But NASA's officials believed that without Boeing's financial participation they would not be able to defend the program before the OMB. OMB's position had consistently been that the corporate partners had to contribute to the program financially. Without Boeing, the airframe technology portion at the least would not survive, but even the propulsion program was at risk, since it would be difficult to justify an engine program without a corresponding aircraft. This meeting ended with an agreement to replan Phase IIA over the next few months as a propulsion research program.

The replanned Phase IIA shrank to a $500-million program that included engine component testing, not a full engine demonstration, a much-reduced airframe technology program, and flight experiments to demonstrate the low-boom technologies from HSR's Phase I. But Boeing's unwillingness to commit to an airplane had sparked a major debate within the administration about the basic goal of supersonic commercial flight and its purported benefits of competitiveness, time savings, productivity improvements, etc. During the previous year, the United States' air transportation system had suffered near gridlock again, prompting the administration to establish a national review commission to make recommendations on what to do about it. Chaired by former congressman Norman Mineta, who had long served in the House of Representatives on the Transportation Committee, the committee had issued a report in December 1997 calling for a shift in national technology investment away from "competitiveness" and toward the "public good."[82] "Public good" issues included greater aviation safety research, investment in new air traffic management technologies, and noise and emissions reduction; they did not include supersonic transportation. In working out its FY 2000 budget request, NASA's Office of Aeronautics had therefore eliminated the Advanced Subsonic Technology program and realigned its funding to environmental compatibility issues and an "Aviation Systems Capacity" program, while shrinking Phase IIA and reprogramming the majority of the deleted funds to non-Aeronautics programs.

In a telephone conference on 23 October, OMB officials laid out the ongoing high-level debate for HSR's top managers. The previous expected outcome of HSR, an industry-led product launch, had been the justification for the government's investment in the technologies. Boeing's lack of interest in a product launch had reopened old debates about whether government should fund technologies aimed at new product development, what magnitude of cost sharing the government should demand, and whether government should be at all involved in near-term research such as HSR. The OMB would therefore no longer support a product-oriented program, although it *might* support a reframed propulsion technology program.

A week after the OMB teleconference, Boeing's Malcolm MacKinnon was quoted in the *Seattle Post-Intelligencer* as saying the company "didn't have an airplane to hang its hat on."[83] Boeing would be scaling back its HSCT effort from three hundred to fifty engineers, and it would not be building an HSCT until 2020 at the earliest.[84] All of the foregoing controversies had occurred without a hint reaching the press, and a brief flurry of articles in newspapers and trade magazines appeared in the aftermath of MacKinnon's statement. MacKinnon, who had taken over as program manager after Henderson was promoted, had been told by his superiors to make the announcement, apparently believing that it had already made clear its intentions to NASA and that it was time HSR's redirection became public. But NASA had not told Congress yet, and the "unfortunate interview," as one NASA official put it, earned the administrator's office some undesired inquiries. The Office of Aeronautics cobbled together a "best face forward" set of responses to questions, but the interview was not taken well.[85] Wally Sawyer later noted that HSR's level of funding had been predicated on an early HSCT launch date; as that date moved further into the remote future, the need for a large, focused technology program vanished. The 2020 date in Boeing's announcement was far from the 2007 or so date that Phase IIA had been predicated on. Hence as far as the OMB was concerned, without Boeing's participation and a near-term product launch, the money could be better used for other priorities.[86] Late in the month, NASA received its annual passback guidance from the OMB recommending that HSR be terminated.

The combination of space station overruns and Boeing's unwillingness to provide further financial support to HSR caused headquarters to agree with the OMB's assessment. While administrator Goldin could have fought the OMB recommendation, he chose to fight for the beleaguered space station instead. The day before Thanksgiving, deputy administrator Sam Armstrong held a teleconference with the center directors in which he announced that HSR Phase II and IIA would not continue in any form. Glenn's Joe Shaw recalled that the announcement had been a shock. At that point, he had believed that his propulsion program would still be going ahead.[87] Instead, Phase II and IIA would be closed down by the start of FY 2000.

Shortly before Christmas, Goldin wrote to the chairs of the various congressional committees responsible for NASA. After thanking them for their support in fully funding HSR in the FY 1999 budget, he explained that Boeing's decision to withdraw had caused him to put the program on "hold" until March, while his staff considered what to do with it. In the meantime, he was reprogramming $4.3 million from HSR to ERAST, the high-altitude environmental unpiloted research aircraft program, and $5 million to the "Bantam Booster," a next-generation lightweight rocket engine program.[88]

In January, headquarters prepared a budget for FY 2000 that did not contain HSR. The remaining funding for HSR II and IIA was reprogrammed, with most of it going into space transportation accounts. NASA's Sam Armstrong drew the unhappy task of explaining the decision in early March to Representative Dana Rohrabacher, devout space enthusiast, NASA critic, and chairman of the House Subcommittee on Space and Astronautics. "The High Speed Research program has always been dependent on an active partnership between the government and our multiple industry partners," he said. "Dramatic technology advances were made against the original HSR program goals; however, the program will be discontinued at the end of FY 1999 based on the lack of financial participation by the major aircraft manufacturer."[89] The HSR project office was directed to close down by the end of September 1999.

For the remainder of 1999, the HSR staff worked to shift its technologies into other programs. An "ultra-efficient engine program" absorbed some of the propulsion materials technologies, while the airframe materials technologies went back into the R&T "base program," which in essence meant they were shelved. The airframe materials, unlike the propulsion materials, were so specific to the HSCT mission that they had no other use. The mixer-ejector nozzle technologies were also wrapped up and shelved, since they too served no other potential purpose. At Glenn, the engine cycle and nozzle teams worked out projections of how far their technologies could be taken in the future, in case HSR came back some day, and they archived them under the rubric "ultimate mixed-flow turbofan" and "ultimate mixer-ejector nozzle."

HSR had started with a set of goals for itself that included measures of vehicle performance, and during the closeout process the technology integration team finally defined a configuration that might meet those original goals. Their baseline concept's major remaining challenge was the high weight produced by the flutter problem, and in their "2015 Technology Configuration," they resolved the flutter problem by moving the engines inboard to a "dual-podded" arrangement like that used on Concorde and changing the inlet to a new, potentially very lightweight "waverider" concept. This configuration appeared able to meet HSR's 1989 noise goal at an acceptable weight, although it would not meet Boeing's more stringent goal of "blending in."

But cancellation had the effect of freeing the Technology Integration team completely from the straitjacket their TCA-like vehicle configurations had become, and the technology integration and aerodynamic configuration teams started to seriously investigate radically different aircraft and propulsion concepts that might meet Boeing's target. They looked at about a dozen configurations briefly, and three in some detail, but the concept that rapidly became the team favorite was a Boeing configuration known as "2154" or HiS-CAT—"Highly Integrated Supersonic Cruise Aircraft Technology." HiSCAT retained the arrow-wing planform and the mixer-ejector nozzles, but repackaged them in a very different configuration. The engines were dual-podded and attached behind the wing, not on it, resolving the flutter problem. The narrowbody-style fuselage was replaced with a widebody, and the aft tail was deleted and replaced with a canard. This concept appeared able to meet Boeing's noise demands at an economical gross weight without a radical new propulsion system, although it was not subjected to a rigorous analysis before HSR closed down in September. Rather defiantly, the Office of Aeronautics replaced the old TCA on the cover of its 1999 annual report with a "sanitized" version of HiSCAT. (See figure 8.7.) The agency's goals still included supersonic travel, but that future no longer had a timetable.

✦ ✦ ✦

Langley's Bill Gilbert reminisced that HSR's demise resulted from naïveté. The program's founders had settled upon what they had believed to be a stable noise standard, and had chosen a set of technologies that they believed

FIG. 8.7. Sanitized HiSCAT. Note that while the planform is similar to the earlier TCA, propulsion integration is very different. The dual-podded engines exhaust into pairs of mixer-ejector nozzles like those developed for HSR. The flat, open area behind the nozzles allows sound to reflect upward, but not downward, reducing noise at ground level—a technique borrowed from military "stealth" programs. [Courtesy NASA]

would meet that standard. But the noise standard was not stable. Stage 3 had been seen not as the final step in noise regulation but as a step toward reducing aircraft noise to background levels, which NASA itself was rendering possible. Hence by the time HSR got started in 1989, the not quite fully implemented Stage 3 standard was already threatened by the agency's own tremendous successes at fostering quieter subsonic aircraft. And because the technology suite HSR had chosen was only just able to meet the Stage 3 standard, with a small margin to compensate for "optimism" in the projections, it could not meet a more stringent standard. HSR had undermined itself by choosing too low a target.

HSR's demise also has to be understood in the context of the international politics of aircraft. The competitive dynamic of the global aircraft market ensured that HSR's goals were too weak. The no-holds-barred competition between Boeing and Airbus (and thus the U.S. and the EU) for aircraft sales practically guaranteed that noise would be used as a discriminating factor, and as the magnitude of that factor became clear during the 1990s, Boeing made the only reasonable judgment that it could have, given the HSCT's huge cost and risk: to maximize an HSCT's chances of market success, it had to be environmentally indistinguishable from all the other airliners. Boeing's decision to require a "blending in" goal for Phase IIA was fatal to the program, but it was a sound one. Nothing less would immunize the vehicle from the politics of the international aircraft trade.

HSR was nonetheless technologically very successful. As this book was being completed, the turbine blade and disc materials were being validated by Pratt and GE for use in their subsonic and military product lines, and the two companies were also planning to use the nozzle program's chevron mixers on future subsonic aircraft to further reduce noise. Boeing's Bob Welge reflected that HSR played a major role in getting computational fluid dynamics accepted as a useful design tool by the manufacturers by demonstrating that it was dependably accurate. HSR also developed a multidisciplinary optimization tool capable of laying out and optimizing complete aircraft configurations from a handful of initial inputs. A new powder-metal process for fabricating nozzle components was adopted for the U.S. Joint Strike Fighter program, and was a candidate for heat shielding on NASA's next-generation reusable launch vehicle project. Hence despite HSR's inability to define an HSCT that met all of its partners' goals, it was technologically very productive.

Similarly, the Atmospheric Effects of Stratospheric Aviation component of HSR had been highly successful. The program had funded several new scientific instruments that became standard equipment on future ozone campaigns. It had substantially improved knowledge about the chemical dynamics of the stratosphere. The information AESA produced eventually led the United Nation's Intergovernmental Panel on Climate Change to issue a special report, *Aviation and the Global Atmosphere,* making the aviation industry the first to be examined in detail for its long-term climate impact.[90]

HSR's story, finally, illustrates a fundamental conflict that exists within the American political system regarding national support for aviation technologies. NASA is pressured by political leaders to work on "visionary," revolutionary technologies, and it is expected to obtain corporate support for them. But aerospace companies are not interested in "revolutionary" technologies that, by definition, are far beyond a rational investor's time horizon in terms of return on investment. Companies will only give substantive support to "evolutionary" technologies with relatively near-term horizons. Requiring corporate support for NASA programs thus means limiting those programs to incremental changes in the state of the art. Such changes will almost certainly never bring about a viable HSCT. Nor will any program that does not proceed through full-scale validation of the technologies, another constraint generally imposed on the agency. The financial risk of a new airplane is so large that nothing less will satisfy investors.

CONCLUSION

The supersonic era finally came to an end in October 2003 with the retirement of the Anglo-French Concordes. After a dramatic, fatal, and televised crash in Paris in 2000, several subsequent nonfatal (but well-publicized) incidents of flutter damage to the aircrafts' vertical stabilizers, and the terrorist attacks in Concorde's only destination of New York in September 2001, the luxury passengers Concorde depended upon began to choose lower-profile, apparently safer conventional aircraft for their overseas voyages. Air France donated a Concorde to the American National Air and Space Museum; an anti-France campaign by the Republican Party over French refusal to support an American invasion of Iraq helped lead to the gift of a British Concorde to New York's USS *Intrepid* memorial.[1]

Concorde had always been a political bird, and thus this end was entirely fitting. To the warriors of the American right, France's gift at this moment was embarrassing. The Air France Concorde quite literally flew in the face of Republican efforts to demonize the nation in pursuit of its own parochial political goals. Placement of the British Airways Concorde at New York City's USS *Intrepid* memorial, within sight of the World Trade Center location, would help balance the symbolism of one of France's other gifts to America, the Statue of Liberty. British Airways Concorde G-BOAD went into retirement 10 November 2003 aboard a barge in New York Harbor. Even in retirement, Concorde was a powerful symbol.[2]

The tiff over Concorde signifies the continuing hold supersonic speeds have on the American imagination. Concorde's now-ancient technology still represents "advanced" technology to aerospace enthusiasts, and one that the

United States unwisely, in their view, abandoned. One finds that futurist presentations in otherwise sober publications promote supersonic and even hypersonic airliners in 2003, not at all coincidently the centennial anniversary of the Wright brothers' first flight.[3] That aerospace engineers and marketing departments ruled the world too small for hypersonic airliners in the late 1980s seems to have already been forgotten. Speed remains linked to progress in the minds of millions of Americans, and supersonic and hypersonic fantasies are still mainstream.

These dreams of speed and power are used by politicians to promote particular visions of America. Like space technologies, they symbolize an America, and a world, without limits to its growth and to its technological progress. They are, in an important political sense, in opposition to the fundamental tenets of environmentalism—that the world is limited, that its resources are limited, and that there will be, ultimately, a limit to growth. Seen in this light, the Coalition against the SST's campaign in 1970 and the Nixon administration's efforts to save the SST concept represent a deep divide in American political thought. A principal goal of the environmental movement that started in the 1960s lay in convincing the public to accept the notion of limits; the election of Reagan represented rejection of it. Reagan deployed the symbolic power of "advanced" aerospace technology to reconstruct the nation's self-image along lines of growth and expansion after the painful 1970s, while simultaneously attacking environmental regulation, particularly those limiting development of wilderness areas. The promotion of supersonic and hypersonic technologies while rejecting development limits are of a piece. They represent a politics of expansion and dominance.

Efforts to devise a "greener" SST, then, represent attempts to reconcile incompatible beliefs through engineering. Aircraft designers are used to having to accommodate multiple conflicting demands in their work, and the community of aerospace engineers has developed design tools to help in what they refer to as "multidisciplinary optimization." These specialized software packages aid in balancing the trade-offs necessary in aircraft design: noise versus weight, internal volume versus aerodynamic drag, etc. In digital code, they embody the myriad conflicts in design and produce pragmatic (and, one hopes, workable!) solutions. These programs also exemplify engineers' efforts to achieve "value neutral" solutions—one cannot accuse nonsentient computers of bias.

Yet reconciliation is only partially possible. Any SST, or hypersonic vehicle for that matter, will be energy inefficient compared to subsonic vehicles. A supersonic future will hasten the end of the fossil fuel era. This will be true even if the next-generation SST is the world's first "hydrogen airliner," as some proponents hope to make it (most of the liquid hydrogen used in the United States is manufactured from coal). Hence while it appears possible to make future SSTs meet the noise rules and to reduce their emissions to the point of ozone neutrality, they will remain "un-green" in their energy use. They will not fully reconcile the conflict between the politics of expansion and those of limits. Instead, should SSTs ever be built again, they will represent a victory for the nineteenth-century version of progress—endless growth and acceleration.

✦ ✦ ✦

For NASA, the quest for an SST led to some unexpected ramifications. The public "discovery" of the ozone layer, the human threat to it, and the intense controversy over it, fueled the resurrection of a line of high-altitude research that had been moribund for decades and that had not been of much interest to the space agency. The SST-ozone layer controversy reversed a declining trend in atmospheric research at NASA, and the agency's involvement in the ozone controversy helped make it a major force in current environmental research aimed at the larger issue of global climate change. How the *space* agency was reoriented toward the *earth* sciences in general is a subject that needs more investigation, however. As of 1999, 10 percent of the agency's budget went to its "Earth Sciences Enterprise," more than the National Science Foundation spends in the area. This is also a larger sum than NASA spends on aeronautical research (the first "A" in its name). It seems unlikely that the ozone controversy by itself could overcome the well-known effects of bureaucratic resistance to change, a game of institutional prerogatives that NASA is otherwise quite good at playing. Witness, for example, its dogged pursuit of the "Large Space Telescope" and permanently crewed space stations despite year after year of "No!" from Congress and from presidents. There is a larger story in the space agency's adoption of what Henry Lambright had called "policy-relevant science" that needs to be told, and the SST is only a piece of it.[4] That story will involve both internal NASA politics and the changing vagaries of national politics, but it is beyond the frame of this book.

If NASA's research agenda has changed over the past five decades, so too has its relationship to the aircraft industry. The agency's role has shifted from provider of basic research to industry toward supplier of development capital. As the NACA, it generated numerous innovations that made supersonic flight feasible and improved its efficiency, although it remained solidly on the "research" end of the R&D spectrum, while the Air Force handled development. In its NASA guise, the agency has moved slowly toward a developmental role. Two reasons for this seem apparent. First, national and international politics have acted to drive the agency toward support for aircraft and engine product development, without allowing that policy to become explicit. NASA and the NACA before it had always served industry's needs in the past through targeting aviation problems for research, of course, with its extensive role in icing research perhaps the best example.[5] And in the past it had assisted manufacturers in overcoming specific problems with their designs. But the launch of HSR in 1989 reflected a strong shift toward financial support for new commercial products. While rhetorically it sought to maintain a distinction between its "research and technology" effort and the resulting "privately financed" HSCT, NASA was able to assemble the HSR partnership only by explicitly pitching the end product, *not* the knowledge that HSR would generate. As we saw in chapter 8, the OMB lost interest when the product, an American HSCT, vanished.

A second factor in NASA's gradual movement toward product development has been the virtual abandonment of commercially relevant technology development by the U.S. Air Force. The subsonic airliner revolution in the United States was driven by technologies developed at great public expense by the Air Force and then commercialized, but since the late 1960s the handful of large Air Force aircraft developed have not been of a nature that made them candidates for commercialization. Indeed, the technology flow has reversed in some ways, with both the Boeing C-17 military transport and the Northrop-Grumman B-2 bomber relying on commercial engines. In the absence of new Air Force-derived technologies, the manufacturers have had to rely on NASA, but as we have seen NASA's aid comes with significant constraints: the agency's funding is much smaller, and it has generally not been allowed to carry technologies into the integration and large-scale testing stages as the Air Force routinely had. NASA's aid tends to cut off just at the most expensive part of the R&D process, making the agency an ineffective substitute for the Air Force's former role as patron of revolutionary technologies. The business risk of truly revolutionary technologies is simply too high for private industry. Nonetheless, the agency has tried to serve a development role within the constraints imposed upon it by national politics. The result has been successful fostering of relatively low-risk "evolutionary" technologies—quieter, more fuel efficient turbofan engines—while repeatedly failing to move much higher risk (and thus more expensive) "revolutionary" technologies into production.

The inability to bring about an SST, then, is at least as much a product of American market fundamentalism as it is a result of environmental constraints. A "revolutionary" aircraft, be it an advanced SST or a hypersonic transport aircraft, will not occur without significant state backing—a reality precluded by the current American secular religion of "free markets." Markets have shown themselves to be excellent instruments for choosing and refining state-originated technologies for commercial deployment, but they have not generally been the originator of revolutionary technological change. Only the state has the ability to fund the grand techno-dreams of aerospace enthusiasts, be they supersonic/hypersonic transports, air-breathing launch vehicles for routine space access, giant orbiting solar-power stations, or the like.

Recognition of the state's role in fostering technological change during the 1960s led environmental activists to demand redirection of state funding toward more environmentally friendly technologies. As we have seen, they were relatively effective at accomplishing that goal in the case of aviation. They also inspired efforts to proactively assess technologies for their environmental impacts, resulting in an Office of Technology Assessment that served as an advisor to the United States Congress from 1972 until its dismantlement by the 1994 Congress. Beyond these successes, however, it is not clear whether environmentalists were effective at championing a redirection of technological change. Declining automotive fuel economy, for example, suggests a larger failure to inculcate an "environmental ethic" of eco-friendly technologizing. Indeed, most of the small body of studies at the intersection of technology and environmental change are declensionist, delineating repeated failure to reverse environmental damage.[6] More research is in order to understand where and under what conditions environmental politics has been successful at fostering redirection, and where and when it has not.

Finally, the long SST saga suggests the ongoing power of the old notion of "technological progress." As late as 1994 dreams of speed could still attract substantial sums of money from politicians. For the full length of the twentieth century, then, speed retained political currency. It could be deployed as a symbol of advancement, representing national commitment to high technology and thus "progress." But it was no longer the unalloyed virtue that it had been in the nineteenth century. Instead, speed was only politically suitable if it came with competitive economics and acceptable environmental impacts.

NOTES

ABBREVIATIONS

AIAA	American Institute for Aeronautics and Astronautics
ANP	Aviation Nuclear Power
ARDC	Air Research and Development Command
BAC	Boeing Aircraft Company
CASTS	composites for advanced space transportation systems
CLASB	Citizen's League against the Sonic Boom
ISABE	International Symposium on Air Breathing Engines
LaRC	Langley Research Center
NARA	National Archives and Records Administration
NACA	National Advisory Committee for Aeronautics
NA/CP	National Archives, College Park, Md.
NA/P	National Archives, Philadelphia
NAA	North American Aviation
NAS	National Academy of Science
NRC	National Research Council
PRO	Public Records Office, Kew Gardens, London
RAE	Royal Aircraft Establishment
RPI	Rensellear Polytechnic Institute
RPL	Ronald Reagan Presidential Library
SAE	Society of Automotive Engineers
SRI	Stanford Research Institute
STAC	Supersonic Transport Advisory Committee (UK)
USAF	United States Air Force
WHCF	White House Central Files (Nixon Papers)
WHSF	White House Special Files (Nixon Papers)
WHO	White House Office (Kennedy Library)

1. John Newhouse, *The Sporty Game* (New York: Knopf, 1982).

2. Keith Hayward, *The British Aircraft Industry* (New York: St. Martin's, 1989); David Weldon Thornton, *Airbus Industrie: The Politics of an International Industrial Collaboration* (New York: St. Martin's, 1995).

3. Author's calculation from data in Aerospace Industries Association of America, *Aerospace Facts and Figures 2001–2002* (Washington, DC: 2001), 32. For earlier data see National Research Council, *The Competitive Status of the U.S. Civil Aviation Manufacturing Industry* (Washington, DC: National Academies Press, 1983).

4. Christian Gelzer, "The Quest for Speed: An American Virtue, 1825–1930" (Ph.D. diss., Auburn University, 1998).

5. "Recent Economic Changes in the United States: Report of the Committee on Recent Economic Changes of the President's Conference on Unemployment" (Washington, DC: GPO, 1929), 1.

6. William M. Leary, *Aerial Pioneers: The U.S. Airmail Service, 1918–1927* (Washington, DC: Smithsonian Institution Press, 1985), 205.

7. Alex Roland, *Model Research: The National Advisory Committee for Aeronautics, 1915–1958*, vol. 1 (Washington, DC: NASA, 1985).

8. Tom Wolfe, *The Right Stuff* (New York: Bantam, 2001 reprint ed.); Laurence K. Loftin, Jr., *Quest for Performance: The Evolution of Modern Aircraft*, NASA SP-468 (Washington, DC: 1985); Richard Hallion, *On the Frontier: Flight Research at Dryden, 1946–1981*, NASA SP-4303 (Washington, DC: 1984).

9. Roland, *Model Research*, 204–5.

10. On the politics of Concorde, see Kenneth Owen, *Concorde and the Americans: International Politics of the Supersonic Transport* (Washington, DC: 1997); the extant history of the Soviet SST program is Howard Moon, *Soviet SST: The Technopolitics of the TU-144* (New York: Orion Books, 1989).

11. Owen, *Concorde and the Americans*.

12. Moon, *Soviet SST*, 6.

13. Mel Horwitch, *Clipped Wings: The American SST Conflict* (Cambridge, MA: MIT Press, 1982).

14. Ibid., 325.

15. James Hansen, "What Went Wrong? Some New Insights into the Cancellation of the American SST Program," in *From Airships to Airbus: The History of Civil and Commercial Aviation*, vol. 1, ed. William M. Leary (Washington, DC: Smithsonian Institution Press, 1995), 168–89.

16. Walter McDougall, review of Horwitch, *Clipped Wings*, in *Technology and Culture* 25 (April 1984): 367–69.

17. Divergent histories of environmental politics in the United States exist. The mainstream interpretation locates the beginnings of modern environmentalism with middle-class responses to Rachel Carson's *Silent Spring*, published in 1962. See Samuel P. Hayes, *Beauty, Health and Permanence: Environmental Politics in the United States, 1955–1985* (New York: Cambridge University Press, 1987). For an alternative approach that locates environmentalism in the radical counterculture, see Robert Gottlieb, *Forcing the Spring:*

The Transformation of the American Environmental Movement (Washington, DC: Island Press, 1993).

18. Richard J. Kent, *Safe, Separated and Soaring: A History of Federal Civil Aviation Policy, 1961–1972* (Washington, DC: FAA, 1980).

19. The environmental campaign against the SST is the subject of Mel Horwitch's history of the SST program. See Horwitch, *Clipped Wings*.

20. Michael Henderson interview with author, 25 March 2002, Langley Research Center historical archives, Hampton, Virginia (hereafter LaRC archives).

21. Joseph J. Corn, *The Winged Gospel: America's Romance with Aviation, 1900–1950* (New York: Oxford University Press, 1983), 3–27.

22. Peter Fritzsche, *A Nation of Fliers: German Aviation and the Popular Imagination* (Cambridge: Harvard University Press, 1992), 3–5.

23. Michael S. Sherry, *The Rise of American Air Power: The Creation of Armageddon* (New Haven: Yale University Press, 1985); Paul Boyer, *By the Bomb's Early Light: American Thought and Culture at the Dawn of the Atomic Age* (Chapel Hill: University of North Carolina Press, 1985).

24. Gabrielle Hecht, *The Radiance of France: Nuclear Power and National Identity after World War II* (Cambridge: MIT Press, 2000), 13.

25. On the technological sublime, see David E. Nye, *American Technological Sublime* (Cambridge: MIT Press, 1996), esp. chap. 9.

1 ✦ CONSTRUCTING THE SUPERSONIC AGE

1. Richard Hallion, *On the Frontier: Flight Research at Dryden, 1946–1981,* NASA SP-4303 (Washington, DC: 1984).

2. Henry Petrosky, *To Engineer Is Human: The Role of Failure in Successful Design* (New York: Barnes and Noble, 1994).

3. In his excellent analysis of U.S. strategic bomber procurement programs, Michael Brown argues that the USAF ran far ahead of the science of supersonic flight. See Michael E. Brown, *Flying Blind: The Politics of the U.S. Strategic Bomber Program* (Ithaca, NY: Cornell University Press, 1992), 162, 226–27.

4. Walter McDougall, . . . *the Heavens and the Earth* (New York: Basic Books, 1985), 141–56.

5. James R. Hansen, *Engineer in Charge: A History of the Langley Aeronautical Laboratory, 1917–1958,* NASA SP-4305 (Washington, DC: 1987), 271–309.

6. Walter G. Vincenti, "Engineering Theory in the Making: Aerodynamic Calculation 'Breaks the Sound Barrier,'" *Technology and Culture* 38:4 (October 1997): 819–51.

7. Hansen, *Engineer in Charge,* 271–73.

8. Ibid., 273–74.

9. Ibid., 277. John Becker, one of Langley's premier hypersonic aerodynamicists, claimed in a later memoir that the difference proved not to matter, because the phenomenon that the Lab wanted to study (the deep shock stall, caused by formation of the shockwave on the wing's upper surface) occurred at a lower speed than anticipated, well below the point at which the wind tunnel data needed correlation. See John V. Becker, *The High Speed Frontier,* NASA SP-445 (Washington, DC: 1980), 97–98.

10. Hansen, *Engineer in Charge,* 278–79.

11. Ibid., 290.

12. Ibid., 299.

13. For a full accounting of Yeager's flight see Richard Hallion, *On the Frontier: Flight Research at Dryden, 1946–1981,* NASA SP-4304 (Washington, DC: 1984), 3–22.

14. Hansen, *Engineer in Charge,* 316–20.

15. See, for example, Steve Pace, *X Fighters: USAF Experimental and Prototype Fighters, XP-59 to YF-23* (Osceola, WI: Motorbooks International, 1991); and Arthur Pearcy, *Flying the Frontiers: NACA and NASA Experimental Aircraft* (Annapolis, MD: Naval Institute Press, 1993).

16. John Becker interview with the author, 5 June 2000.

17. Brown, *Flying Blind.*

18. Pace, *X Fighters,* 89–90.

19. Ibid., 91.

20. Ibid., 92.

21. Ibid., 120.

22. Vincenti, "Engineering Theory in the Making," 819–51.

23. Hansen, *Engineer in Charge,* 337. The NACA report is Richard T. Whitcomb, "A Study of the Zero Lift Drag Characteristics of Wing-Body Combinations Near the Speed of Sound," NACA RM-L52H08, September 1952.

24. Richard D. Thomas, *History of the Development of the B-58 Bomber,* vol. 1 (Dayton, OH: USAF Aeronautical Systems Division, Air Force Systems Command, November 1965), 54–60.

25. Carl F. Damberg to director, NACA, 1 March 1951, box 374, NACA correspondence series, RG 255, National Archives and Records Administration (hereafter NARA), College Park, Maryland. See also Richard D. Thomas, *History of the Development of the B-58 Bomber,* vol. 1 (Dayton, OH: USAF Aeronautical Systems Division, Air Force Systems Command, November 1965), chap. 3.

26. John Swihart interview with author, 5 June 2000.

27. AD-33 to AD-3, "Problems of AD-33 in evaluation of performance of current and future aircraft," n.d. but c. 1958, box 224, file A173-5, LaRC series 2, RG 255, National Archives, Philadelphia (hereafter NA/P); Laurence K. Loftin, Jr., telephone conversation with author, 15 May 2000.

28. Richard D. Thomas, *History of the Development of the B-58 Bomber,* vol. 1 (Dayton, OH: USAF Aeronautical Systems Division, Air Force Systems Command, November 1965), 169.

29. The B-58's subsonic cruise L/D was estimated by NACA at 11.3. See Laurence K. Loftin, Jr., *Quest for Performance: The Evolution of Modern Aircraft,* NASA SP-468 (Washington, DC: 1985), 494.

30. Robert E. Hage and Richard D. Fitzsimmons, "Economic Aspects of the Supersonic Jet Transport," *Aeronautical Engineering Review* 11:9 (September 1952): 42–45.

31. Robert E. Hage, *Jet Propulsion in Commercial Air Transportation* (Princeton: Princeton University Press, 1948).

32. Author's estimate derived from Hage and Fitzsimmons, "Economic Aspects," p. 44, fig. 8.

33. Ibid., 45.

34. Ibid.

35. Kent A. Mitchell, "Nuclear-Powered Aircraft Program," *AAHS Journal* 42:2 (Summer 1997): 142.

36. The earliest example is L. V. Humble, W. W. Wachtl, and R. B. Doyle, "Preliminary Analysis of Three Cycles for Nuclear Propulsion of Aircraft," NACA RM-E50H24, 1950. This study examined air-cooled, liquid-metal cooled, and helium-cooled reactors.

37. Mitchell, "Nuclear-Powered Aircraft Program," 143.

38. ANP Project collection, file "GE-ANP Program November 1951–June 1952," LaRC archives.

39. Air Research and Development Command, Historical Division, "History of the Air Research and Development Command, 1 January–30 June 1959," vol. 2, "The B-70 Story" (Dayton, OH: Wright-Patterson Air Force Base, 1959), 4.

40. On WS-110 as a "hedge" against failure of the nuclear-powered bomber, see ibid., 5.

41. Ibid., 11.

42. This version of the story differs substantially from that given in the open literature. All published sources agree that Curtis LeMay hated both Boeing and North American Aviation designs because of the floating wing tips. However, the internal "Secret" level study done by ARDC in 1959 makes clear that as of 1956, SAC had accepted them. I have chosen to rely on the classified history. The relevant published sources on the B-70 are Steve Pace, *North American Valkyrie XB-70A* (Blue Ridge Summit, PA: Aero, 1984), 12–13, and Jeannette Remak and Joe Ventolo, *XB-70 Valkyrie: The Ride to Valhalla* (Osceolo, WI: MBI Publishing, 1998), 30. The formerly classified study is Air Research and Development Command, Historical Division, "History of the Air Research and Development Command," 14–16. They have most likely drawn their conclusion from the Senate investigation of the B-70 program, published as U.S. Senate, Preparedness Investigating Subcommittee of the Armed Services Committee, "The B-70 Program," 86th Cong., 2nd sess., July 1960, p. 4.

43. Historical Division, Air Research and Development Command, "History of the Air Research and Development Command," 17.

44. Richard T. Whitcomb and Thomas L. Fischetti, "Development of a Supersonic Area Rule and an Application to the Design of a Wing-Body Combination Having High Lift-to-Drag Ratios," NACA-RM-L53H31a, 1953.

45. Robert T. Jones, "Theory of Wing-Body Drag at Supersonic Speeds," NACA-TR-1284, 1956.

46. Eggers interview with author.

47. A. J. Eggers, Jr., and Clarence A. Syvertson, "Aircraft Configurations Developing High Lift-Drag Ratios at High Supersonic Speeds," NACA-RM-A55L05, 5 March 1956.

48. Summarized from ibid., 6–7.

49. Robert Schlaifer and S. D. Heron, *Development of Aircraft Engines* (Boston: Harvard University Press, 1950), 424–28.

50. The earliest NACA investigation of turbine cooling was published as a multivolume study entitled "Cooling of Gas Turbines," NACA-RM-E7B11, 1947. The lab published a comprehensive review of turbine cooling in 1955: Jack B. Esgar and Robert R. Ziemer, "Review of Status, Methods, and Potentials of Gas-Turbine Air-Cooling," NACA-RM-E54I23, February 1955.

51. Esgar and Ziemer, "Review of Status," esp. fig. 1, p. 25.

52. Ibid., 11.

53. J. Reeman and R. W. A. Buswell, "An Experimental Single-Stage Air-Cooled Turbine: Part I: Design of the Turbine and Manufacture of Some Experimental Internally Cooled Noz-

zles and Blades," *Proceedings Part A, Institution of Mechanical Engineers* 167:4 (1953): 341–50; D. G. Ainley, "Research on the Performance of a Type of Internally Air-Cooled Turbine Blade," *Proceedings Part A, Institution of Mechanical Engineers* 167:4 (1953): 351–62, and Discussion 363–70.

54. Telephone conversation with Marv Stibich, Wright-Patterson AFB, 2 June 2000. See also Virginia Dawson, *Engines and Innovations,* NASA SP-4306 (Washington, DC: 1991), 127–44.

55. Quoted in Senate, Preparedness Investigating Subcommittee of the Armed Services Committee, 86th Cong., 2nd sess., July 1960, pp. 28–29.

56. Historical Division, Air Research and Development Command, "History of the Air Research," 16–17.

57. Eugene C. Draley, "Status of Langley Research Program for High-Performance Supersonic Aircraft," 22 October 1957, box 223, file A173-5, LaRC series 2, RG 255, NA/P.

58. Donald D. Baals, Thomas A. Toll, and Owen G. Morris, "Airplane Configurations for Cruise at a Mach Number of 3," NACA Conference on High-Speed Aerodynamics, March 18–20, 1958, 521–41. See also Draley, "Status of Langley Research Program."

59. "F-108 and B-70 Program Analysis 1955–1959," doc. nr. K243.04-12, Air Force Historical Research Agency, Maxwell AFB. NAA saw the F-108 and B-70 programs as very complementary, and the company had timed the two programs so that its design engineers would phase off the F-108 work as that program was completed onto the B-70, permitting easy transfer of expertise from one to the next.

60. NAA's initial proposal cited an L/D of 6.90 in November 1957. See Historical Division, Air Research and Development Command, "History of the Air Research and Development Command," 35.

61. Ibid., 33.

62. E. E. Weismantel and R. H. Kuhn, "Electrochemical Machining of Aerospace and Commercial Fabrications," 1968, preprint courtesy of the General Electric Aircraft Engines Archives, Cincinnati.

63. All data from "WS-110A Highlights," 13 August 1957, BAC document no. D2-2116, box 5845, Boeing Archives, Everett, Washington.

64. John Swihart interview.

65. Historical Division, Air Research and Development Command, "History of the Air Research and Development Command," 32.

66. Holden R. Withington interview with author, 7 May 2001.

67. Historical Division, Air Research and Development Command, "History of the Air Research and Development Command," 19.

68. George W. Rathjens to Dr. Kistiakowsky, 1 November 1960, DOD-1960, box 6, Security Affairs series, Eisenhower Library, Abilene, Kansas.

69. Richard Fitzsimmons interview with the author, 7 February 2000.

70. William Cook, *The Road to the 707* (Bellevue, WA: TYC Publishing, 1991), 221.

71. Boeing Co. SST chronology, Boeing Archives.

72. John Swihart interview.

73. Kenneth Owen, *New Shape in the Sky* (London: Janes, 1982), 36. The original article is E. C. Maskell and D. Küchemann, "Controlled Separation in Aerodynamic Design" (RAE Technical Memo Aeronautics #463, 1956).

74. This was the "Mutual Weapons Development Program." France also participated. The best-known result of this collaboration is the AV-8 "Harrier" V/STOL fighter.

75. Owen, *New Shape,* 12.

76. T. E. Blackall, *Concorde: The Story, the Facts, and the Figures* (Oxfordshire: G. T. Foulis, 1969), 13.

77. McDougall, . . . *the Heavens and the Earth,* 141–53.

78. ARDC to contractors in SR-169, 6 December 1957, box 225, LaRC series 2, RG 255, NA/P.

79. "USAF Considering SST," *American Aviation* 21:21 (10 March 1958): 14.

80. Convair's transport is detailed in Hearings, House, Special Investigating Subcommittee of the Committee on Science and Astronautics, "Supersonic Air Transports," 86th Cong., 2nd sess., 17 May 1960, pp. 126–32.

81. Robert Hotz, "Supersonic Transport Era," *Aviation Week,* 26 May 1958, p. 21.

82. This phenomenon had been a subject of NACA research from 1955. See the Minutes of the Subcommittee on High-Speed Aerodynamics, 3 October 1956, Committee and Subcommittee alpha-numeric file 1948–1958, box 10, file "Minutes 1955–1956," RG 255, NARA, College Park, Md.

83. Hotz, "Supersonic Transport Era," 21.

84. "America Revisited," *The Aeroplane,* 27 June 1958, pp. 887–90.

85. The Boeing Company Annual Reports for 1958, 1959, and 1960, copies in author's collection; Douglas Aircraft Company Annual Reports for 1958, 1960, 1961, and 1962, copies in author's collection; Testimony of William M. Allen, 25 October 1963, Senate, Committee on Interstate and Foreign Commerce; transcript from W. M. Allen Papers, box 31, file "SST materials 1963," Boeing Archives.

86. "America Revisited," 887–90.

87. McDougall, . . . *the Heavens and the Earth,* 162–63.

88. A White House decision in December 1956 effectively terminated the program, but the termination effort ran into 1958. See Historical Division, Air Research and Development Command, "History of the Air Research and Development Command," 19.

89. A. J. Goodpaster, Brig. Gen. U.S.A., "Memorandum of Conference with the President," 2 December 1959 and Staff Notes November 1959, box 45, Dwight D. Eisenhower Diary Series, Eisenhower Library.

90. Ibid.

91. Air Force arguments defending the B-70 are concisely summarized in A. J. Goodpaster, Brig. Gen. U.S.A., "Memorandum of Conference with President," 20 January 1960, JCS September 1959–May 1960, White House, Office of Staff Secretary, box 4, Records 1952–1961, Eisenhower Library. The conference this memo reports was actually held 18 November 1959.

92. Historical Division, Air Research and Development Command, "History of the Air Research and Development Command," 22.

93. Goodpaster, "Memorandum of Conference with the President," 2 December 1959.

94. Ibid. See also George B. Kistiakowsky, *A Scientist in the White House* (Boston: Harvard University Press, 1976), 161–62.

95. Stephen Ambrose, *Eisenhower: Soldier and President* (New York: Simon and Schuster, 1990), 495.

96. Ibid., 501.

97. "Reds Build Transport Plane Similar to Shelved U.S. B-70," *Washington Post,* 25 January 1960, p. 9. Rumors of a Soviet SST program had circulated within the aeronautical engineering community during 1958–1959. See, for example, R. W. Rummel, "Comments on R. C. Sebold's Paper 'Commercial Air Transportation beyond the Subsonic Jets,'" *Supersonic Transports (Proceedings),* Institute of the Aeronautical Sciences, 26–29 January 1959, p. 10. Rummel was TWA's VP-engineering; R. C. Sebold was VP-engineering at Convair.

98. Testimony of Ira Abbott, House Hearings, "Supersonic Air Transports," 11.

99. Ibid., 53. Quesada also admitted that the Sputnik crisis had been irrational.

100. Ibid., 42.

101. Ibid., 52.

102. Senate, Committee on Armed Services, Preparedness Investigating Subcommittee, "The B-70 Program," 86th Cong., 2nd sess., July 1960, pp. 30, 37.

103. For the 1960 campaign's aerospace relevance see McDougall, . . . *the Heavens and the Earth,* chap. 10.

104. A Langley researcher had demonstrated in 1959 that the "supersonic wedge principle" did, in fact, produce enough additional drag to make the extra lift not worth having. See Lowell Hasel, "An Experimental Pressure-Distribution Investigation of Interference Effects Produced at a Mach number of 3.11 by Wedge-Shaped Bodies Located under a Triangular Wing," NASA TM-X-76, October 1959. Wind tunnel results that challenged an accepted theory, however, had to be verified in flight to become "settled knowledge." The XB-70 did not fly until 1964.

105. See Hearings before House Committee on Science and Astronautics, "Boron High Energy Fuels," 86th Cong., 1st sess., August 1959, pp. 53–86.

106. Douglas Aircraft Company doc. #SM-23378, Air Force Historical Research Agency, Maxwell AFB, Montgomery AL.

107. "Supersonic Transport Studies," BAC doc. D6-3617, folder 11, box 2816, Boeing Archives.

108. Glenn Garrison, "Supersonic Transport May Aim at Mach 3," *Aviation Week,* 2 February 1959, p. 39.

109. McDougall, . . . *the Heavens and the Earth,* 4–13, 303–6.

110. Ibid., 158–76; Alex Roland, *Model Research: The National Advisory Committee for Aeronautics, 1915–1958,* vol. 1, NASA SP-4103 (Washington, DC: 1985), 296–300.

2 ✦ TECHNOLOGICAL RIVALRY AND THE COLD WAR

1. The Federal Aviation Agency, constructed out of the Civil Aeronautics Administration in 1958, was renamed the Federal Aviation Administration (FAA) in 1966, when the Department of Transportation Act made it a subunit of the new Department of Transportation. See Richard J. Kent, Jr., *Safe, Separated, and Soaring: A History of Federal Civil Aviation Policy* (Washington, DC: FAA, 1980), 167–80; Federal Aviation Administration, *FAA Historical Chronology* (Washington, DC: FAA, 1998).

2. The lab's efforts were measured in "professional man-years" assigned to each type of research. See John Becker et al., "Review of Langley Research Effort," 31 January 1958, file 33, Floyd Thompson collection, LaRC archives.

3. James R. Hansen, *The Spaceflight Revolution: NASA Langley Research Center from Sputnik*

to Apollo, SP-4308 (Washington, DC: 1995), 15–22, 69; and Virginia Dawson, *Engines and Innovation: Lewis Laboratory and American Propulsion Technology,* SP-4306 (Washington, DC: NASA, 1991), 179.

4. J. D. Hunley, ed., *The Birth of NASA: The Diary of T. Keith Glennan,* NASA SP-4105 (Washington, DC: 1993), 188–89.

5. Stuart I. Rochester, *Takeoff at Mid-Century: Federal Civil Aviation Policy in the Eisenhower Years, 1953–1961* (Washington, DC: FAA, 1976), 189–220; Mel Horwitch, *Clipped Wings: The American SST Conflict* (Cambridge, MA: MIT Press, 1982), 21.

6. Robert Gilpin, *France in the Age of the Scientific State* (Princeton: Princeton University Press, 1968); Gabrielle Hecht, *The Radiance of France: Nuclear Power and National Identity after World War II* (Cambridge, MA: MIT Press, 1998).

7. Langley Research Center, "Summary of NACA/NASA Technical Support of the TFX (F-111) Variable Sweep Aircraft Program," 15 February 1963, file "TFX 1961–1963," box 103, LaRC series 1, RG 255, NA/P; Charles Donlan and William Sleeman, "Low-Speed wind-tunnel investigation of the longitudinal stability characteristics of a model equipped with a variable-sweep wing," NACA RM-L9B18, 1949.

8. Jay Miller, *The X-Planes* (Arlington, TX: Aerofax, 1988), 59.

9. On the flight testing see Richard Hallion, *On the Frontier: Flight Research at Dryden, 1946–1981* (Washington, DC: NASA, 1984), 51–54. Langley also carried out an extensive wind tunnel correlation program resulting in nineteen technical publications. See Langley Research Center, "Summary of NACA/NASA Technical Support."

10. Langley Research Center, "Summary of NACA/NASA Technical Support."

11. Langley Research Center, "Summary of NACA/NASA Technical Support"; see also W. J. Alford, Jr., "Exploratory investigation of the low-speed aerodynamic characteristics of variable-wing-sweep configurations," NASA TMX-142, 1959; W. J. Alford, Jr., "Wind-tunnel studies at subsonic and transonic speeds of a multiple mission variable wing-sweep airplane configuration," NASA TMX-206, 1959. Alford and Polhamus were awarded a patent on the concept on 11 September 1962.

12. John Stack to director, "MWDP Project Trip, June–July 1959," John Stack Collection, file 68, LaRC archives.

13. A detailed, if Langley-centered, chronology appears in Langley Research Center, "Summary of NACA/NASA Technical Support."

14. This program became extremely controversial, but that controversy has no bearing on this book. For a history of the TFX program, see Robert J. Art, *The TFX Decision: McNamara and the Military* (Boston: Little, Brown, 1968.)

15. Laurence K. Loftin interview with author, 23 November 1999; Don Baals and Lowell Hasel interview with author, 5 September 2000; John Swihart interview with author, 5 June 2000.

16. F. Edward McLean, *Supersonic Cruise Technology,* NASA SP-472 (Washington, DC: 1985), 42.

17. Clinton E. Brown and L. K. Hargrave, "Investigation of minimum drag and maximum lift-drag ratios of several wing-body combinations including a cambered triangular wing at low Reynolds numbers and at supersonic speeds," NACA-RM-L51E11, 1951.

18. Richard T. Whitcomb interview with author, 22 September 2000.

19. Ibid. See also Neil Driver interview with James Hansen, 21 July 1989, LaRC archives.

20. Horwitch, *Clipped Wings*, 9–10. See also the incomplete, unpublished FAA history of the SST program: "FAA SST History part I," SST collection, box 30-4, FAA History Office.

21. "FAA SST History part I," SST collection, box 30-4, FAA History Office, Washington, DC, and "The Supersonic Transport: A Technical Summary prepared for the Federal Aviation Agency," 11 December 1959, p. 1.

22. NASA, "The Supersonic Transport: A Technical Summary."

23. Ibid.

24. Ibid.

25. Ibid.

26. "United States Leadership in World Aviation," John Stack Collection, file 84, LaRC archives.

27. Glennan did not wish to "manage and fund this kind of an activity" despite considerable pressure from people inside NASA. See J. D. Hunley, ed., *The Birth of NASA: The Diary of T. Keith Glennan*, NASA SP-4105 (Washington, DC: 1993), 188–89.

28. Horwitch, *Clipped Wings*, 16.

29. J. G. Borger et al., "The Commercial Supersonic Transport: Some Operational Considerations," Society of Automotive Engineers (SAE), paper 341A, 1961.

30. Richard Rhode to Langley, 23 February 1954, A313-1, box 456, LaRC series 2, RG 255, NA/P.

31. T. L. K. Smull to H. A. Sweet, 12 November 1953, box 224, LaRC series 2, RG 255, NA/P.

32. "Sonic Bangs: A Qualitative Explanation," RAE Technical Note Aero 2192, September 1952.

33. Lina Lindsay to associate director, 16 October 1953, A313, box 456, LaRC series 2, RG 255, NA/P.

34. Handwritten note initialed "FJB," on Lina Lindsay to associate director, 16 October 1953, A313, box 456, LaRC series 2, RG 255, NA/P. The note states: "I believe this problem is going to get real big—have great military potential and should be investigated by us—pronto!" Other members of the Langley staff recall that John Stack believed this as well. See Laurence K. Loftin interview with author, 24 November 1999.

35. Studies of the boom phenomenon done in the U.S. during the early 1960s credit G. B. Whitham for the basic theory and F. Walkden, of the University of Manchester, for the addition of lift effects to Whitham's theory. See Harvey H. Hubbard, Dominic Maglieri, Vera Huckel, and David Hilton, "Ground Measurements of Sonic-Boom Pressures for the Altitude Range of 10,000 to 75,000 feet," NASA TR R-198, July 1964, and E. L. Crosthawait, "Sonic Boom Theory and the B-58," Convair Report FZA-4-405, 28 April 1961, Langley Research Center Library.

36. Dominic Maglieri and Harvey Hubbard, "Ground Measurements of the Shock Wave Noise from Supersonic Bomber Airplanes in the Altitude Range From 30,000 to 50,000 Feet," NASA TN-D-880, July 1961. See also Crosthawait, "Sonic Boom Theory and the B-58."

37. Harry W. Carlson, "An Investigation of the Influence of Lift on Sonic-Boom Intensity by Means of Wind-Tunnel Measurements of the Pressure Fields of Several Wing-Body Combinations at a Mach Number of 2.01," NASA TN-D-881, 1961. He correlated the wind tunnel measurements with the High Speed Flight Station's data in Carlson, "Wind-Tunnel Measurements of the Sonic-Boom Characteristics of a Supersonic Bomber Model and a Correlation with Flight-Test Ground Measurements," NASA TM-X-700, 1962.

38. For an example of congressional interest see W. B. Wilmot to Office of Congressional Liaison, "Congressional Inquiry—Sonic Boom, Lower Carolina—Upper Dorchester Counties" (Congressman T. F. Johnson), 25 March 1959, Committee Series, A313-1, box 455, LaRC series 2, RG 255, NA/P; the 1960 meeting is recounted in Harvey Hubbard to assistant director, "Visit to WADC Jan 11, 1960, to participate in a meeting to discuss various aspects of the problem of sonic boom induced damage to building structures," 21 January 1960, file 1959, A313-1, box 455, LaRC series 2, RG 255, NA/P.

39. See Domenic J. Maglieri, Vera Huckel, and Tony L. Parrott, "Ground Measurements of Shock-Wave Pressure for Fighter Airplanes Flying at Very Low Altitudes and Comments on Associated Response Phenomena," NASA TN D-3443 (July 1966).

40. Don Dwiggins, *The SST: Here It Comes, Ready or Not* (Garden City, NY: Doubleday, 1968), 66.

41. Harvey Hubbard to John Stack, "Community response phase of the sonic boom program," 15 November 1961, A313-1 1962, box 454, LaRC series 2, RG 255, NA/P.

42. Charles W. Nixon and Harvey Hubbard, "Results of the USAF-NASA-FAA Flight Program to study Community Responses to Sonic Booms in the Greater St. Louis Area," NASA TN D-2705, May 1965.

43. Ibid., 14.

44. Convair's marketing plan is contained in "Convair/SAC B-58 Cooperative Orientation Program," n.d. (c. 1958), Convair Records, San Diego Aerospace Museum.

45. Quoted in Horwitch, *Clipped Wings*, 44.

46. Jack Brewer, "NASA Supersonic Transport Research Committee (5th meeting)," 25 April 1962, Roll 144, LaRC series 3, RG 255, NA/P.

47. W. H. Statler to W. M. Hawkins, "NASA Research on Aircraft," 15 August 1962, Box 214, LaRC series 2, RG 255, NA/P.

48. Ralph Rudd to Floyd Thompson, 23 January 1963, "SST October 1963," LaRC series 2, RG 255, NA/P.

49. See "Proceedings of NASA Conference on Supersonic Transport Feasibility Studies and Supporting Research, September 17–19, 1963," NASA TM-X-905, December 1963.

50. Lockheed's numbers. See R. Richard Heppe and Jim Hong, "Summary of Lockheed SCAT Feasibility Studies," in *Proceedings of NASA Conference on Supersonic Transport Feasibility Studies and Supporting Research, September 17–19, 1963*, NASA TM-X-905, December 1963, pp. 21–92.

51. Both Lockheed and Boeing argued that the NASA "D" engine, a duct-burning, 1.5 bypass ratio turbofan engine with a turbine-inlet temperature of 2,700R (2,241F) was the minimum "technology level" that would make an SST feasible. Use of the "state of the art" engine B (1,900°F turbine inlet temperature) resulted in an approximate 50,000-pound weight penalty for SCAT 17, based on Boeing's analysis. Lloyd Goodmanson, William T. Hamilton, and Maynard Pennell, "Summary of Boeing SCAT Feasibility Studies," *Proceedings of NASA Conference on Supersonic Transport Feasibility Studies and Supporting Research, September 17–19, 1963*, NASA TM-X-905, December 1963, p. 128.

52. This was 2.0 psf for acceleration, 1.5 psf for cruise. Essentially, the higher the acceleration altitude, the lower the boom. However, higher altitudes required a higher mass-flow engine (or larger amounts of afterburner) to perform the transonic acceleration. Hence while a subsonic aircraft's engine size was determined by takeoff performance, the SST's

engine size was largely determined by the altitude at which transonic acceleration was to occur.

53. Goodmanson, Hamilton, and Pennell, "Summary of Boeing SCAT Feasibility Studies," 93–138.

54. I take this payload fraction estimate from Boeing's George Schairer. Schairer made this estimate much later, but it is consonant with the tone of both Lockheed and Boeing presentations. See George Schairer to Holden Withington, 9 June 1967, Allen Papers, box 31, file "SST #3," Boeing Archives.

55. John Stack, 6 April 1961, file "SST 1960–1962," Stack collection, LaRC.

56. There is a direct correlation between jet engine performance and altitude caused by the reduction in air density as altitude increases. To maintain a constant airflow rate (and thus constant efficiency) as one increases altitude, one must also increase speed (so the engine literally "scoops in" more air). Hence a Mach 2 aircraft cruises most efficiently around 50,000 feet, while a Mach 3 aircraft cruises most efficiently between 60,000 and 70,000 feet.

57. Kenneth Owen, *Concorde and the Americans: International Politics of the Supersonic Transport* (Washington, DC: Smithsonian Institution Press, 1997), 37. The "smoking gun" document is "Conclusions of a Meeting of the Cabinet Held at Admiralty House 6 November 1962," PRO/CAB 36.

58. The chief secretary to the Treasury was quite explicit in his condemnation of the program's economics. See "Supersonic Airliner: Memorandum by the Chief Secretary to the Treasury," 22 June 1962, PRO/CAB 134/1585. His belief in the uneconomic nature of the program is reflected in the cabinet's decision to launch the program. See "Conclusions of a Meeting of the Cabinet Held at Admiralty House 6 November 1962," PRO/CAB 36.

59. Robert Gilpin, *France in the Age of the Scientific State* (Princeton: Princeton University Press, 1968), chap. 1; Walter A. McDougall, "Space Age Europe: Gaullism, Euro-Gaullism, and the American Dilemma," *Technology and Culture* 26 (1985): 179–203; see also Walter A. McDougall, "Technocracy and Statecraft in the Space Age," *American Historical Review* 87:4 (October 1982): 1010–40.

60. Quoted in Kenneth Owen, *Concorde: New Shape in the Sky* (London: Jane's, 1982), 34.

61. Ibid., 35.

62. E. C. Maskell and D. Küchemann, "Controlled Separation in Aerodynamic Design," RAE Technical Memo Aero 463, March 1956.

63. Andrew Nahum, "The Royal Aircraft Establishment from 1945 to Concorde," in *Cold War, Hot Science,* ed. Robert Bud and Philip Gummet (London: Harwood Academic Publishers, 1999), 29–58.

64. M. B. Morgan to Air Chief Marshal Sir Claude Pelly, 9 March 1959, reproduced in Owen, *New Shape,* 255–56.

65. Harold Watkinson to prime minister, 22 July 1959, PRO/AVIA 63/30.

66. Royal Aircraft Establishment, "Report of the Supersonic Transport Aircraft Committee," reproduced in Owen, *New Shape,* 258–59.

67. Cabinet approval is contained in Cabinet Conclusion (59) 52nd, minute 3, 18 September 1953, PRO/AVIA 128/32.

68. Owen, *Concorde and the Americans,* 24–25.

69. T. E. Blackall, *Concorde: The Story, the Facts, and the Figures* (Oxfordshire: G. T. Foulis, 1969).

70. Sud's management believed that the Caravelle had succeeded because it was aimed at

a portion of the market that the big American manufacturers had ignored—small, short-to-medium-haul aircraft. They tried at first to convince British authorities that this was the best way to avoid competing with the U.S. manufacturers in the SST market, too, but could not overcome the fundamental problem that higher speeds do not produce much time savings over a short route.

71. Owen, *New Shape*, 46.

72. Ibid., 46–51.

73. Treasury, "Supersonic Airliner," 22 June 1962, PRO/CAB 134/1585. The TSR.2 program, which aimed at producing a Mach 2–capable tactical fighter, had seen its estimates soar from £25–50 million in 1959 to £137 million in March 1962, causing Treasury to seek its cancellation and to expect the same to happen on the SST. See John Law and Michel Callon, "The Life and Death of an Aircraft: A Network Analysis of Technical Change," in *Shaping Technology / Building Society*, ed. Wiebe E. Bijker and John Law (Cambridge, MA: MIT Press, 1992), 36.

74. First Secretary of State, "The Supersonic Airliner," 25 October 1962, PRO/CAB 134/1585.

75. Thorneycroft was moved to Defence in a cabinet shakeout during July, but continued to be involved in the SST campaign. See Owen, *Concorde and the Americans*, 36.

76. "Minutes of a Meeting of the Committee held in Conference Room 'C', Cabinet Office, Thursday, 27 September 1962," PRO/CAB 134/1585.

77. "Conclusions of a Cabinet Meeting held at Admiralty House, 6 November 1962," PRO/CAB 128/62.

78. Owen, *Concorde and the Americans*, 37–38.

79. Horwitch, *Clipped Wings*, 21.

80. Najeeb Halaby, *Crosswinds: An Airman's Memoir* (Garden City, NY: Doubleday, 1978), 182.

81. The legislation stated: "The Administrator is empowered and directed to encourage and foster the development of civil aeronautics and air commerce in the United States and abroad," in section 305, and "The Administrator is empowered to undertake or supervise such development work and service testing as tends to the creation of improved aircraft, aircraft engines, propellers and appliances," section 312 (b). See Federal Aviation Act of 1958.

82. This assessment of the FAA incapacity to manage a large development program was widespread at the time. See: Horwitch, *Clipped Wings*, 28–29; Allen to Gilpatrick, 18 June 1962, file "SST Correspondence 1961–1963," box 74, acc. # 255-77-0677, Suitland Federal Records Center, Suitland, MD.

83. "FAA SST History part I," typescript, SST collection, box 30-4, FAA History Office.

84. Horwitch, *Clipped Wings*, 24.

85. Quoted in ibid., 25.

86. J. G. Borger, L. H. Allen, N. D. Folling, and W. F. Hibbs, "The Commercial Supersonic Transport—Some Operational Considerations," SAE Paper 341A, 1961, p. 9. Pan Am's analysis was based on a substantial part of its international route structure and was run on an IBM 705 computer. A major reason for the SST's poor showing was operating restrictions imposed by noise regulations barring operations after 10 P.M. These restrictions reduced the SST's productivity by 50 percent, which translated into an increase in direct costs of 30 percent. Pan Am's analysis used a 400,000-pound Boeing SST design as its basis.

87. Quoted in Horwitch, *Clipped Wings*, 33.

88. The IATA's "Ten Commandments" are summarized by Kenneth Owen in *Concorde:*

Story of a Supersonic Pioneer (London: Science Museum, 2002), 203–4; see also International Air Transport Association, *Symposium on Supersonic Air Transport* (Montreal: IATA, 1961).

89. In his memoir, Halaby reports that LBJ pulled him aside after the meeting and told him who his enemies were: McNamara, Kermit Gordon (the budget director), and the Commerce Department. Halaby, *Crosswinds*, 194.

90. McNamara believed that the U.S. Air Force (and its budget) were "out of control" and he did not intend to give them any path to restoring either the B-70 bomber (which the SST enthusiasts wanted) or the Air Force's internal drive toward a military supersonic transport.

91. Kennedy appointed LBJ to chair the National Aeronautics and Space Council. See Robert Dallek, *Flawed Giant: Lyndon Johnson and His Times, 1961–1973* (New York: Oxford University Press, 1998), 11.

92. Washington (U.K. Embassy) to Foreign Office, 20 October 1964, PRO/AIR 8/2296.

93. Najeeb Halaby to the president, 15 November 1962, "SST 1962," box 6, SST collection, FAA History Office.

94. Halaby, *Crosswinds*, 192. See also Najeeb Halaby, "Intensification of Supersonic Program," 31 January 1963, SST collection, box 6, file "SST 1963," FAA History Office.

95. Halaby, *Crosswinds*, 189.

96. Supersonic Transport Advisory Group, "Report to the Chairman, Supersonic Transport Steering Group," 15 November 1962.

97. The "balance of payments" affected the value of the dollar. During the 1960s, the balance was negative, meaning that the United States imported more goods than it exported. This forced the treasury to sell gold to sustain the dollar's value. This problem was a central economic issue for both the Kennedy and the Johnson presidencies because the gold supply was limited and thus a negative balance of payments was not sustainable over a long period of time. Numerous programs were promoted as "solutions" to the balance of payments problem, including the SST. The "solution," ultimately, was President Nixon's decision to take the United States off the gold standard.

98. Supersonic Transport Advisory Group, Supplementary Report to the Supersonic Transport Steering Group Supersonic Transport Program Planning, 14 May 1963, FAA History Office.

99. In his detailed examination of SRI's report, he noted that it did not state the source of the market data upon which the SST market predictions were based. Further, the economic model used by SRI's analysts did not include the possibility that the market for the SST could be highly sensitive to fares, and this could trigger an airline demand for operating subsidies. See Walter Heller to the vice president, "Supersonic Transport," 12 March 1963, microfilm #30, Kennedy Library, Boston, Massachusetts.

100. Department of Commerce, "Summary Statement of Position on the Supersonic Transport Problem," 24 May 1963, microfilm #30, Kennedy Library.

101. See Horwitch, *Clipped Wings*, 51–52. See also Robert McNamara's very carefully worded expression of support: Robert S. McNamara to Halaby, 1 June 1963, file "SST 1963," box 6, SST collection, FAA History Office.

102. Horwitch, *Clipped Wings*, 52–53.

103. Halaby rejects the importance of the Pan Am order in his memoir, but Mel Horwitch's analysis of Pan Am's role is more compelling, given Pan Am's role in the introduction of subsonic jets. See ibid., 53.

104. This is explained in John Costello and Terry Hughes, *The Concorde Conspiracy* (New York: Scribner's, 1976), 64.

105. Trippe did not make clear the difference between his agreement and a typical airline order in his conversations with Halaby and Johnson, perhaps to strengthen his lobbying position. It also was not reported in the trade magazines, which published alarmist articles on the implications of Pan Am's order.

106. Horwitch, *Clipped Wings,* 53.

107. Quoted from ibid., 54.

108. Halaby made this comment on JFK's attitude on 31 January 1963, two weeks after de Gaulle's rebuff to Britain's Common Market application. See Administrator to Mr. Shank, "Intensification of the Supersonic Transport Program," 31 January 1963, file "SST 1963," box 6, SST collection, FAA History Office.

109. Washington (U.K. Embassy) to Foreign Office, 20 October 1964, PRO/AIR 8/2296.

110. Gabrielle Hecht, *The Radiance of France: Nuclear Power and National Identity after World War II* (Cambridge, MA: MIT Press, 1998).

3 ✦ ENGINEERING THE NATIONAL CHAMPION

1. See, for example, Najeeb Halaby to William Boeing, telephone conversation, 12 July 1963, file "SST #1," box 31, Allen Papers, Boeing Archives.

2. Mel Horwitch follows the cost-sharing debate in considerable detail. See Mel Horwitch, *Clipped Wings: The American SST Conflict* (Cambridge: MIT Press, 1982), 64–67.

3. W. Henry Lambright, *Presidential Management of Science and Technology: The Johnson Presidency* (Austin: University of Texas Press, 1985), 117–26.

4. Kenneth Owen, *Concorde and the Americans: International Politics of the Supersonic Transport* (Washington, DC: Smithsonian Institution Press, 1997), 53.

5. McNamara's accomplishments at the Pentagon are summarized at www.defenselink .mil/specials/secdef_histories/bios/mcnamara.htm. Copy in author's collection. See also James M. Roherty, *Decisions of Robert McNamara: A Study of the Role of the Secretary of Defense* (Coral Gables: University of Miami Press, 1970), 65–88.

6. Only rockets and rocket planes had reached Mach 3. The XB-70, which was supposed to fly at this speed, was not yet complete. The CIA's A-12 had begun flying but had not yet gotten much above Mach 2.

7. Donald Douglas, Jr., to Najeeb Halaby, 26 August 1963, "FAA Supersonic Transport 1963 documents," box 78, President's Office Files, John F. Kennedy Library, Boston, Massachusetts.

8. From Richard Shevell interview with author, February 2000.

9. Douglas Aircraft Company, "Douglas Supersonic Transport System," document M014875, McDonnell-Douglas Archives, Long Beach, California.

10. E. F. Barnes to J. A. Gorgenson, "S.R.I. and DACo forecasts of the SST market," 19 July 1963, document M017502-63, McDonnell-Douglas Archives.

11. W. D. Smith, "Summary Comments from the viewpoint of Market Research," 18 July 1963, document M017502-63, McDonnell-Douglas Archives.

12. *New York Times,* 10 September 1963, p. 79.

13. Halaby to president, "Douglas Aircraft Company will not Participate in the Super-

sonic Transport Development Program," 26 August 1963, Box 78, Pres. Office Files, Kennedy Library.

14. Horwitch, *Clipped Wings*, 68.

15. Eugene Black and Stanley Osborne, "Report to the President on the SST," 19 December 1963, "Report to the President on the SST," box 170, RG 200, Robert F. McNamara Papers, RG 200, NA/CP.

16. See Horwitch, *Clipped Wings*, 68–72.

17. McNamara, who was often accused of wanting to "steal" the program from FAA, was opposed to placing it within his own organization. He believed that the U.S. Air Force would attempt to use it as justification for a military SST, which he saw no use for. For McNamara's lack of interest in "owning" the SST, see Roswell Gilpatric to Vice President Johnson, "Comments on FAA Supersonic Transport Recommendation," 22 May 1963, box 171, "Material for SST meeting 30 May 1963," McNamara Papers.

18. Don Baals to associate director, 5 August 1963, A173-5, box 13, LaRC series 2.

19. FAA, "Supersonic Transport Evaluation Report," 1 April 1964, pp. 1, 49, box 56, PAC/SST collection, Johnson Presidential Library, Austin, TX.

20. Robert Serling, *Legacy and Legend: The Story of Boeing and Its People* (New York: St. Martin's, 1992), 270.

21. Aircraft that climb faster radiate less noise into areas underneath their flight path because their greater altitude allows more of the noise to be absorbed and dispersed in the atmosphere.

22. FAA, "Supersonic Transport Evaluation Report," 4–5.

23. Ibid., 68.

24. Ibid., 68–71.

25. Ibid., 15.

26. Stated expressly in Russell G. Robinson to the director, "Supersonic transport engine and airframe proposals in response to FAA request," 24 January 1964, box CF-53, file "E16-75," Ames records, NARA, San Bruno.

27. Louis R. Eltscher and Edward M. Young, *Curtiss-Wright: Greatness and Decline* (New York: Twayne Books, 1998), 160.

28. A very nice summary of the evaluation independent of the FAA exists in the Ames lab records. See Russell G. Robinson to the director, 21 April 1964, box CF-53, file "E16-75," Ames records, NARA, San Bruno. Here the vote against Curtiss-Wright is clear: "Cumulative risk of numerous innovations too great. Success not very probable." But the author is also clear that this was the most advanced engine proposed.

29. FAA, "Supersonic Transport Evaluation Report," 15.

30. Emanuel Boxer, "Advanced Research and Development for the Supersonic Transport Propulsion System," 13 November 1963, box 213, A173-5, LaRC series 2, RG 255, NA/P. The context of the memo was a meeting at Wright-Patterson at which the members of the joint NASA/DOD SST propulsion committee had agreed that current technologies would be at best marginal for the SST, and that whoever was selected as the manufacturer for the SST development program would be too busy to conduct the advanced research necessary for a superior "second generation" engine that would make the SST a truly competitive airplane. They expected that either GE or Pratt would be that manufacturer, as each already had working supersonic engines and thus could claim lower development risk. Hence the group wanted

Curtiss-Wright kept under contract to develop its advancements free from the pressure of manufacturing deadlines.

31. FAA, "Supersonic Transport Evaluation Report," 17–18.

32. Gordon Bain, "Supersonic Transport Program Recommendation," 1 April 1964, "SST Memos, April–November 1964," box 171, McNamara Papers.

33. "Draft Proceedings of the President's Advisory Committee on the Supersonic Transport Plane," 13 April 1964, "Minutes of PAC/SST v. 1, 13–18 April 196," box 168, McNamara Papers.

34. John E. Keller and Lt. Col. Robert E. Pursley, "Review of Report by the Administrator, FAA dated 13 April 1964 and entitled 'Major Issues—Supersonic Transport Development Program,'" 16 April 1964, "SST Memos April and November 1964," box 171, pp. 6–8, McNamara Papers.

35. Ibid., 9.

36. "Draft Proceedings of the President's Advisory Committee on the Supersonic Transport Plane," 13 April 1964, "Minutes of PAC/SST v. 1, 13–18 April 1964," box 168, pp. 27, McNamara Papers.

37. Joseph Califano, "Draft Proceedings of the President's Advisory Committee on the Supersonic Transport Plane," 18 April 1964, "Minutes of PAC/SST v. 1, 13–18 April 1964," box 168, McNamara Papers.

38. Ibid.

39. Ibid. The letter is: The President to Najeeb Halaby, 23 April 1964, "SST Memos, Apr–Nov 1964," box 171, McNamara Papers.

40. Pratt & Whitney presentation, "Minutes of the PAC SST Meeting 1 May 1964,"box 168, pp. 4–5, McNamara Papers.

41. The difference in military and commercial feasibility is simply the engine's "time between overhaul," or TBO. Military requirements typically specified certification by the manufacturer of 150 hours TBO. Airlines would not touch an engine rated less than 500 hours TBO—and FAA's SST economic projections were based upon 2,000 hours TBO. At this time, after several million flight hours worth of experience, the J57 engine powering Pan Am's 707s had a certified TBO of only 1600 hours.

42. "Draft Proceedings of the President's Committee on the Supersonic Transport," 8 May 1964, "Minutes of PAC/SST for 8 May and 30 Mar 1965," box 168, p. 15, McNamara Papers.

43. Ibid.

44. Ibid. The DeHavilland Comet, the world's first commercial jet airliner, suffered several catastrophic structural failures in flight. These were traced to metal fatigue and an inadequate structural design. The accidents destroyed confidence in the aircraft, and also destroyed its market.

45. Ben Rich and Leo Janos, *Skunk Works: A Personal Memoir of My Years at Lockheed* (Boston: Little, Brown, 1994), 220–25. Rich designed the inlets for the aircraft.

46. Supersonic inlets provide thrust because they increase the pressure of the incoming air. On the A-12/SR-71, the inlet's internal pressure is about 40 psi, while the outside air at the plane's cruising altitude is about 1 psi. This pressure increase occurs as the incoming supersonic air passes through a shockwave inside the inlet and is compressed. In an "unstart," the shockwave is ejected from the inlet due to a mismatch between the position of the "spike" that controls the inlet geometry and the incoming airflow, and the compression effect ceases.

On the inlet programming issue, see Paul F. Crickmore, *Lockheed SR-71 Blackbird* (London: Osprey Press, 1986), 26.

47. Ibid., 29.

48. "Draft Proceedings of the President's Committee on the Supersonic Transport," 8 May 1964, "Minutes of PAC/SST for 8 May and 30 Mar 1965," box 168, p. 63, McNamara Papers.

49. E. Eminton and W. T. Lord, "Note on the Numerical Evaluation of the Wave Drag of Smooth Slender Bodies Using Optimum Area Distribution for Minimum Wave Drag," *Journal of the Royal Aeronautical Society* 60:541 (January 1956): 61–63.

50. Harry W. Carlson and D. Middleton, "A Numerical Method for the Design of Camber Surfaces of Supersonic Wings with Arbitrary Planforms," NASA TN D-2341, 1964.

51. Roy Harris, Jr., interview with author, 7 June 2000.

52. Donald D. Baals, A. Warner Robins, and Roy V. Harris, Jr., "Aerodynamic Design Integration of Supersonic Aircraft," AIAA paper No. 68-1018, October 1968.

53. Najeeb Halaby to the president, 23 December 1964, "SST Reference Book tabs 1–14," box 169, McNamara Papers.

54. On propulsion integration see Mark R. Nichols, "Aerodynamics of Airframe-Engine Integration of Supersonic Aircraft," NASA TN D-3390, 1966; A. Warner Robins, Odell A. Morris, and Roy V. Harris, Jr., "Recent Research Results in the Aerodynamics of Supersonic Vehicles," AIAA paper no. 65-717, November 1965.

55. The Boeing Company, "Phase IIA Summary," Boeing doc. D6-8680-1, pp. 33–34.

56. "Memorandum on the FAA evaluation of the Manufacturers' November 1, 1964 Phase IIA Supersonic Transport Proposals," 16 December 1964, "Report on FAA Evaluation of SST Proposals," box 172, McNamara Papers.

57. Horwitch, *Clipped Wings,* 98–101, covers this issue in more detail.

58. Robert S. McNamara to the president, 21 November 1964, and Lyndon B. Johnson to Najeeb Halaby, 3 December 1964, "SST Reference Book tabs 1–14," box 169, McNamara Papers.

59. Halaby to the president, 23 December 1964, box 169, "SST Reference Book tabs 1–14," McNamara Papers.

60. LBJ to administrator, FAA, 21 January 1965, "SST Reference Book tabs 1–14," box 169, McNamara Papers. The drafter's marks on this copy show the memo was written for LBJ by McNamara.

61. "Draft Proceedings of the President's Committee on the Supersonic Transport," 30 March 1965, "Minutes of PAC/SST for 8 May and 30 March 1965," box 168, McNamara Papers.

62. Ibid., 21–22.

63. Richard J. Kent, Jr., *Safe, Separated, and Soaring: A History of Federal Civil Aviation Policy, 1961–1972* (Washington, DC: FAA, 1980), 136–39; Horwitch, *Clipped Wings,* 109–10.

64. Horwitch, *Clipped Wings,* 109.

65. Ibid., 113.

66. Willard E. Foss to associate director, 21 April 1966, "Phase IIC interim assessment of the Boeing 733-390," A173-5, box 207, LaRC series 2, RG 255, NA/P.

67. Ibid.

68. The industry reaction to McNamara's delay was so vehement that the PAC discussed several ways to accelerate the program, with an earlier-than-scheduled prototype decision being the best choice. *Aviation Week and Space Technology* was the most violently anti-

McNamara publication. See Robert B. Hotz, "Flogging with Feathers," *Aviation Week and Space Technology,* 5 July 1965, p. 7.

69. Ibid.

70. Designing the flight control system so that it will not allow the aircraft to reach a high angle of attack can eliminate the deep stall problem. Since there was no reason for an airliner to operate at high AOAs—the flight attendants, drink carts, etc. would be thrown into a heap in the tail if they did—this was a perfectly reasonable solution.

71. Howard Phelps to "T" (Thornton Wilson), 1 April 1966, file "SST #2," box 31, Allen Papers, Boeing Archives.

72. Laurence K. Loftin, interview with James R. Hansen, August 1989, LaRC.

73. Horwitch, *Clipped Wings,* 159.

74. Ibid., 162.

75. Supersonic Transport Division, Boeing Co., "Phase III Proposal Summary," Boeing doc. V1-B2707-1. Courtesy Lowell Hasel.

76. Neil Driver interview, 28 September 1999.

77. The further away from an aircraft's aerodynamic center the control surfaces are, the more effect they have.

78. A. E. Raymond to McKee, 5 December 1966, "Personal Views on the SST Evaluation following review by Bisplinghoff Group, Dec. 2–5," "1966 Evaluation Docs," Box 30-5, SST History Collection, FAA History Office.

79. "Basis for selection of Boeing Supersonic Transport Airframe," 17 January 1967, "1966 Evaluation Docs," Box 30-5, SST History Collection, FAA History Office. This document was written three weeks after the selection, suggesting an after-the-fact defense.

80. Charles W. Harper to Laurence K. Loftin, A173-5, box 208, LaRC series 2, RG 255, NA/P .

81. Nick A. Komons, SST program history, chap. 11, "The Design Change," unpublished manuscript courtesy the FAA. Copy in author's collection.

82. Charles J. V. Murphy, "Boeing's Ordeal with the SST," *Fortune* 78 (October 1968): 129–33, 161–64.

83. George Schairer to H. Withington, 9 June 1967, file "SST #3," box 31, Allen Papers, Boeing Archives.

84. Murphy, "Boeing's Ordeal," 133.

85. Ibid., 191.

86. Komons, "The Design Change," 8–10.

87. Wells and Allen quoted in Murphy, "Boeing's Ordeal," 193.

88. Ibid., 192.

89. Neil Driver interview, 28 September 1999.

90. A. Warner Robins interview with author, September 2000. The "sandbagged" quote is attributed to Edgar Cortwright, the Langley lab's director at the time.

91. A. Warner Robins interview with author, September 2000.

92. John Swihart interview, June 2000.

93. Owen, *New Shape in the Sky,* 72–73.

94. Howard Moon, *Soviet SST: The Techno-Politics of the TU-144* (New York: Orion, 1989), 75.

95. Ibid., 76.

96. Horwitch, *Clipped Wings,* 180.

1. Scott Hamilton Dewey, *Don't Breathe the Air!: Air Pollution and U.S. Environmental Politics, 1945–1970* (College Station, TX: Texas A&M University Press, 2000), 3–17. Dewey's argument differs significantly from Samuel P. Hays, *Beauty, Health, and Permanence: Environmental Politics in the United States, 1955–1985* (Cambridge: Cambridge University Press, 1987), the classic work in the field. Hays argues that environmentalism was primarily a middle-class movement, while Dewey finds activist groups spanning the economic spectrum and he contends that these efforts were the root of modern environmentalism.

2. On Nixon's environmental record, see J. Brooks Flippen, *Nixon and the Environment* (Albuquerque: University of New Mexico Press, 2000).

3. Langdon Winner, *Autonomous Technology: Technics-out-of-Control as a Theme in Political Thought* (Cambridge: MIT Press, 1978), 2–12, 14.

4. See, for example, Charles A. Reich, *The Greening of America* (New York: Crown, 1970); Theodore Roszak, *The Making of a Counterculture* (Garden City, NY: Doubleday, 1969).

5. Mel Horwitch, *Clipped Wings: The American SST Conflict* (Cambridge, MA: MIT Press, 1982), 220.

6. Bo Lundberg, "The Menace of the Sonic Boom to Society and Civil Aviation," *Noise and Smog News* 14:1–4 (1966): 77–91.

7. Monroney had written the FAA's founding legislation in 1957 and was commercial aviation's biggest supporter in the Senate. On the founding of FAA, see John R. M. Wilson, *Turbulence Aloft: The Civil Aeronautics Administration in War and Rumors of War, 1938–1958* (Washington, DC: FAA, 1980).

8. Horwitch, *Clipped Wings*, 76.

9. R. W. Robberson to A. S. Monroney, 24 July 1964, file "Sonic Boom program 1963–1964," box 4, ACC 255-73-663, NASA HQ Records, Suitland Federal Records Center; Monroney to Robberson, 28 July 1964, ibid. See also Don Dwiggins, *The SST: Here It Comes, Ready or Not* (New York, 1968), 72–73.

10. Robert S. McNamara et al., "First Interim Report of the President's Advisory on Supersonic Transport," 14 May 1964.

11. Horwitch, *Clipped Wings*, 79.

12. Stanford Research Institute, *Sonic Boom Experiments at Edwards Air Force Base* (Arlington, VA: National Sonic Boom Evaluation Office, 1967).

13. "Draft Proceedings of the President's Advisory Committee on Supersonic Transport," 21 May 1965, box 168, "Minutes of PAC/SST for 8 May and 30 Mar 1965," McNamara Papers.

14. Horwitch, *Clipped Wings*, 84.

15. Ibid., 84–85.

16. Bo Lundberg, "The Menace of the Sonic Boom to Society and Civil Aviation," *Noise and Smog News* 14:1–4 (1966): 77–91.

17. Bo Lundberg to William Vogt, 9 November 1966, box 4, Citizen's League against the Sonic Boom collection, Massachusetts Institute of Technology Archives, Cambridge, MA (hereafter CLASB).

18. Ibid.; Lundberg to Senator William Proxmire, 22 December 1966 and 4 January 1967, box 4, CLASB.

19. Horwitch, *Clipped Wings*, 150–51.

20. During 1967, the FAA was made part of the new Department of Transportation, which also assumed all sonic boom and aircraft noise authority. The DOT's leadership was also pro-SST, and actively supported FAA's attempts to cast the boom in a favorable light. See Horwitch, *Clipped Wings*, 156–57.

21. "Suppose They Ban the Boom," *The Economist 221* (3 December 1966): 1046–47.

22. Lundberg to Proxmire, 4 January 1967, box 4, CLASB.

23. Michael Lumb, "Concorde Cracks along at $2000 per Hour," *The Aeroplane*, 12 January 1967, pp. 4–6. Shurcliff's annotated copy is in box 4, CLASB.

24. Lundberg to Shurcliff, 13 January 1967, box 4, CLASB.

25. Horwitch, *Clipped Wings*, 223.

26. Ibid.

27. K. H. Hohenemser, "SST," *Bulletin of the Atomic Scientists* 22 (December 1966): 8–12.

28. John E. Gibson, "The Case against the SST," *Harper's* 233 (July 1966): 76–80.

29. *Wall Street Journal*, 28 July 1966, p. 14.

30. R. G. O'Lone, "Guthrie Urges SST Economics Scrutiny," *Aviation Week and Space Technology*, 19 September 1966, pp. 38–39.

31. Horwitch, *Clipped Wings*, 237.

32. Dwiggins, *The SST: Here It Comes*, 66; Horwitch, *Clipped Wings*, 219.

33. Bruce Welch, "SST: Coming Threat to Wilderness," *National Parks Magazine* 42 (March 1968): 9–11.

34. Thomas Wellock, *Critical Masses: Opposition to Nuclear Power in California, 1958–1978* (Madison: University of Wisconsin Press, 1998), 90–91.

35. Wellock, *Critical Masses*, 94–95.

36. Ibid., 92–94.

37. William A. Shurcliff, *S/S/T and Sonic Boom Handbook* (New York: Ballantine Books, 1970).

38. Michael McCloskey interview, 5 December 2000.

39. Brock Evans interview with Ann Lage, "Building the Sierra Club's National Lobbying Program," Bancroft Library, University of California, Berkeley; also Brock Evans interview with author, 27 March 2001.

40. James J. Kilpatrick, "Nixon Blunders on Supersonic Transport," *Washington Star*, 28 September 1969; see also David Dietch, "Can Boeing Co. Handle the SST?," *Boston Globe*, 9 November 1969; "Mr. Nixon's SST Says Much on Priorities," *Nashville Tennessean*, 30 September 1969.

41. Horwitch, *Clipped Wings*, 272; "Report of the SST Ad Hoc Review Committee," *Congressional Record*, 31 October 1969, H10432-46.

42. "Long-Suppressed Report by President Nixon's SST Review Committee," file 429, box 9, CLASB records, MIT Archives.

43. Elizabeth Levy, *The People Lobby* (New York: Delacorte, 1973), 17.

44. Horwitch, *Clipped Wings*, 262.

45. Levy, *The People Lobby*, 18–19; George W. Liebmann, "Memorandum of Meeting on the SST, March 24, 1970," March 1970, box 6, CLASB, MIT Archives.

46. Horwitch, *Clipped Wings*, 271. The "boom ban" was mentioned by Secretary of Transportation John Volpe, who announced that no SST would be allowed to fly supersonically over land unless noise levels were "acceptable," which, of course, meant nothing.

47. This was the Rolls-Royce "Conway" engine.

48. The propulsion efficiency equation works out to be:

$$\text{efficiency} = 2 * V_{\text{flight}} / (V_{\text{exhaust}} + V_{\text{flight}}).$$

49. George E. Smith and David A. Mindell, "The Emergence of the Turbofan Engine," in *Atmospheric Flight in the Twentieth Century,* ed. Peter Galison and Alex Roland (Boston: Kluwer Academic Publishers, 2000), 136.

50. William Cook, *The Road to the 707* (Bellevue, WA: TYC Publishing, 1991), 260.

51. The bypass ratio is the ratio of "bypass" air to air that passes through the turbine, or the ratio of "unburned air" to "burned air."

52. Robert V. Garvin, *Starting Something Big: The Commercial Emergence of GE Aircraft Engines* (Reston, VA: AIAA, 1998), 35–37.

53. Allen telephone conversation with Jack Parker, 11 October 1965, transcript in box 33, Allen Papers; see also Ken Holtby interview by Bob Serling, 9 April 1990; T. A. Wilson interview by George Schairer, 1985. All courtesy of the Boeing Company Archives.

54. Robert Serling, *Legend and Legacy: The Story of Boeing and Its People* (New York: St. Martin's, 1992), 283.

55. John Newhouse, *The Sporty Game* (New York: Knopf, 1982), 116–17; Serling, *Legend and Legacy,* 286–87.

56. John Newhouse reports that the airlines who thought it too large included United, American, Delta, and Eastern, all major airlines at the time. See Newhouse, *The Sporty Game,* 121–22.

57. Ibid., 143.

58. The case was *Griggs vs. Allegheny County,* 5 March 1962. Richard J. Kent, *Safe, Separated and Soaring: A History of Federal Civil Aviation Policy, 1961–1972* (Washington, DC: FAA, 1980), 158.

59. Daggett Howard to administrator, "Recent Decision of the Supreme Court in Griggs vs. Allegheny County," 3 September 1962, "Noise Abatement 1962–1964" file, FAA History Office; Philip Swatek to administrator, "National Airport Problems," 23 March 1962, Noise Abatement file, 1962–1964, FAA History Office. Daggett was the agency's general counsel, and Swatek head of its PR department.

60. The members of Hornig's panel were: himself, physicist Richard Garwin, acoustician Karl Kryter, Nicholas Golovin, NASA's Harvey Hubbard, American Airlines' Frank Kolk, Air Force acoustician Herman von Gierke, and Captain Charles Ruby from the Air Line Pilots Association.

61. Office of Science and Technology, "Alleviation of Jet Aircraft Noise Near Airports: A Report of the Jet Aircraft Noise Panel," March 1966, file "Noise Abatement, 1966," FAA History Office.

62. Kent, *Safe, Separated and Soaring,* 160; Lyndon B. Johnson, "Special Message to the Congress on Transportation," *Public Papers of the Presidents of the United States, 1966* (Washington, DC: 1967), 250–63.

63. Kent, *Safe, Separated and Soaring,* 283.

64. The measurement points were: sideline, 1,500 feet to the side at the point of brake release; takeoff, 3.5 n.m. from brake release on the extended runway centerline; approach, 1 n.m. from the threshold, also on the extended centerline. The 108 EPNdB standard applied to aircraft over 600,000 pounds gross weight. Smaller aircraft had stricter standards. See "Ini-

tial FAA Noise Proposal Hits Mainly New Aircraft," *Aviation Week and Space Technology*, 13 (January 1969), p. 24.

65. Newell D. Sanders, "Supersonic Transport: Assessment of the Engine Noise Problem," 9 April 1969, "Background to May Interim Report," SST-Sonic Boom Committee collection, National Academy of Sciences Archives.

66. Jack Kerrebrock to John R. Dunning, "Letter Report, Committee on SST-Sonic Boom, Subcommittee on Aircraft Noise Research," 7 May 1969, file "Subcommittee on Acft Noise Research, Letter Report," Committee on SST-Sonic Boom collection, NAS Archives. The fundamental problem the GE4 engine faced in meeting the sideline standard was its use of afterburners on takeoff. The afterburners heated the engine's turbine exhaust, dramatically increasing the exhaust's velocity. Velocity essentially determined the noise level, of course. Because what mattered to the aircraft was not the exhaust velocity but the exhaust's *energy*, the alternative to using afterburners was to increase the exhaust's *mass*. That meant increasing the engine's airflow, which in turn meant increasing the engine's diameter and, probably, its compression ratio. (Recall that kinetic energy is equal to the product of mass and the square of velocity. Mass and velocity are thus interchangeable.)

67. This engine is also described in Sanders's letter of 9 April 1969.

68. S. J. Converse to chief of technical operations, "Boeing Noise Control Program," 1 August 1969, FAA History Office, box 30-3, "Engine Noise," SST collection, FAA History Office.

69. T. J. O'Brien to chief, Propulsion Branch, "TBD Noise Control Program," 15 August 1969, box 30-3, "Engine Noise," SST collection, FAA History Office.

70. General Maxwell had rotated back to the Air Force at the end of his normal three- year tour of duty during mid-1969, and several months elapsed before Secretary Volpe selected Magruder. Magruder took over as SST program director in April 1970. Horwitch, *Clipped Wings*, 280.

71. Serling, *Legend and Legacy*, 275; author interview with Bob Withington, 6 December 2000.

72. Ibid. Withington's memory corroborates that of James Beggs, then the deputy secretary of transportation. James Beggs interview with author, 29 May 2001.

73. "DOT/GE/TBC Production Airplane/Engine Planning Meeting, Office of SST Development, Washington, DC," 4 September 1970, box 30-3, "Engine Noise," SST collection, FAA History Office.

74. T. Wilson to John Volpe, 21 January 1971, Allen Papers, Boeing Archives.

75. On the Keep California Green campaign see Nick Kotz, *Wild Blue Yonder: Money, Politics and the B-1 Bomber* (New York: Pantheon, 1988), 100–103.

76. Horwitch, *Clipped Wings*, 281.

77. Ibid., 282; Andrew F. Blake and Michael Kenney, "13 Arrested at Logan Jet Protest," *Boston Globe*, 23 April 1970, p. 1.

78. Brent Blackwelder interview with author, 25 March 2000.

79. Horwitch, *Clipped Wings*, 275. "Open War" is the title of Horwitch's chapter on the anti-SST campaign.

80. Richard Garwin et al., "Report of the Ad Hoc Supersonic Transport Review Committee of the Office of Science and Technology, March 30, 1969," CA-8, "Aircraft Development 1969–1970, 1 of 3," White House Central Files, Nixon Presidential Materials Project, College Park, MD.

81. Lee DuBridge to the president, 2 April 1969, CA6, "Aircraft Development, 1969–1970, 1 of 3," White House Central Files (WHCF), Richard M. Nixon Presidential Materials Project, College Park, MD.

82. Richard L. Garwin, "Presidential Science Advising," in *Science Advice to the President*, ed. William T. Golden (Washington, DC: AAAS Press, 1993), 189.

83. Greg Herken, *Cardinal Choices: Presidential Science Advising from the Atomic Bomb to SDI* (Oxford: Oxford University Press, 1992), 178–79.

84. Joint Economic Committee, Subcommittee on Economy in Government, *Hearings*, 91st Cong., 2nd sess., 7 May 1970, pp. 904–20.

85. Magruder's testimony appears in the transcript of the Joint Economic Committee *Hearings*, 977.

86. Ralph Pinnes to chief, engineering division, 11 May 1971, "Dr. Garwin's Comments on SST Sideline Noise of the SST," box 30-3, "Engine Noise," SST collection, FAA History Office. The calculation is based on the definition of the decibel. Decibels are additive logarithmically, such that two identical noise sources produce 3 decibels more noise than a single source. Since the SST's sideline noise signature was 17 decibels higher than the 747's, one needed fifty 747s to produce the same noise level.

87. Joint Economic Committee, *Hearings*, 998.

88. Ibid., 999.

89. Ibid., 1000.

90. Horwitch, *Clipped Wings*, 287.

91. Massachusetts Institute of Technology, *Man's Impact on the Global Environment: Report of the Study of Critical Environmental Problems* (Cambridge, MA: MIT Press, 1970), 17.

92. Michael McCloskey interview with author, 5 December 2000; Lloyd Tupling interview with author, 8 December 2000.

93. All fifteen letters were placed in the *Congressional Record* by Fulbright. See *Congressional Record*, 91st Cong., 2nd sess., 15 September 1970, S15397-405.

94. "Downwind from the SST," *New York Times*, 2 December 1970, p. 4c.

95. "DOT Report Concedes SST Might Harm the Environment," *Congressional Record*, 17 September 1970, S15844-50.

96. William W. Prochnau, "'New Politics,' Emotion Win in Showdown," *Seattle Times*, 4 December 1970.

97. Robert B. Hotz, "Supersonic Shock Wave," *Aviation Week and Space Technology*, 14 December 1970, p. 11.

98. T. Wilson to Bill Magruder, Senator Warren Magnuson, and Senator Henry Jackson, "SST Appropriation," 8 December 1970, file "SST #4," box 31, Allen Papers, Boeing Archives.

99. This report was the most significant issue in Richard Garwin's decision to testify against the government in 1970. Boeing had been required to submit it in 1968 and had done so, but the SST office had not liked what it had to say and had "buried it." Garwin's committee saw the report, and when Transportation testified that Boeing had not submitted it, Garwin realized that the agency would say whatever was necessary to protect the program, including making claims that were outright lies. He could not testify about this particular report, because he knew about it only through his position on the ad hoc committee and he had agreed to testify only about matters that were in the open record. But to him, the incident was decisive.

100. Senator Henry Jackson to William Allen, phone call, transcript date 12 December 1970, file "SST #4," box 31, Allen Papers, Boeing Archives.

101. Wilson to Bill Magruder, 11 January 1971, "Materials for SST meeting at the White House," Allen Papers; Wilson to Secretary Volpe, 21 January 1971, "Materials for SST meeting at the White House," Box 31, Allen Papers.

102. E. E. Hood to Bill Magruder, 14 January 1971, "SST #4," box 31, Allen Papers.

103. Horwitch, *Clipped Wings*, 308.

104. Beranek testimony, Senate Appropriations Committee, *Commercial Supersonic Aircraft Development (SST)*, 92nd Cong., 1st sess., 10 March 1971, p. 284.

105. Beranek's notes indicate that both GE and Pratt & Whitney had presented several new engines, from which the collected group selected a GE proposal. His congressional testimony discusses only a single engine, however.

106. See Richard G. O'Lone, "SST Engine, Wing Changes Weighed," *Aviation Week and Space Technology*, 15 March 1971, pp. 36–38.

107. The definition of a true intercontinental airliner was New York to Rome because any aircraft that could cover this distance unrefuelled could fly all of the world's major commercial routes with at most a single fuel stop. In an odd sense, this makes Rome the center of the airline world.

108. Withington interview with author, 6 December 2000; James M. Beggs interview with author, 29 May 2001.

109. On Nixon's antienvironmental campaign, see J. Brooks Flippen, *Nixon and the Environment* (Albuquerque: University of New Mexico Press, 2000), 129–58.

110. Ibid., 135.

111. Bill Magruder to George Low, 8 February 1971, box 52, file 7, Low Papers, Rensellear Polytechnic Institute (hereafter RPI).

112. T. A. Heppenheimer discusses the 1970 aerospace recession in *The Space Shuttle Decision: NASA's Search for a Reusable Space Vehicle* (Washington, DC: NASA, 1999), 291–330.

113. Boeing's actions during the recession are well described in Eugene Rodgers, *Flying High: The Story of Boeing and the Rise of the Jetliner Industry* (New York: Atlantic Monthly Press, 1996), 290–304.

114. Christopher Lydon, "Drive with Ads Planned," *New York Times*, 23 February 1971, p. 73. See also George Alderson, "Status of the SST Campaign," 28 February 1971, folder 27, box 151, Sierra Club collection, Bancroft Library.

115. House Committee on Appropriations, *Civil Supersonic Aircraft Development (SST)*, 92nd Cong., 1st sess., March 1971, pp. 32, 48.

116. Senate SST hearings, 10 March 1972, p. 211.

117. Ibid.

118. Ibid., 218. Ruppenthal had written several anti-SST articles and also been an airline captain for twenty-six years.

119. Andrew Wilson, "Airline Bombshell for the Government—BOAC Can't Afford the Concorde," *The Observer*, 21 February 1971, reprinted in House SST hearings, 1 March 1971, 500–501.

120. "Senate 51–46 vote dooms federal financing of the SST," press release from the Sierra Club, March 1971, folder 27, box 151, Sierra Club records.

121. Horwitch, *Clipped Wings*, 326.

122. On the restart movement, see ibid., 326–27; and J. R. Haldeman, *The Haldeman Diaries: Inside the Nixon White House* (New York: Putnam's, 1994), 246, 260, 285, 288.

123. Nixon quoted in Stephen E. Ambrose, *Nixon: Triumph of a Politician,* vol. 2 (New York: Simon and Schuster, 1989), 433.

124. Robert B. Hotz, "Toward a Technological Appalachia," *Aviation Week and Space Technology,* 29 March 1971, p. 9.

125. Haldeman, *The Haldeman Diaries,* 288.

126. Rodgers, *Flying High,* 302.

127. John R. Wheeler to Richard Shurcliff, 22 March 1972, box 6, CLASB, MIT Archives.

128. Kotz calls the B-1 the "Born Again Bomber." See Kotz, *Wild Blue Yonder,* 6–9, 87.

129. Gregory C. Kunkle, "New Challenge of the Past Revisited? The Office of Technology Assessment in Historical Context," *Technology in Society* 17:2 (1995): 175–96.

130. On corporate mobilization, see Patrick J. Akard, "Corporate Mobilization and Political Power: The Transformation of U.S. Economic Policy in the 1970s," *American Sociological Review* 57 (1992): 597–615. While it does not address the SST in any way, this article is an important starting point for understanding the corporate role in the conservative revival of the 1980s.

131. The rise of conservative antienvironmentalism is discussed in Hays, *Beauty, Health and Permanence,* 287–328 and 491–526.

132. Thomas Wellock, *Critical Masses,* 92–113.

5 ⚛ OF OZONE, THE CONCORDE, AND SSTs

1. R. M. Nixon to John Ehrlichman, 1 June 1971, CA-8, WHCF, Nixon Papers.

2. Peter G. Peterson to John Ehrlichman, 6 August 1971, CA-8, WHCF, Nixon Papers.

3. Harlan B. Byrne, "GE Head Never Gets to Fiscal Forecast as War, Dirt Protesters Roil Meeting," *Wall Street Journal,* 23 April 1970, p. 8.

4. J. M. Huntsman to John Ehrlichman, 10 August 1971, "Aircraft Development 1971–1974," box 4, CA-8, White House Special Files (WHSF), Nixon Papers.

5. Nixon attempted to establish a "Department of Economic Affairs" that was to contain an "Office of Technology Development" to foster high-technology research. The "New Technology Initiative" was supposed to be a forerunner of this department.

6. William M. Magruder to John Ehrlichman, 10 January 1972, "Aircraft Development 1971–1974," box 4, CA-8, WHSF, Nixon Papers.

7. Magruder to Tod R. Hullin, 24 August 1971, "Aircraft Development 1971–1974," box 4, CA-8, WHSF, Nixon Papers.

8. William S. Aiken to deputy associate administrator for programs, 17 May 1972, "SST 1972," NASA HQ History Office.

9. "Policy Matter for Consideration: Advanced Supersonic Transport Technology Program," n.d. (but c. September 1972), file "SST 1972," NASA HQ History Office.

10. Albert Karr, "Try, Try Again: Die-Hard Supporters of Supersonic Airliner Plot to Revive Project," *Wall Street Journal,* 1 November 1972, p. 1.

11. See Kenneth Owen, *Concorde: Story of a Supersonic Pioneer* (London: Science Museum, 2002), 210; "Slowdown of Concorde Production Seen," *Aviation Week and Space Technology,* 5 February 1973, pp. 30–31.

12. "Descent of the SST," *Washington Evening Star,* 8 February 1973.

13. "Bleak Arithmetic of Concorde: Cancelled Orders," *Business Week,* 10 February 1973, p. 29.

14. See, for example, Roy Jackson testimony, "Hearings on FY 1974 NASA Authorization, pt. 4," House Science and Astronautics Committee, 93rd Cong., 1st sess., March 1973.

15. Quoted in Gerald J. Mossinghoff, "Memo for the Record," 27 September 1972, file 4, box 50, Low Papers, RPI.

16. This criticism exists in every House Science and Astronautics Committee budget hearing through 1980, with the sole exception of 1976. See, for example, Budget Operations Division, "Chronological History of the Fiscal Year 1973 Budget Submission," 5 September 1972; Budget Operations Division, "Chronological History of the FY 1974 Budget Submission," 26 June 1974; Office of Budget Operations, "Chronological History of the Fiscal Year 1975 Budget Submission," 13 August 1975; Office of Budget Operations, "Chronological History Fiscal Year 1977 Budget Submission," 25 August 1976, all NASA HQ History Office.

17. George Cherry testimony, "Advanced Supersonic Technology: Hearings before the Science and Astronautics Committee, 22 February 1974," Senate, 93rd Cong., 2nd sess., February 1974, p. 186.

18. For a summary of SCAR/VCE, see F. Edward McLean, *Supersonic Cruise Technology,* NASA SP-472 (Washington, DC: 1985), 101–14. For a complete listing of the program's publications, see Sherwood Hoffman, "Supersonic Cruise Aircraft Research (SCAR) program bibliography, July 1972 through June 1976," NASA TM X-73950, 1976; S. Hoffman, "Supersonic Cruise Research (SCR) Program Publications for FY 1977 through FY 1979 (preliminary)," NASA TM-80184, 1979; Sherwood Hoffman, "Bibliography of Supersonic Cruise Research Program from 1980–1983," NASA RP-1117, 1984.

19. See, for example, Paul Crutzen and Veerabhadran Ramanathan, "The Ascent of Atmospheric Sciences," *Science,* 13 October 2000, pp. 299–304.

20. Summarized from David E. Newton, *The Ozone Dilemma* (Denver: ABC Clio, 1995), 25–26.

21. MIT, *Man's Impact on the Global Environment: Report of the Study of Critical Environmental Problems* (Cambridge, MA: MIT Press, 1970).

22. Ibid., 100–106.

23. Halstead Harrison, "Stratospheric Ozone with Added Water Vapor: Influence of High-Altitude Aircraft," *Science,* 13 November 1970, pp. 734–36. Harrison told his own story in an online document: Halstead Harrison, "Boeing Adventures, with Digressions," 5 March 2003, copy in author's collection.

24. McDonald cited F. Urbach, ed., *The Biologic Effects of Ultraviolet Radiation* (New York: Pergamon Press, 1969), A. Hollaender, ed., *Radiation Biology,* vol. 2 (New York: McGraw Hill, 1965), and H. F. Blum, *Carcinogenesis by Ultraviolet Light* (Princeton: Princeton University Press, 1959), as his sources for the UV–skin cancer link.

25. The *New York Times* reported the story, *sans* byline but complete with space aliens, and buried the piece far in the back. See "Scientist calls SST skin cancer hazard," *New York Times,* 3 March 1971, p. 87.

26. Lydia Dotto and Harold Schiff, *The Ozone War* (Garden City, NY: Doubleday, 1978), 45.

27. Ibid., 45.

28. See ibid., 39–68.

29. The paper is Paul Crutzen, "The Influence of Nitrogen Oxides on the Atmospheric Ozone Content," *Quarterly Journal of the Royal Meteorological Society* 96 (1970): 320.

30. Two separate accounts of this sequence exist. The first is Dotto and Schiff, *The Ozone War,* 59–68. In a retrospective article published in 1992, Johnston gives a somewhat different account from his voluminous notes: Harold S. Johnston, "Atmospheric Ozone," *Annual Review of Physical Chemistry* (1992), 1–32. I have synthesized this account from both sources, relying on Johnston where they differ.

31. Sen. Clinton P. Anderson to James Fletcher, 10 June 1971, file "NASA's Advanced SST Program, Catalytic Destruction of Ozone," box 52, acc. 255-77-0677, Washington National Records Center, Suitland, MD.

32. Herbert Friedman to Homer Newell, 1 September 1971, file "NASA's Advanced SST Program, Catalytic Destruction of Ozone," box 52, acc. 255-77-0677, Suitland. The author of this letter was a researcher at the Naval Research Laboratory and chair of the Ad Hoc Panel on the NOx-Ozone Problem, formed by the National Research Council on an administration request.

33. Harold Johnston, "Reduction of Stratospheric Ozone by Nitrogen Oxide Catalysts from Supersonic Transport Exhaust," *Science,* 6 August 1971, pp. 517–22.

34. Harold Johnston, "Atmospheric Ozone," *Annual Review of Physical Chemistry* (1992), 23.

35. A. J. Grobecker, S. C. Coroniti, and R. H. Cannon, Jr., "The Effects of Stratospheric Pollution by Aircraft" (Washington, DC: DOT, December 1974), vi.

36. Bill Grose interview, 3 April 2001, LaRC archives.

37. This was almost certainly prompted by the advocacy of three scientists within the agency who believed the shuttle might have a problem: Robert Hudson of the Johnson Space Flight Center, James King at the Jet Propulsion Lab, and I. G. Poppoff at the Ames lab. See Dotto and Schiff, *The Ozone War,* 127. Hudson managed the contract with Stolarski and Cicerone, according to Stolarski. The shuttle assessment is R. J. Cicerone, D. H. Stedman, R. S. Stolarski, A. N. Dingle, and R. A. Cellarius, "Assessment of Possible Environmental Effects of Space Shuttle Operations," NASA CR-129003, 3 June 1973.

38. Author interview with Richard Stolarski, 26 April 2001, LaRC Historical Archives; Johnston, "Atmospheric Ozone," 26–27.

39. Harvey Herring, "Research on Stratospheric Pollution," 13 February 1974, file 13638, Low Papers, NASA HQ History Office.

40. Ibid.

41. James Fletcher to H. Guyford Stever, 20 January 1975, "Committees: NAS/NAE Climatic Impact Committee," box 9, acc. 80-0608, RG 255, Suitland; and Hearings before the Subcommittee on the Upper Atmosphere, Committee on Aeronautical Space Sciences, "Stratospheric Ozone Research and Effects," Senate, 94th Cong., 2nd sess., 25 February 1976.

42. Senate Hearings, "Stratospheric Ozone Research and Effects," 25.

43. Stolarski interview, 26 April 2001.

44. Senate Hearings, "Stratospheric Ozone Research and Effects," 54–55.

45. Low to Wilford Rommel, 19 March 1971, box 18, file 17, Low Papers; William S. Aiken, "Memorandum for the Record: Notes on AST Program Planning," 17 March 1972, "SST 1972," NASA HQ History Office.

46. Driver interview, 25 May 2001.

47. F. Edward McLean, "SCAR Program Overview," *Proceedings of the SCAR Conference part 1*, NASA CP-001, November 1976, p. 2.

48. Vincent R. Mascitti, "SCR Program Overview," *Supersonic Cruise Research '79*, NASA CP-2108, pp. 2–3.

49. Author interview with Garry Klees, 2 May 2001.

50. Laurence H. Fishbach and Michael J. Caddy, "NNEP—The Navy NASA Engine Program," NASA TM X-71857, December 1975, p. 1.

51. A "normal" turbofan engine produced a hot, high-speed exhaust from the engine core, and the bypass air stream was cold and slow.

52. L. H. Fishbach et al, "NASA Research in Supersonic Propulsion—A Decade of Progress," NASA TM-82862, 1982, p. 4.

53. Fishbach et al., "NASA Research in Supersonic Propulsion," 9.

54. C. C. Gleason and R. W. Niedzwiecki, "Results of the NASA/General Electric Experimental Clean Combustor Program," AIAA 76-763, p. 1.

55. See Thomas Donahue, "The SST and Ozone Depletion," *Science* 28 (March 1975): 4185.

56. A. J. Grobecker, S. C. Coroniti, and R. H. Cannon, Jr., "Report of Findings: The Effects of Stratospheric Pollution by Aircraft," DOT-TST-75-50, December 1974, xvii–xix. The CIAP executive summary has long been controversial because it placed a positive "spin" on the results—it was based upon the implicit assumption that propulsion engineers *could* achieve emissions reductions of this magnitude in ten to fifteen years, and thus represented what many critics in the environmental movement and quite a few scientists considered a very blind faith in "techno-fixes." Furthermore, the summary was released in a way that caused most press outlets to claim that the report exonerated the SST—which it very clearly did *not* do. On the release controversy, see Dotto and Schiff, *The Ozone War*, 68–89. The "spinning" of this report caused an infuriated exchange of letters in the journal *Science* between Grobecker and the study's chairman. See Thomas Donahue, "The SST and Ozone Depletion," *Science*, 28 March 1975, p. 4185, and response: A. J. Grobecker, "The SST and Ozone Depletion," *Science*, 28 March 1975, p. 4185.

57. The grants were to UC Berkeley, MIT, Cornell University, and the University of Michigan. See Larry A. Diehl, Gregory M. Reck, Cecil J. Marek, and Andrew J. Szniszlo, "Stratospheric Cruise Emission Reduction Program," NASA TN 78-11063, pp. 357–91.

58. Combustion of liquid droplets occurs at "stoichiometric" flame temperatures. By definition, "stoichiometric" conditions are the exact amounts of fuel and air necessary to result in complete chemical reaction of both air and fuel masses. Combustion under these conditions produces the maximum possible flame temperature. Internal combustion engines provide far more air than is necessary to fully consume the fuel, but only the air that is in contact with a droplet can participate in the combustion process. Hence the surface of a droplet replicates stoichiometric conditions.

59. "Proceedings of the SCAR Conference," NASA CP-001, November 1976.

60. See chapter 2.

61. Total operating costs are the sum of *indirect operating costs,* which include maintenance, crew, and other ground costs, and *direct operating costs,* which include fuel and other consumables and the airplane's purchase cost. Higher speeds reduce indirect costs by reduc-

ing crew costs. Regulations prevent crews from flying more than eight hours per day, which means that aircraft on missions longer than eight hours carry multiple crews—which significantly raises costs. A Mach 2 aircraft would nearly eliminate the need to double-crew long flights. McDonnell-Douglas believed that the decrease in crew costs would outweigh the increase in fuel costs, producing a net reduction in total costs.

62. This was the Woodward program, described in F. A. Woodward, J. W. Larsen, and E. N. Tinoco, "Analysis and Design of Supersonic Wing Body Combinations," NASA CR-73106, 1967. See R. L. Radkey, H. R. Welge, and R. L. Roensch, "Aerodynamic Design of a Mach 2.2 Supersonic Cruise Aircraft," AIAA paper 76-955, September 1976.

63. See Robert L. Roensch, "Aerodynamic Validation of a SCAR Design," *Proceedings of the SCAR Conference,* NASA CP-001, November 1976, pp. 155–68.

64. Details of the theoretical correction are in R. L. Roensch and G. S. Page, "Analytic Development of an Improved Supersonic Cruise Aircraft Based on Wind Tunnel Data," *Proceedings of the SCAR Conference,* NASA CP-001, November 1976, pp. 205–12.

65. Ibid., 205–28.

66. R. L. Roensch et al., "Results of a Low-Speed Wind Tunnel Test of the MDC 2.2M Supersonic Cruise Aircraft Configuration," *Supersonic Cruise Research '79,* NASA CP-2108, November 1979, pp. 35–57.

67. Paul L. Coe, Jr., et al., "Overview of the Langley Subsonic Research effort on SCR configurations," *Supersonic Cruise Research '79,* NASA CP-2108, November 1979, pp. 13–33; Dhanvada M. Rao, "Exploratory Subsonic Investigation of Vortex-Flap Concept on Arrow Wing Configuration," *Supersonic Cruise Research '79,* NASA CP-2108, November 1979, pp. 117–30; L. James Runyan et al., "Wind Tunnel Test Results of a New Leading Edge Flap Design for Highly Swept Wings—A Vortex Flap," *Supersonic Cruise Research '79,* NASA CP-2108, November 1979, pp. 131–48.

68. J. E. Fischler, "Opportunities for Structural Improvements for an Advanced Supersonic Transport Vehicle," *Supersonic Cruise Research 1979,* NASA CP-2108 pt. 2, November 1979, p. 595; William T. Rowe, "Technology Development Status at McDonnell Douglas," *Supersonic Cruise Research 1979,* NASA CP-2108 pt. 2, November 1979, p. 879.

69. ATLAS, an integrated structural analysis and design program, was supported by a Langley contract, while FLEXSTAB, a program designed to analyze aerodynamic loads, was designed under an Ames contract.

70. Cited in Richard Heldenfels, "Automating the Design Process: Progress, Problems, Prospects, Potential," AIAA Paper 73-410, 20 March 1973.

71. William T. Rowe, "Technology Status for an Advanced Supersonic Transport," SAE paper 820955, 16 August 1982, p. 2.

72. One must note the inherent bias in this number. The AST had 33 percent more seats than the DC-10-30, meaning that its appropriate comparison was not the DC-10 but the 350-to-400-seat Boeing 747.

73. Lockheed also believed this was true, and stated what Douglas did not: the substantial loss of revenue on the subsonic fleet would force the airlines to raise fares on the subsonic jets as well. Hence the first-order effect of a large-scale SST program would be to raise everyone's airfares. See R. S. Claus et al., "The Impact of Materials Technology and Operational Constraints on the Economics of Cruise Speed Selection," *Supersonic Cruise Research '79,* NASA CP-2108 pt. 2, November 1979, pp. 909–35. The second-order effect that neither company explained in its published analyses was that increased airfares would significantly reduce the

total size of the market by putting air travel beyond the economic reach of some portion of the traveling public.

74. The estimate was based on 3,600 hours per year utilization, a block speed of 1,265 mph, a 55 percent load factor, and no discount fares. The ratio of AST to DC-10-30 was $184 million to $43 million. W. T. Rowe, H. R. Welge, E. S. Johnson, and I. S. Rochte, "Advanced Supersonic Transport Propulsion and Configuration Technology Improvements," AIAA paper 81-1595, July 1981, p. 10.

75. R. E. G. Davies, *Supersonic (Airliner) Nonsense: A Case Study in Applied Market Research* (McLean, VA: Paladwr Press, 1998), 42–44.

76. Due to the anti-SST movement's efforts and the ozone problem, Concorde had been the subject of the first environmental impact statement ever written on an aircraft. See Horwitch, *Clipped Wings,* 336–37; Kenneth Owen, *Concorde and the Americans: International Politics of the Supersonic Transport* (Washington, DC: Smithsonian Institution Press, 1997), 92–93.

77. Owen, *Concorde and the Americans,* 104–5.

78. Ibid., 110.

79. This decision emerged from a complex set of four separate cases: two original and two appellate. For a full explanation, see Owen, *Concorde and the Americans,* chap. 10.

80. The surcharge was established by the International Air Transport Association, which at the time regulated international air fares. Events quickly showed that the surcharge was not high enough to cover Concorde's operating costs, but until the IATA's regulatory scheme was dismantled in the early 1980s, British Airways and Air France could not impose the 100 percent or so surcharge over standard first class that Concorde required for profitability.

81. Richard Fitzsimmons, "The Advanced Supersonic Transport: What It Is and How It Compares," *Acta Astronautica* 4 (1977): 131–43.

82. J. C. Brizendine to administrator, 23 June 1977, file "Aeronautical Research and Technology," box 9, Acc. 92-0644, RG 255, Suitland.

83. On the three manufacturers' ideas see Craig Covault, "SST Technology Readiness Studied," *Aviation Week and Space Technology,* 17 October 1977, pp. 33–34.

84. The idea behind this assessment is simply that having more design teams increased the chance that one of them would produce necessary "breakthroughs." Further, NASA's engineers believed that the existence of competition would "stimulate" the design teams' efforts and force management to participate more actively in the program. Hence technical risk was minimized by maximizing the number of qualified participants in any given research program. See Covault, "SST Technology Readiness Studied," 33.

85. On Lovelace's response, see ibid. The committee's response is summarized in NASA Office of Budget Operations, Chronological History of the FY 1979 Budget Submission, 17 November 1978, NASA HQ History Office.

86. Cited in Jeffrey Ethell, *Fuel Economy in Aviation,* NASA SP-462 (Washington, DC: 1983), 29.

87. The program's basic structure was much like the VCE program, with the Lewis Research Center providing contracts to Pratt & Whitney and GE while also conducting some in-house research of its own. GE published a self-congratulatory article in 1984 in its in-house organ, praising itself for a demonstrated 13.5 percent SFC reduction and a projected 18.5 percent reduction with minor modifications. It also claimed a five-decibel reduction in noise over the existing standards. See "Keeping a Bold Promise," *The Leading Edge,* Summer 1984, pp. 15–17. Eventually, GE used the technology in its 1990 GE90 engine.

88. See Craig Covault, "Interest in Supersonic Aircraft Surges," *Aviation Week and Space Technology,* 11 September 1978, pp. 110–12; and Office of Budget Operations, "Chronological History of the FY 1979 Budget Submission," 17 November 1978, p. 30.

89. R. D. Fitzsimmons to Robert Frosch, 2 March 1978, file "Aeronautical Research and Technology," box 9, Acc. 92-0644, RG 255, Suitland; and Robert Frosch to R. D. Fitzsimmons, 4 April 1978, Acc. 920644, box 9, file "Aeronautical Research and Technology," RG 255, Suitland.

90. See NASA Office of Budget Operations, "Chronological History of the FY 1980 Budget Submission," NASA HQ History Office.

91. Robert Hotz, "New Look at Supersonics," *Aviation Week and Space Technology,* 30 April 1979, p. 21.

92. Frank Melville, "The Concorde's Disastrous Economics," *Fortune,* 30 January 1978, p. 70.

93. "British, French Begin Process of Ending Concorde Program," *Aviation Week and Space Technology,* 1 October 1979, p. 28.

94. Howard Moon, *Soviet SST: The Technopolitics of the TU-144* (New York: Orion Books, 1989), 192.

95. Office of Technology Assessment, *Impact of Advanced Air Transport Technology: Part I, Advanced High-Speed Aircraft* (Washington, DC: 1980).

96. Office of Budget Operations, "Chronological History of the FY 1980 Budget Submission," 9 October 1980, NASA HQ History Office.

97. George Low, who had left NASA to become president of Rensellear Polytechnic University, headed the NASA transition team. Low responded to the demand for budget cuts by contending that Carter had underfunded NASA and opposed the 2 percent cut he had been asked to find. See Low to Richard Fairbanks, 19 December 1980, Transition Files, Low Report, NASA HQ History Office; Low to Richard Fairbanks, 18 December 1980, folder 8, box 181, Low Papers, RPI; Low, "NASA Budget Cuts," 2 December 1980, folder 8, box 141, Low Papers.

98. Office of Budget Operations, "Chronological History of the FY 1981 Budget Submission," 14 August 1981, NASA HQ History Office.

99. See Sherwood Hoffman, "Supersonic Cruise Aircraft Research (SCAR) Program Bibliography, July 1972 through June 1976," NASA TM X-73950, 1976; S. Hoffman, "Supersonic Cruise Research (SCR) Program Publications for FY 1977 through FY 1979 (preliminary)," NASA TM-80184, 1979; Sherwood Hoffman, "Bibliography of Supersonic Cruise Research Program from 1980–1983," NASA RP-1117, 1984.

100. Charles W. Harper to Edgar Cortwright, "NASA's role in SST development," 12 March 1969, box 472, B10-1 (DOT), LaRC series 2, RG 255, NA/P.

6 ⊹ THE AIRBUS, THE ORIENT EXPRESS, AND THE RENAISSANCE OF SPEED

1. The classic discussion of the emergence of science in the United States is A. Hunter Dupree, *Science in the Federal Government: A History of Policies and Activities to 1940* (Cambridge, MA: Harvard University Press, 1957). On the arguments surrounding civilian support for scientific research immediately following World War II, see Daniel J. Kevles, "The National Science Foundation and the Debate over Postwar Research Policy, 1942–1945," *Isis* 68 (1977): 5–26; J. Merton England, *A Patron for Pure Science: The National Science Foundation's Formative Years, 1945–1957* (Washington, DC: NSF, 1983); Jessica Wang, "Liberals, the Progressive Left,

and the Political Economy of Postwar American Science: The National Science Foundation Debate Revisited," *Historical Studies of the Physical and Biological Sciences* 26:1 (1995): 139–66. For the impact of NSF, see Toby A. Appel, *Shaping Biology: The National Science Foundation and American Biological Research* (Baltimore, MD: Johns Hopkins University Press, 2000); Dian Olson Belanger, *Enabling American Innovation: Engineering and the National Science Foundation* (West Lafayette, IN: Purdue University Press, 1998).

2. On the conservative resurgence, see Maurice Isserman and Michael Kazin, *America Divided: The Civil War of the 1960s* (New York: Oxford University Press, 2000), 205–20.

3. On Reagan, see Lou Cannon, *President Reagan: Role of a Lifetime* (New York: Public Affairs Press, 2000); Haynes Johnson, *Sleepwalking through History: America in the Reagan Years* (New York: Doubleday, 1991).

4. Stockman's intentions are spelled out in his bitter 1985 memoir: David A. Stockman, *The Triumph of Politics: Why the Reagan Revolution Failed* (New York: Harper and Row, 1987).

5. George Soros, "The Capitalist Threat," *Atlantic Monthly* 279 (1997): 45–58. There are harsher indictments of market fundamentalism than Soros's, but stemming from theology: David R. Loy, "The Religion of the Market," *Journal of the American Academy of Religion* 65:2 (Summer 1997): 275–90.

6. Robert H. Nelson, *Economics as Religion: From Samuelson to Chicago and Beyond* (University Park: Pennsylvania State University Press, 2001), xv–xxvi.

7. Merrit Roe Smith, *Harper's Ferry Armory and the New Technology* (Ithaca, NY: Cornell University Press, 1977).

8. See Donald Worster, *Rivers of Empire: Water, Aridity, and the Growth of the American West* (New York: Oxford University Press, 1985); Marc Reisener, *Cadillac Desert: The American West and Its Disappearing Water* (New York: Viking, 1986).

9. On the "Great Inflation," see Bruce Shulman, *The Seventies* (New York: Free Press, 2001), 121–44.

10. On airline deregulation and its aftermath, see Thomas Petzinger, *Hard Landings* (New York: Times Books, 1997).

11. John M. Logsdon and Craig Reed, "Commercializing Space Transportation," in *Exploring the Unknown*, vol. 4: *Accessing Space*, NASA SP-4407 (Washington, DC: 1999), 405–22.

12. Stockman's take on NASA is brief but telling: Stockman, *The Triumph of Politics*, 150–51.

13. The best book on Airbus Industrie's foundation is David Weldon Thornton, *Airbus Industrie: The Politics of an International Industrial Collaboration* (New York: St. Martin's, 1995).

14. Ibid., 74–75.

15. John Newhouse, *The Sporty Game* (New York: Knopf, 1982), 28.

16. National Research Council, "NASA's Role in Aeronautics: A Workshop, volume 1" (Washington, DC: National Academy Press, 1981).

17. National Research Council, "NASA's Role in Aeronautics: A Workshop, volume III," (Washington, DC: National Academy Press, 1981), 1.

18. Ibid., 5.

19. Ibid., 6–7.

20. The group had some trouble defining "vehicle class" technologies. This category "concentrated on specific vehicle classes and on the preparation of the unique data base required to improve the design and development of certain classes of aircraft." Ibid., 30.

21. Ibid., 20.

22. Ibid.

23. Ronald Reagan to Clifton von Kann, 11 September 1980, Low Papers, box 142, folder 8, RPI.

24. Office of Budget Operations, "Chronological History of the FY 1981 Budget Submission," 14 August 1981, NASA HQ History Office.

25. National Research Council, "Aeronautics Research and Technology: A Review of Proposed Reductions in the FY 1983 NASA Program," July 1982, p. 65.

26. Edward P. Boland and Jake Garn to James M. Beggs, 8 March 1982, reprinted in ibid., 59.

27. George A. Keyworth, "Talking Paper for 12 March Steering Group Meeting," file "Aeronautical Policy Review Committee [6 of 6]," box 17, Keyworth collection, Reagan Presidential Library, Simi Valley, California (hereafter RPL).

28. The appropriated budget for aeronautics was $280 million, a $4 million decrease from the previous year. See Budget Operations Division, "Chronological History of the Fiscal Year 1983 Budget Submission," 8 August 1983, NASA HQ History Office.

29. NRC, "Aeronautics Research and Technology: A Review of Proposed Reductions in the FY 1983 NASA Program," National Academy Press, July 1982, pp. 1–3.

30. Ibid., 2.

31. Office of Science and Technology Policy, "Aeronautical Research and Technology Policy vol. 1," November 1982.

32. Ibid., 17–20.

33. Ibid., 21–22.

34. Ibid., 23.

35. James M. Beggs and Richard D. DeLauer to George A. Keyworth, 7 April 1983, file "Aeronautics Policy Review Committee [4 of 6]," box 17, Keyworth collection, RPL.

36. Assessment based on the DOD nominee's list and cover letter. NASA recommendations were not contained in the file. See James P. Wade to George Keyworth, 31 March 1983, file "Aeronautics Policy Review Committee [1 of 6]," box 17, Keyworth collection, RPL.

37. "First Annual Report: Aeronautical Policy Review Committee," November 1983, file "Aeronautics Policy Review Committee [1 of 6]," box 17, Keyworth collection, RPL.

38. Ibid.

39. James P. Wade, Jr., to George A. Keyworth, 15 May 1984, file "Aeronautics Policy Review Committee [1 of 6]," box 17, Keyworth collection, RPL.

40. See "DoD response to OSTP Aeronautical Policy Review Committee Report Recommendations," attachment to James P. Wade, Jr., to George A. Keyworth, 15 May 1984, file "Aeronautics Policy Review Committee [1 of 6]," box 17, Keyworth collection, RPL; "NASA Response to November 1983 Recommendations of the OSTP Aeronautical Policy Review Committee," 24 April 1984, attachment to James M. Beggs, to G. A Keyworth, 22 May 1984, file "Aeronautics Policy Review Committee [1 of 6]," box 17, Keyworth collection, RPL; National Research Council, "Aeronautics Technology Possibilities for 2000: Report of a Workshop," National Academy Press, 1984, p. xiii.

41. National Research Council, "Aeronautics Technology Possibilities for 2000: Report of a Workshop," National Academy Press, 1984; NRC, "Aeronautical Technology 2000: A Projection of Advanced Vehicle Concepts," National Academy Press, 1985.

42. "Second Annual Report: Aeronautical Policy Review Committee," 28 December 1984, file "Aeronautical R&T Policy [5 of 5]," box 17, Keyworth collection, RPL.

43. John E. Steiner to George A. Keyworth II, 28 December 1984, file "Aeronautical R&T Policy [5 of 5]," box 17, Keyworth collection, RPL.

44. "Second Annual Report: Aeronautical Policy Review Committee," 5.

45. See Rob Williams to G. A. Keyworth, 10 January 1985, file "Aeronautical R&T Policy [5 of 5]," box 17, Keyworth collection, RPL.

46. The second edition of the "goals" report is dated 15 February 1985. See "Second Annual Report: Aeronautical Policy Review Committee."

47. Ibid.

48. Ibid.; "National Aeronautics Research Goals implementation Plan," 3 April 1985, file "Aeronautical R&T Policy [3 of 5]," box 17, Keyworth collection, RPL.

49. G. A. Keyworth II to James Beggs, 21 March 1985, Keyworth collection, box 17, file "Aeronautical R&T Policy [4 of 5]," RPL.

50. NASA made one significant change in the process of constructing the road maps through the addition of a rotorcraft goal. The Ames Research Center had specialized in rotorcraft, and would have been left out of the research agenda otherwise. So too would the Army, which was primarily interested in helicopters. See "NASA Response to the Office of Science and Technology Policy: Technology Roadmaps for the National Aeronautical R&D Goals," 15 May 1985, "Aeronautical R&T Policy [2 of 5]," box 17, Keyworth collection, file RPL.

51. On the HRE project see Earl H. Andrews and Ernest A. Mackley, "NASA's Hypersonic Research Engine Project—A Review," NASA TM-107759, 1994.

52. Ibid.

53. Robert Jones interview with author, 25 June 2001.

54. See Dave Stone to Dr. Keyworth, 15 November 1985, file "Aerospace [2 of 2]," box 17, Keyworth collection, RPL; British Aerospace brochure titled "HOTOL," n.d., file "Aerospace [2 of 2]," box 17, Keyworth collection, RPL; and Russel J. Hannigan, *Spaceflight in the Era of Aero-Space Planes* (Malabar, FL: Kreiger, 1994), 101–4.

55. On Keyworth's enthusiasm for Copper Canyon, see Maurice Roesch III, in *High Speed Commercial Flight: The Coming Era,* ed. James P. Loomis (Columbus, OH: Battelle, 1987), 165.

56. G. A. Keyworth to James M. Beggs, 30 July 1985, Keyworth collection, box 17, "Aeronautics-Hypersonics," RPL; G. A. Keyworth to Robert McFarlane, 29 July 1985, "Aeronautics-Hypersonics," box 17, Keyworth collection, RPL.

57. These were NSDD-144, NSDD-261, and NSSD-685.

58. Keyworth testimony, hearings entitled "High Speed Aeronautics," House, Committee on Science and Technology, 99th Cong., 1st sess., 24 July 1985, pp. 20, 22.

59. Paul Czysz interview with author, 17 July 2001.

60. Quoted from Hannigan, *Spaceflight in the Era of Aero-Space Planes*, 81.

61. Ibid., 82.

62. Czysz interview; Hannigan, *Spaceflight in the Era of Aero-Space Planes,* 82.

63. Rohrabacher was elected to the House of Representatives in 1988.

64. The most incisive analysis is Robert Dallek, *Ronald Reagan: The Politics of Symbolism,* 2nd ed. (Cambridge, MA: Harvard University Press, 1984). See also Michael Schaller, *Reckoning with Reagan* (New York: Oxford University Press, 1992), 56–57.

65. Eliot Marshall, "NASA and Military Press for a Spaceplane," *Science,* 10 January 1986, pp. 105–7.

66. Loomis, ed., *High Speed Commercial Flight*, vii.

67. Clarence J. Brown, keynote address, in *High Speed Commercial Flight,* ed. Loomis, 9.

68. Ibid., 12.

69. John F. Horn, in *High Speed Commercial Flight,* ed. Loomis, 28.

70. Frederick Smith, in *High Speed Commercial Flight,* ed. Loomis, 52.

71. John Swihart, "Practical Applications of Hypersonic Flight," in *High Speed Commercial Flight,* ed. Loomis, 208–18.

72. Takamasa Ryuzaki, "Economic Aspects of High Speed Commercial Transport Operations," in *High Speed Commercial Flight,* ed. Loomis , 137–45.

73. Klaus Nittinger, in *High Speed Commercial Flight,* ed. Loomis, 117–36.

74. Donald P. Schenk, in *High Speed Commercial Flight,* ed. Loomis, 100–104.

75. Sam Dollyhigh interview with author, LaRC.

76. A. Warner Robbins et al., "Concept Development of a Mach 3.0 High-Speed Civil Transport," NASA TM-4058, September 1988.

77. Boeing Commercial Airplanes, "High Speed Civil Transport Study," NASA CR-4233, 1989, pp. 6–11.

78. Malcolm MacKinnon interview, 25 February 2002.

79. Douglas Aircraft Company, "Study of High Speed Civil Transports," NASA CR-4235, 1989, p. 120.

80. Bob Welge interview with author, 9 February 2000.

81. Boeing Commercial Airplanes, "High Speed Civil Transport Study," NASA CR-4233, 1989, p. 103.

82. Douglas Aircraft Company, "Study of High Speed Civil Transports," 145–47; Boeing Commercial Airplanes, "High Speed Civil Transport Study," 81–84.

83. The ozone hole was discovered in 1982 but the Survey did not publish their findings until 1985. See J. C. Farman, B. G. Gardiner, and J. D. Shanklin, "Large Losses of Total Ozone in Antarctica Reveal Seasonal ClOx/NOx Interaction," *Nature,* 16 May 1985, pp. 207–10.

84. Loomis, ed., *High Speed Commercial Flight,* 2–3.

85. Congressional Research Service, "Commercial High-Speed Aircraft: Opportunities and Issues," March 1989, p. 6.

86. Hearings on the National Aeronautics and Space Administration Budget, Senate, Committee on Commerce, Science, and Transportation, Subcommittee on Science, Technology, and Space, 101st Cong., 1st sess., 4 April 1989, p. 226.

87. See James Shoch, *Trading Blows: Party Competition and U.S. Trade Policy in a Globalizing Era* (Chapel Hill: University of North Carolina Press, 2001), esp. chap. 6.

88. Ben Rich and Leo Janos, *Skunk Works* (Boston: Little, Brown, 1994), 301. Rich pulled no punches in his memoir—the NASP was, he said, a "simple-minded boondoggle from start to finish." Others have proposed that NASP was a cover for a black-world military project to develop a hypersonic spy plane code named "Aurora" and was designed to fail. See Czysz interview with author, 17 July 2001. For a skeptical summary of the "Aurora mythology," see Curtis Peebles, *Dark Eagles* (Navato, CA: Presidio Press, 1995), 251–80.

7 ✦ TOWARD A GREEN SST

1. See notes on "High-Speed Transport Aircraft Research" briefing, author's collection; author interview with Jerry Hefner, 24 September 2001.

2. For the "gull wing," see D. Bushnell, "Supersonic Aircraft Drag Reduction," AIAA Paper

90-1596, June 1990. The "oblique wing" derives from an idea of Robert T. Jones in the 1950s. NASA built and flew a subsonic oblique-wing experimental aircraft called the AD-1 during the early 1980s. The Ames Research Center had the oblique wing analyzed as a supersonic transport in 1989 by Stanford University researchers. See Alexander J. M. Van Der Velden and Ilan Kroo, "The Aerodynamic Design of the Oblique Flying Wing Supersonic Transport," NASA CR-177552, June 1990.

3. Michael Henderson, "Boeing Commercial Airplane Group Overview," *Third Annual NASA/Industry High-Speed Research Program Review,* 18 May 1993, p. 7. Also see National Research Council, *U.S. Supersonic Commercial Aircraft: Assessing NASA's High Speed Research Program* (Washington, DC: National Academy Press, 1997), 27.

4. Lou Williams interview with author, 22 October 2001; Jerry Hefner interview with author, 6 July 2001.

5. This figure is given in constant 1989 dollars. The best published accounting for the government portion of HSR is NRC, *U.S. Supersonic Commercial Aircraft,* p. 14.

6. Office of Aeronautics, Exploration, and Technology, "High Speed Research Program Plan," March 1990, p. 48.

7. Karen Crandall and Gary Golinski, "Teaming with the Enemy: A New Approach to Strategic Alliances" (MBA thesis, Cleveland State University, 1992).

8. The program's goals were defined initially as -1 PNdB from the stage-three "sideline" measurement, -3 PNdB from the stage-three "community" measurement, and -1 PNdB from the stage-three approach noise measurement—and abbreviated -1 / -3 / -1. The 747-400's certificated noise level at the "community" measurement point was stage-three -3 PNdB, which the program management considered the most important for "blending in" to the existing traffic picture.

9. J. C. Farman, B. G. Gardiner, and J. D. Shanklin, "Large Losses of Total Ozone in Antarctica Reveal Seasonal ClOx/NOx Interaction," *Nature* 315 (16 May 1985): 207–10.

10. See R. S. Stolarski, "Nimbus-7 Satellite Measurements of the Springtime Antarctic Ozone Decrease," *Nature* 322 (28 August 1986):. 808–11.

11. There is a huge body of literature on the Montreal Protocol. I used Karen T. Litfin, *Ozone Discourses: Science and Politics in Global Environmental Cooperation* (New York: Columbia University Press, 1994), and Richard Elliot Benedick, *Ozone Diplomacy: New Directions in Safeguarding the Planet* (Washington, DC: World Wildlife Fund, 1991).

12. Paul Crutzen and Frank Arnold, "Nitric acid cloud formation in the cold Antarctic stratosphere: a major cause for the springtime 'ozone hole,'" *Nature,* 324 (25 December 1986): 651–55; Mario J. Molina et al., "Antarctic Stratospheric Chemistry of Chlorine Nitrate, Hydrogen Chloride, and Ice: Release of Active Chlorine," *Science,* 238 (27 November 1987): 1253–57; Susan Soloman, "The Mystery of the Antarctic Ozone 'Hole,'" *Review of Geophysics* 26:1 (February 1988): 131–48. A good overview of the community's understanding as of 1987 is Ralph Cicerone, "Changes in Stratospheric Ozone," *Science* 237 (3 July 1987): 35–41.

13. On AAOE see Sharon Roan, *Ozone Crisis: The Fifteen-Year Evolution of a Sudden Global Emergency* (New York: John Wiley, 1989), 180–88, 212–24; Maureen Christie, *The Ozone Layer* (New York: Cambridge University Press, 2000), 60–69.

14. Michael Prather interview with author, 15 December 2002.

15. Michael Oppenheimer interview with author, 15 December 2002. See also Howard Wesoky interview with author, 27 June 2001.

16. Altitude of emission has a large effect on the severity of depletion. Higher cruise alti-

tudes, and thus higher Mach numbers, produce significantly more depletion than lower altitudes. Johnston's 1989 study showed that a Mach 1.6 SST produced the least depletion, .9 percent, while the Mach 2.4 SST that was HSR's focus might cause 10.4 percent depletion. See National Research Council, *The Atmospheric Effects of Stratospheric Aircraft Project: An Interim Review of Science and Progress* (Washington, DC: National Academy Press, 1998), 25.

17. Harold S. Johnston, "Topical Review of Stratospheric Aircraft and Global Ozone (1989)," in author's collection; later revised and published as Johnston et al., *The Atmospheric Effects of Stratospheric Aircraft: A Topical Review*, NASA RP-1250, January 1991.

18. The *Journal of Geophysical Research* published a special issue on the AAOE campaign results: *Journal of Geophysical Research*, 94(D9), 30 August 1989. See also Susan Solomon, "Progress towards a quantitative understanding of Antarctic ozone depletion," *Nature*, 347 (27 September 1990): 347–54. The polar stratospheric cloud phenomenon is described in M. P. McCormick et al., "Polar Stratospheric Cloud Sightings by SAM II," *Journal of the Atmospheric Sciences* 39 (June 1982), 1387–97.

19. Wesoky interview, 27 June 2001.

20. Howard Wesoky to Thomas E. Graedel, 18 June 1993. Copy in author's collection.

21. See Steven L. Baughcum et al., *Stratospheric Emissions Effects Database Development*, NASA CF-4592, July 1994.

22. Z. H. Landau et al., *Jet Aircraft Engine Exhaust Emissions Database Development—Year 1990 and 2015 Scenarios*, NASA CR-4613, July 1994.

23. Daniel L. Albritton et al., *The Atmospheric Effects of Stratospheric Aircraft: Interim Assessment Report of the NASA High-Speed Research Program* (June 1993).

24. Ibid., 12. The potential impact of sulfate aerosols was first suggested by David Hofman and Susan Solomon in "Ozone Destruction through Heterogeneous Chemistry Following the Eruption of El Chichon," *Journal of Geophysical Research* 94:D4 (20 April 1989): 5029–41.

25. Hofman and Solomon, "Ozone Destruction through Heterogeneous Chemistry." I'm indebted to Richard Stolarski for explaining this process.

26. See, for example, A. R. Ravishankara, "Chemical Processing: Homogeneous and Heterogeneous Reactions," in *The Atmospheric Effects of Stratospheric Aircraft: Interim Assessment Report of the NASA High-Speed Research Program*, ed. Daniel L. Albritton et al., June 1993, pp. 55–74; Richard Stolarski, Robert Watson, and Dan Albritton, "Executive Summary," in *Atmospheric Effects of Stratospheric Aircraft*, ed. Albritton et al., 1–8.

27. National Research Council, *Atmospheric Effects of Stratospheric Aircraft: An Evaluation of NASA's Interim Assessment* (Washington, DC: National Academy of Sciences, 1994), 5.

28. These were all primarily funded by the Upper Atmosphere Research Program office, with contributions from AESA/HSR.

29. Ozone was known to be a strong "greenhouse gas," which absorbed infrared radiation that would otherwise escape into space. The model results had suggested that HSCTs would increase the troposphere's ozone density while decreasing the stratosphere's. While this produced very little *net* change in the atmosphere's average ozone density, and therefore prevented a significant increase in ultraviolet radiation at the Earth's surface, an alteration in the relative proportion of ozone between the two might cause tropospheric heating and stratospheric cooling.

30. NRC, *Atmospheric Effects of Stratospheric Aircraft*, 6–7.

31. Statement of Thomas E. Graedel, Subcommittee on Technology, Environment, and

Aviation, Committee on Science, Space, and Technology, House of Representatives, 103rd Cong., 2nd sess., 10 February 1994.

32. Howard Wesoky and Richard Stolarski, "Atmospheric Effects of Stratospheric Aircraft: Budget Augmentation Request," 24 February 1994, author's collection.

33. Richard Stolarski interview with author, 22 October 2001.

34. Howard Wesoky interview with author, 11 September 2001. Skepticism in the German research community also came out at the 7th European Aerospace Conference, held in Toulouse in 1994. See Ulrich Shumann, "Impact of Emissions from Aircraft on the Atmosphere," in *The Supersonic Transport of Second Generation: Proceedings of the 7th European Aerospace Conference* (1994), 71–80.

35. The European "Concorde II" effort had no public funding until 1994, when it received $15 million per year. It was primarily driven by Aerospatiale. For more details, see Kieran Daly, "BAe pleads for more supersonic funding," *Flight International*, 15 December 1993, p. 10; Anon., "Hasty Decisions," *Flight International*, 15 December 1993, p. 3; Anon., "Europe: Supersonic Aircraft Research Program launched," *S&T Perspectives*, 31 May 1994, pp. 17–18; Anon., "European Effort at Critical Point," *Aviation Week and Space Technology*, 21 November 1994, p. 72; Pierre Sparaco et al., "France UK Promote European HSCT Effort," *Aviation Week and Space Technology*, 21 November 1994, p. 62.

36. Richard Stolarski et al., *1995 Scientific Assessment of the Atmospheric Effects of Stratospheric Aircraft*, NASA RP-1381, 1995; NRC, *The Atmospheric Effects of Stratospheric Aircraft*, 40–41.

37. The accepted measurement for nitrogen oxides emissions is grams of NOx produced per kilogram of fuel burned, and is abbreviated "g NOx/kg."

38. R. W. Niedzwiecki interview with author, 22 May 2001.

39. Ibid.

40. Bill Strack interview with author, 19 July 2002. See also Allen H. Whitehead, Jr., "Impact of Environmental Issues on the High Speed Civil Transport," paper presented at the Von Karman Institute for Fluid Mechanics, Rhode-Saint-Genese, Belgium, 25–29 May 1998, p. 3.

41. Bernie Blaha interview with author, 23 October 2001.

42. Ibid.

43. Ibid.

44. Bill Strack interview with author, 19 July 2002.

45. Ibid.

46. See Rudolph J. Swigart, "An Experimental Study of the Validity of the Heat-Field Concept for Sonic-Boom Alleviation," NASA CR-2381, March 1974.

47. R. Seebass and R. George, "Sonic Boom Minimization," *Journal of the Acoustical Society of America*, vol. 52, no. 2, pt. 3 (February 1972): 686–94.

48. Office of Aeronautics, Exploration, and Technology, "High Speed Research Program Plan," March 1990, p. 32.

49. On the unrestricted market, see Munir Metwally, "Economic Benefits of Supersonic Overland Operation," in *High Speed Research: Sonic Boom v. 2*, comp. Christine Darden, NASA CP-3173, 1992, pp. 1–15.

50. Kevin Shepherd interview with author, 28 September 2001.

51. Summarized from Jack D. Leatherwood et al., "Summary of Recent NASA Studies of

Human Response to Sonic Booms," *Journal of the Acoustical Society of America* 111:1, pt. 2 (2002): 586–98.

52. Jack D. Leatherwood and Brenda Sullivan, "Subjective Loudness Response to Simulated Sonic Booms," in *High Speed Research: Sonic Boom v. 2,* comp. Darden, 151–70.

53. See, for example, John Morgenstern, "MDC Low Boom HSCT Design Status," in *High Speed Research: Sonic Boom v. 2,* ed. Thomas A. Edwards, NASA CP-10133, 1994, pp. 1–16; Christine Darden, "Status of Sonic Boom Program," in *Third Annual NASA/Industry High-Speed Research Program Review,* 18–19 May 1993.

54. Susan E. Cliff, "On the Design and Analysis of Low Boom Configurations," in *High Speed Research: Sonic Boom v. 2,* ed. Thomas A. Edwards, NASA CP-10133, 1994, pp. 37–80; George T. Haglund, "Low Sonic Boom Studies at Boeing," in *High Speed Research: Sonic Boom v. 2,* ed. Thomas A. Edwards, NASA CP-10133, 1994, pp. 81–94; Daniel Baize and Peter G. Coen, "A Mach 2.0/1.6 Low Sonic Boom High-Speed Civil Transport Concept," in *High Speed Research: Sonic Boom v. 2,* ed. Edwards, 125–42.

55. Kevin Shepherd interview with author, 25 March 2002; Daniel Baize interview with author, 4 December 2001; Peter G. Coen interview with author, 25 September 2001.

56. Malcolm MacKinnon interview with author, 25 February 2002.

57. Shepherd interview, 28 September 2001.

58. See James M. Fields, *Reactions of Residents to Long-Term Sonic Boom Noise Environments,* NASA CR-201704, June 1997.

59. Whitehead, "Impact of Environmental Issues on the High-Speed Civil Transport," 7.

60. Bruce Bunin, "The Market for High Speed Civil Transport," paper presented to the American Association for the Advancement of Science, 1997, copy in author's collection.

61. Michael Henderson interview with author, 25 March 2002.

62. Shepherd interview, 28 September 2001; Whitehead, "Impact of Environmental Issues on the High Speed Civil Transport," 7.

63. O. L. Austin, Jr., W. B. Robertson, Jr., and G. E. Woolfender, "Mass Hatching Failure in Dry Tortugas Sooty Terns," *Proceedings of the International Ornithological Congress* 15:627 (1970).

64. Shepherd interview, 28 September 2001.

65. Ann Bowles and William C. Cummings, "Effects of Sonic Booms on Marine Mammals: Problem Review and Recommended Research," in *1995 NASA High Speed Research Program Sonic Boom Workshop,* vol. 1, ed. Daniel G. Baize, NASA CP-3335, July 1996; see also Whitehead, "Impact of Environmental Issues on the High-Speed Civil Transport," 8.

66. Ming Tang interview with author, 3 December 1999.

67. James Shoch, *Trading Blows: Party Competition and U.S. Trade Policy in a Globalizing Era* (Chapel Hill: University of North Carolina Press, 2001), 161. Initially, the administration intended to push through a large package of technology investments, but that strategy failed. The administration turned to an export strategy instead. The HSR program fit well into both strategies, since around two-thirds of all HSCTs sold would be sold overseas.

68. David Moore, *Reinventing NASA* (Washington, DC: Congressional Budget Office, March 1994).

69. Ray Hook to associate administrator (Aeronautics), 12 November 1992, Goldin Papers, file 74540, NASA HQ History Office, and attachments. See also Lou Williams (handwritten notes), 23 October 1992, Goldin Papers, file 73529. Copies in author's collection.

70. Ibid.

71. Ibid.

72. Quoted from Lou Williams (handwritten notes), 23 October 1992, Goldin Papers, file 73529.

73. Robert Whitehead interview with author, 17 July 2002.

74. Goldin's comment, of course, was political speech and should not be taken as a true reflection of the agency's priorities. Instead, it was a statement of what he believed the audience wanted to hear. Goldin, and NASA, continued to treat the International Space Station as its top priority even though it clearly was not the top priority of either congressional committee.

75. Senate, Committee on Commerce, Science, and Technology, "Hearings on NASA's Fiscal Year 1994 budget," 103rd Cong., 2nd sess., 20 April 1993, p. 19.

8 SIC TRANSIT HSCT

1. Malcolm MacKinnon interview with author, 4 November 2002.

2. Alan Wilhite interview with author, 19 September 2001.

3. Wilhite attributes the idea to Gilkey; Gilkey rejected credit. See Wilhite interview and Gilkey interview with author, 6 December 2001.

4. I'm indebted to Paul Hergenrother for the simplified explanation of composite manufacturing.

5. John G. Davis, Jr., "Composites for Advanced Space Transportation Systems (CASTS)," in *Graphite/Polyimide Composites*, ed. H. Benson Dexter and John G. Davis, Jr., NASA CP-2079, 1979, pp. 5–16.

6. For a summary of ACEE see Jeffrey L. Ethell, "Fuel Economy in Aviation," NASA SP-462, 1983.

7. Marvin B. Dow, "The ACEE Program and Basic Composites Research at Langley Research Center (1975–1986)," NASA RP-1177, October 1987, p. 14.

8. The polyimide-graphite example they chose was LARC-160, from the Langley Research Center's CASTS effort. The study's authors explain that it was the only suitable material in production. See J. R. Kerr and J. F. Haskins, "Time-Temperature-Stress Capabilities for Composite Materials for Advanced Supersonic Technology Application," NASA CR-178272, May 1987, pp. 3–5 to 3–7.

9. On the 1993 state of the art, see Michael Brunner and Alex Velicki, "Study of Materials and Structures for High-Speed Civil Transport," NASA CR-191434, September 1993, pp. 15–16.

10. Paul Hergenrother, "Development of Composites, Adhesives and Sealants for High-Speed Commercial Airplanes," *SAMPE Journal* 36:1 (January/February 2000): 30–41; Paul Hergenrother interview with author, 20 November 2001, LaRC. See also M. L. Rommel, L. Konopka, and P. M. Hergenrother, "Process Development and Mechanical Properties of IM7/LaRC PETI-5 Composites," Proceedings of the 28th International SAMPE Technical Conference, Seattle, WA, November 1996.

11. Thomas Gates interview with author, 18 December 2002. See also Thomas S. Gates and Karen S. Whitley, "Long-Term Durability of Polymer Matrix Composites under Simulated Supersonic Flight Profile Testing," AIAA 2000-1684, 2000; Thomas S. Gates and Michael A. Grayson, "On the Use of Accelerated Aging Methods for Screening High Temperature Polymeric Composite Materials," AIAA A99-1296, 1999.

12. Paul Hergenrother, "Development of Composites, Adhesives and Sealants for High-Speed Commercial Airplanes," *SAMPE Journal* 36:1 (January/February 2000): 30–41; Hergenrother interview.

13. Ibid.

14. The root cause of turbulence in a wing's airflow is separation of the thin "boundary layer" of air just above the wing's surface from the wing. Boundary layer separation has many specific causes, including roughness or irregularity of the wing surface, shock wave impingement, ice crystal formation, and insect impacts.

15. Albert Braslow, *A History of Suction-Type Laminar-Flow Control with Emphasis on Flight Research* (Washington, DC: NASA, 1999), 10–11.

16. Ibid., 13–26; Roy V. Harris, Jr., and Jerry N. Hefner, "NASA Laminar-Flow Program—Past, Present, Future," *Research in Natural Laminar Flow and Laminar-Flow Control*, NASA CP-2487 pt. 1, 1987, pp. 1–24.

17. D. M. Bushnell and M. R. Malik, "Supersonic Laminar-Flow Control," *Research in Natural Laminar Flow and Laminar-Flow Control*, NASA CP-2487 pt. 3, 1987, p. 924.

18. Boeing Commercial Aircraft Company, "Application of Laminar Flow Control to Supersonic Transport Configurations," NASA CR-181917, July 1990, p. 5. See also A. G. Powell, S. Agrawal, T. R. Lacey, "Feasibility and Benefits of Laminar Flow Control on Supersonic Cruise Airplanes," NASA CR-181817, July 1989.

19. Jerry Hefner interview with author, 6 July 2001. Braslow gives a slightly different version of events: Braslow, *A History*, 32–33.

20. Braslow, *A History*, 33.

21. Jeff Lavell interview with author, 30 October 2001.

22. MacKinnon interview.

23. Joseph Chambers to director for aeronautics, 13 March 1993, HSR collection, Gilbert series, box 5, file "TU-144."

24. Lou Williams interview with author, 27 June 2001.

25. On the Gore-Chernomyrdin agreement see Office of Aeronautics, "NASA-Russian Aeronautical Research," 4 February 1994, TU-144 file, author's collection; on adding the TU-144 to this program see Wesley L. Harris, "NASA-Russian Cooperation in Aeronautical Research: White House Presentation," March 1994, TU-144 file, author's collection.

26. E. R. Beeman, "Feasibility Assessment of a TU-144 Supersonic Laminar Flow Control Flight Test Experiment," 27 October 1994, p. 16; M. Fischer and J. Lavell, "Comments on Rockwell TU-144 SLFC Feasibility Study," n.d.; and Gilbert notes on copy of Beeman, all in TU-144 file, author's collection.

27. Tom Huntington, "Encore for an SST," *Air and Space Smithsonian* 10:4 (October/November 1995): 29. See also Craig Covault, "Russian TU-144 Advances Work toward US SST," *Aviation Week and Space Technology*, 8 September 1997, pp. 50–53.

28. Wilhite interview.

29. Andrew Johnson interview with author, 23 January 2002.

30. Ibid.

31. Paul Bartolotta interview with author, 20 December 2001.

32. Bob Draper, Bob Miller, and Rebecca McKay interview with author, 23 January 2002. The blade coatings could be renewed at the first major scheduled overhaul period, which for commercial aircraft was at 4,500 operating hours. Hence this was the *minimum* lifetime the blades could have.

33. Johnson interview.

34. Ibid.

35. Andrew M. Johnson, Benton J. Bartlett, and William A. Troha, "Material Challenges and Progress toward Meeting the High Speed Civil Transport Propulsion Design Requirements," International Symposium on Air Breathing Engines (ISABE) paper 97-7179, 1997, pp. 1321–28.

36. Ajay Misra interview with author, 1 April 2002; Ajay K. Misra, Andrew M. Johnson, and Benton J. Bartlett, "Progress toward Meeting Material Challenges for High Speed Civil Transport Propulsion," ISABE paper 99-7018, 1999; Robert J. Shaw, Leigh W. Koops, Richard Hines, "Progress toward Meeting the Propulsion Technology Challenges for the High Speed Civil Transport," ISABE paper 99-7005, 1999.

37. R. A. Miller, "Thermal Barrier Coatings for Aircraft Engines: History and Directions," Journal of Thermal Spray Technologies 6:1 (March 1997): 35; Harry E. Eaton et al., "EBC Protection of SiC/SiC Composites in the Gas Turbine Combustion Environment," American Society of Mechanical Engineers paper 2000-GT-631, 2000.

38. Subsonic airliners have this seemingly odd engine cycle because they are significantly overpowered for cruise. Commercial aircraft certification regulations require all airliners to be able to complete a takeoff after a single engine failure during the takeoff run, which in effect requires designers to install twice the thrust an aircraft actually needs in a twin engine aircraft, one-third more in a trijet, or 25 percent more in a four-engine aircraft. For supersonic aircraft, the much higher drag above Mach 1 compensates for the overpowered takeoff condition, forcing the engines to run at full power for most of the aircraft's mission.

39. Draper, McKay, and Miller interview.

40. Ibid.; Shaw, Koops, and Hines, ISABE paper 99-7005, p. 41; Misra, Johnson, and Bartlett, ISABE paper 99-7018.

41. Draper, McKay, and Miller interview.

42. FAA, "FAA Historical Chronology, 1926–1996" (Washington: GPO, 1996), 211.

43. Christopher P. Fotos, "New Law to Permit Head Tax, Phase Out Stage 2 Aircraft," Aviation Week and Space Technology, 5 November 1990, p. 31. The law was "The Airport Noise and Capacity Act." See FAA, "FAA Historical Chronology, 1926–1996" (Washington: GPO, 1996), 270.

44. The five-nation "Nordic Council," consisting of Denmark, Sweden, Finland, Iceland, and Norway, adopted a ban on "Chapter 2" aircraft after 2000 in 1989. See Edward H. Phillips, "EEC Delays Action on Aircraft that Fail to Meet Noise Limits," Aviation Week and Space Technology, 13 February 1989, p. 30.

45. David Hughes, "ICAO Assembly Fails to Agree on Aircraft Noise Restrictions," Aviation Week and Space Technology, 16 October 1989, pp. 82–83.

46. David Hughes, "ICAO Members Set Noise Guidelines for Restricting Chapter 2 Aircraft," Aviation Week and Space Technology, 5 November 1990, pp. 38–39.

47. Philip Buttersworth-Hayes, "Raising the Volume on a Noisy Issue," Aerospace America 31:10 (October 1993): 8–10.

48. The regulation is DOT 1050.1D.

49. Bob Cuthbertson to Mike Henderson et al., "HSR Phase II Noise Certification Requirements and Noise Goals," 7 April 1995, author's collection.

50. Bill Gilbert interview with author, 26 June 2001.

51. Robert Golub interview with author, 12 December 2001.

52. Chester Nelson, "Overview of Technology Integration Activities Related to Configuration Aerodynamics," NASA CP-1999-209692, 1999, pp. 1762, 1763.

53. National Research Council, *U.S. Supersonic Commercial Aircraft: Assessing NASA's High Speed Research Program* (Washington, DC: National Academy Press, 1997), 92. This report was unusually explicit about the challenge posed by aeroelastic considerations on the vehicle: they would have "enormous" impact and could result in "catastrophic pilot-vehicle dynamic effects."

54. Finite element analysis is a means to accurately predict the structural weight of a configuration. Engineers generate a detailed structural model of the vehicle, size the structural members for the static load they must bear (i.e., the stationary load), then "fly" the vehicle through its operating envelope in order to resize the structural members for the loads they carry in flight.

55. Rodney Ricketts interview with author, 27 September 2001.

56. Michael Henderson interview with author, 25 March 2002.

57. The figures cited represent the cumulative total of noise reduction at each of the three required measurement points—the twenty-decibel total might be (-6, -7, -7) at the side-line, takeoff, and approach measurement points. The single number is simply a common shorthand notation that permits less clumsy sentence construction, but bears explanation because it is misleading.

58. MacKinnon interview.

59. Elizabeth Drew, *Showdown: The Struggle between the Gingrich Congress and the Clinton White House* (New York: Simon and Schuster, 1996), 26–38.

60. On the International Space Station as political symbol, see Andrew Lawler, "Onward into Space," *Astronomy* 26:12 (1998): 47–48.

61. See Howard E. McCurdy, *Faster, Better, Cheaper: Low-Cost Innovation in the U.S. Space Program* (Baltimore: Johns Hopkins University Press, 2001), 44.

62. "FY 1996 President's Budget," chart in Goldin Papers, file 71899, NASA HQ History Office; and T. J. Glauthier to Daniel Goldin, 21 November 1994, Goldin Papers, file 71899.

63. Newt Gingrich, *Window of Opportunity: A Blueprint for the Future* (New York: Tomas Doherty & Associates, 1984).

64. The various charges made against NASA are nicely summarized in David Moore, *Reinventing NASA* (Washington, DC: Congressional Budget Office, 1984).

65. Andrew J. Butrica, *Single Stage to Orbit: Politics, Space Technology, and the Quest for Reusable Rocketry* (Baltimore: Johns Hopkins University Press, 2003).

66. Amy J. Boatner, "Consolidation in the Aerospace and Defense Industries: The Effect of the Big Three Mergers in the United States Defense Industry," *Journal of Air Law and Commerce* 64 (Summer 1999): 913–40.

67. See National Academy of Sciences, "Recent Trends in U.S. Aeronautics Research and Technology," 1999, www.nap.edu; copy in author's collection, box 2, file 98, LaRC.

68. Homan Jenkins, "Boeing's Trouble: Not Enough Monopolistic Arrogance," *Wall Street Journal,* 16 December 1998, A23.

69. Frederic M. Biddle, "Boeing Effort to Cushion Itself from Cycles Backfires," *Wall Street Journal,* 24 October 1997; James Wallace, "Boeing Stuns Wall Street with Big Loss: $1.6 Billion Now and More in 1998," *Seattle Post-Intelligencer,* 23 October 1997, A1; Karen West, "Boeing Puts on the Brakes: Production Slowdown Affects 747 and 737," *Seattle Post-Intelligencer,* 4 October 1997, A1.

70. Frederic Biddle, "Boeing Throttles Back Development Drive," *Asian Wall Street Journal,* 4 December 1998, p. 2.

71. James Ott, "Five Aircraft Manufacturers Forming HST Study Group," *Aviation Week and Space Technology,* 14 May 1990, pp. 26–27. On the British Aerospace/Aerospatiale alliance see Carole A. Shifrin, "Britain and France Begin Concorde Follow-On Study," *Aviation Week and Space Technology,* 14 May 1990, pp. 24–26.

72. The United States suffered near-paralyzing airspace congestion during 1999 and 2000 due to the route fragmentation problem and the airlines' purchase of large numbers of small "regional" jets. The airlines blamed the government for not building new runways, while the government blamed the airlines for overscheduling flights (i.e., for buying and operating too many small aircraft). The problem was "solved" temporarily by the onset of a major recession in 2001. See David Bond, "Commercial Aviation on the Ropes," *Aviation Week and Space Technology,* 21 September 2000.

73. Michael Mecham and James T. McKenna, "Cost, Not Size, to Drive Success of Superjumbo," *Aviation Week and Space Technology,* 21 November 1994, pp. 45–46; Pierre Sparaco, "Airbus in Hot Pursuit of A3XX Launch," *Aviation Week and Space Technology,* 1 July 2000.

74. NRC, *U. S. Commercial Supersonic Aircraft,* 2.

75. "FY 1997 OMB Budget Preview," 8 August 1995, Goldin Papers, file "074067," NASA HQ History Office.

76. "HSR Phase III Planning: Report to IPT," 16 May 1996, copy in author's collection.

77. Ibid.

78. "High-Speed Research Program Phase IIA: Technology Demonstration," presentation to Office of Management and Budget, October 1996, copy in author's collection.

79. Malcolm Peterson, "FY 1999 Budget Passback Presentation to the Capital Investment Council," 17 December 1997, Goldin Papers, file "FY 1999 Budget Passback."

80. See "NASA's Fiscal Year 1999 Budget Request, Parts I–IV," Subcommittee on Space and Aeronautics, Science Committee, House of Representatives, 105th Cong., 2nd sess., 5 February 1998.

81. Rich Christiansen, "CIC Meeting: OMB Passback Finalization," 17 December 1997, Goldin Papers, "FY 1999 Budget Passback."

82. National Civil Aviation Review Commission, "Avoiding Aviation Gridlock and Reducing the Accident Rate: A Consensus for Change," 11 December 1997.

83. Bruce Ramsey, "Supersonic Jet Project Postponed," *Seattle Post-Intelligencer,* 3 November 1998.

84. Stanley Homes, "Boeing Backs away from Plans for Supersonic Jet," *Seattle Times,* 3 November 1998.

85. "Response to Query: High Speed Research Program as of 11/4/1998," HSR collection, box 10, file "MacKinnon interview."

86. Wallace Sawyer interview with author, 19 December 2001, LaRC.

87. Robert J. Shaw interview with author, 6 June 2002, LaRC.

88. Daniel Goldin to John McCain, 22 December 1998, Goldin Papers, file 76492.

89. Testimony of Spence Armstrong, Subcommittee on Space and Astronautics, House of Representatives, "Fiscal Year 2000 NASA Authorization Part IV: Aero-Space Technology," 106th Cong., 1st sess., 3 March 1999, p. 422.

90. Joyce E. Penner, David Lister, David J. Griggs, David J. Dokken, and Mack McFarland, *Aviation and the Global Atmosphere* (London: Cambridge University Press, 1999).

1. The anti-France campaign was led by two media organizations, the *New York Post,* which declared France and Germany "weasels" in its headlines repeatedly during March 2003, and by *NewsMax,* which sponsored a boycott of French products. They were aided by a brief effort to pass anti-France legislation in Congress, including a ban on American participation at the Paris Air Show. See "NY Times Publishes NewsMax 'Boycott France' Ad," *NewsMax,* 26 March 2003; John Podhoretz, "Show 'Em—President Makes Weasels Choose," *New York Post,* 7 March 2003, p. 33; "Weasel Whacking," *New York Post,* "It's Showdown Time at U.N. as Powell Takes on Euro-Weasels," *New York Post,* 14 February 2003, p. 1; "Axis of Weasel—Germany and France Wimp Out on Iraq," *New York Post,* 24 January 2003, p. 1; "France Presses Press Complaint," CBSNEWS.com, 15 May 2003 (copy in author's collection).

2. "Concorde quietly heads to Manhattan," CNN.com, 25 November 2003 (copy in author's collection); Edward Wong, "Concorde Jet to Make the Intrepid Its Home," *New York Times,* 31 October 2003, p. B8; "Concorde flies into history," CNN.com, 27 February 2003 (copy in author's collection).

3. Michael Klesius, "The Future of Flying," *National Geographic,* December 2003.

4. For the Hubble complete history, see Robert W. Smith, *The Space Telescope: A Study of NASA, Science, Technology and Politics* (Cambridge: Cambridge University Press, 1989). For the space station quest, consult Howard McCurdy, *The Space Station Decision: Incremental Politics and Technological Choice* (Baltimore: Johns Hopkins University Press, 1990), and W. Henry Lambright, "NASA, Ozone, and Policy-Relevant Science," *Research Policy* 24 (1995): 747–60.

5. On this see William Leary, *We Freeze to Please!* (Washington, DC: NASA, 2001). Another well-known case is flying-qualities research; see Michael H. Gorn, *Expanding the Envelope: Flight Research at NACA and NASA* (Lexington: University Press of Kentucky, 2001).

6. Thanks to my colleagues Maureen McCormick and Sara Pritchard for pointing this out.

ESSAY ON SOURCES

This book is based on extensive research in a number of public and private archives. In addition, I was given access to the office files of a number of individuals at the NASA Langley and NASA Glenn Research Centers for the final two chapters. Although these files are not public, I was able to get a few key documents cleared for release and have placed copies in the project collection held by the NASA History Office.

Due to the time span covered by this book, the availability, nature, and sources of the "archival" documents referenced herein change radically and my research approach had to change in response. For the first six chapters, I was able to rely on historians' traditional approach to research, digging through memos, correspondence, reports, etc. held in discrete, organized archival collections. The archives I used are discussed below. For these chapters, I used oral histories sparingly, to fill in gaps in the record and to flush into the open the sorts of things that do not get written down—in aerospace, at least, a great deal happens that never gets inscribed on paper.

Beginning with chapter 7, roughly 1988 on, the traditional approach was unworkable. The only organized archive from this period, the George Herbert Walker Bush Presidential Library, chose to adopt an extremely broad interpretation of "executive privilege" that effectively prevented me from gaining access to anything dealing with the NASA budget, even communication between low-level staffers.

Without organized records to mine, I had to rely on the office files of several individuals in NASA to reconstruct the HSR program. These records were quite complete, but they differed in their basic nature from the kinds of materials available before 1990. The vast majority of these records are viewgraph presentations made using Microsoft Powerpoint and similar software. These differ from traditional kinds of documents in several important respects. The first is that they were written by and for technically trained people and tend to present important information in charts and graphs instead of words. Most historians are untrained in the art of visual representation of data, especially as practiced in modern engineering, making this material a great challenge to master. The second important difference is lack of context. One often cannot know why a particular presentation was given. Sometimes the reason is obvious—annual and monthly progress reports—but frequently it isn't. Some presentations were generated after "something went wrong" and are thus important bits of evidence, but there is no way to discover that from the documents themselves. The third difference is that presentations do not serve as a record of discussion. Unlike an exchange of memos, they give only the presenter's view. And they only give an incomplete depiction of that view. We cannot know what, exactly, the speaker said during the presentation or how it was said. Electronic mail records were not available to me and were not systematically kept on paper, eliminating this source.

For all of these reasons, I had to approach the last two chapters very differently. I relied much more on oral history, in some cases taking presentations back to their originators and asking questions about them. The result is that the book's final chapters are based on much less traditional "evidence" than the rest of it. One countervailing benefit, however, is that they are backed by a substantial set of oral histories that will be useful to future researchers.

ARCHIVES

Air Force Historical Research Agency, Maxwell Air Force Base, Montgomery, Alabama. Contains material related to the XB-70 and the "Advanced Manned Strategic Aircraft." Much of this is still classified because it deals directly with nuclear strategy, but the staff is willing to review selected documents for prompt declassification.

The Boeing Company. Alone among the major aerospace companies, Boeing maintains an archive that is accessible to researchers. It contains a very large photographic collection, reference files searchable via a database, and a substantial set of oral history interviews. It also has a huge amount of unprocessed material from the former North American Aviation, but this is unfortunately unavailable. Most useful for this project were the papers of William Allen, a former chairman of the company, which contained a substantial amount of SST-related material. Several of the oral histories were useful as well, and the reference collection contained copies of Boeing's various SST proposals.

Langley Research Center archives, Hampton, Virginia (Cited as "LaRC archives" in the notes.) Langley Research Center has a small historical archive in its library building. It contains most of the Research Authorization files generated during the center's days as the NACA's primary laboratory, a number of personal collections, oral histories, and the documents I collected in writing this book. Langley Research Center is a restricted facility, and potential users must obtain permission to visit. The archive also has no staff. Researchers should contact the NASA Headquarters history office to explore options for using this material.

The Massachusetts Institute of Technology Library and Archives. The Citizens' League against the Sonic Boom deposited its papers here when it went out of business. A small but thorough and well-organized collection, this was very useful in reconstructing the league's actions as well as some of the motivations of the anti-SST campaign.

National Academy of Sciences Archive, Washington, D.C. Contains the records of the NAS Committee on SST-Sonic Boom. This body of records is very useful for understanding the lobbying efforts of William Shurcliff, who caused a great deal of trouble within the academy by his attacks on his colleagues. Ultimately, he drove the organization to reform its review process.

National Archives and Records Administration, College Park, Maryland. This is NARA's primary facility, and it houses the twentieth-century records of all Washington, D.C., area agencies. For this project, relevant records were in the records of the National Advisory Committee on Aeronautics and in the records of the Transportation Department. The NACA records consist of the records of the NACA's individual committees; for my purposes, the useful records were those of the High Speed Committee and the Committee on Aircraft Problems (for noise). These records are reasonably well described in the finding aids. The Transportation Department records, however, are completely undocumented. They appear to consist of the Transportation Secretary's official correspondence, and I was able to use them only by convincing a staff archivist to escort me into the closed stacks to scan the box labels for relevant subjects. A small amount of material related to the SST was here, primarily correspondence between DOT and NASA and responses to letters from the public.

The most useful group of records for this project proved to be in RG 200, the "gift collection." This consists of papers donated by former public officials. Robert S. McNamara donated a large body of records to NARA, including several boxes of material related to the SST and to the A-12/SR-71. These records proved to be central to chapter 3.

National Archives and Records Administration, Mid-Atlantic Region, Philadelphia. This location contains the correspondence files of the Langley Research Center, from roughly 1920 through 1980. They are in three series. Series 1 consists of project files organized by name; this is not SST material, but valuable for many other projects the center was involved in. Series 2 contains the center's central files from 1920 to approximately 1968, organized by subject code number. Because there was no specific code for the SST, correspondence about this project was scattered throughout the "airplanes" category—about forty archive boxes of material. Both series 1 and 2 have been very well organized and documented by the NARA staff. Series 3 is a three-hundred-reel microfilm collection containing the central files from 1970 through 1980. There is no meaningful SST material here. This series is very difficult to use because the "finding aid" is simply a very long printout listing the subject code numbers reflected on each reel.

National Archives and Records Administration, San Bruno, California. The San Bruno branch contains the records of the NASA Ames Research Center in Sunnyvale, California. Like the Langley records in Philadelphia, these are organized by subject code. Unlike the Langley records, they are cataloged in a searchable database via the Online Archive of California project. Ames had little involvement with the SST project, and only a small amount of material is present here.

Public Records Office, Kew Gardens, London, England. The PRO is the British equivalent of the U.S. National Archives and Records Administration, but vastly better organized. Computer searchable databases cover the entire twentieth century collection, and records are automatically opened after thirty years, with the exception of personal collections. When I visited in 2000, I was able to glean Concorde records from several ministries, including Transport, Treasury, and Prime Minister's Office, through 1970. The most useful collections were the cabinet collection (CAB) and the Aviation Ministry (AVIA) collection.

Rensselaer Polytechnic Institute Special Collections, Troy, New York. RPI's archive contains the papers of George Low, who was deputy administrator under James Fletcher. His papers are well described in the institute's finding aid and are exhaustive in their coverage of his tenure at NASA. They are essential for understanding the agency's direction (or lack thereof) during the 1970s.

Washington Area National Records Center, Suitland, Maryland. This facility warehouses records turned over to NARA by federal agencies in the Washington, D.C., area that the agencies still maintain control over. Access to these records requires agency permission. NASA headquarters has a substantial volume of material stored here under the control of the agency records manager. Many of the "finding aids," in the guise of SF-135 Records Transfer forms, are held in the NASA History Office. There are a great many of these, sorted by year of transfer but not by office or subject, and therefore searching for possible records is quite time consuming. This is the only way to access NASA records after 1970, however, as the agency has transferred very little to the National Archives since. There is a small volume of SST material scattered through NASA's accessions.

Suitland also contained the very extensive records of the FAA's SST Program Office until 1996, when they were destroyed. The FAA History Office holds a small collection of material gleaned from the project office records before they were destroyed. The most useful material in this collection are two sets of incomplete draft histories, one started when the SST program was still running, and one started by FAA historian Nick Komons before he retired.

PRESIDENTIAL LIBRARIES

George H. W. Bush Library, College Station, Texas. Under Freedom of Information Act request number 2000-0085-F, I requested release of material related to the National Aerospace Plane, the High Speed Research program, and NASA budget formation. All useful material was redacted under an extremely restrictive interpretation of "executive privilege."

Dwight D. Eisenhower Library, Abilene, Kansas. Eisenhower was the first president to address the subject of supersonic transports. Two collections here contain relevant material: the papers of Elwood Quesada, Eisenhower's special assistant for aviation, and the National Security collection, which contained records of Eisenhower's conversations about the SST and the XB-70 bomber program.

Lyndon B. Johnson Library, Austin, Texas. By far the largest collection of surviving SST documents, 118 boxes, is in the Presidential Advisory Committee on the SST collection here. A small portion of this material has not been declassified yet, but most of it is open to researchers. With the destruction of the FAA's project office collection, this is the primary collection for research into the U.S. SST program. Unfortunately, it ends with the Johnson administration in 1969.

John F. Kennedy Library, Boston, Massachusetts. SST-related material available at this facility is limited to a small number of documents in the Federal Aviation Agency collection and a handful of documents in the President's Office File collection. Most of the SST-related material at this facility has not been opened to the public.

Richard M. Nixon Presidential Materials Project, College Park, Maryland. The Nixon project includes SST-related documents scattered through several collections: The White House Central Files, the White House Special Files, the President's Office Files, and the Whitaker Papers. Since I completed the research for chapters 5 and 6, the Nixon Project has released the Oval

Office tapes for 1971, and these contain additional SST material. However, time did not permit me to utilize this resource.

Ronald Reagan Library, Simi Valley, California. The only material useful to this study was contained in the White House Office of Science and Technology collection, and was released under Freedom of Information Act request number F2000-001.

INTERVIEWS

All interviews were conducted by the author, and audio recordings are deposited at the Langley Research Center history archive.

Aiken, William	11 October 1999
Anderson, Robert	16 April 2002
Baals, Donald	5 September 2000
Baize, Daniel	4 December 2001
Bartolotta, Paul	20 December 2001
Becker, John V.	5 June 2000
Beggs, James	29 May 2001
Beranek, Leo	15 December 1999
Blackwelder, Brent	25 March 2000
Blaha, Bernie	23 January 2001
Boxer, Emmanuel	8 June 2000; 18 January 2001
Bromley, D. Allen	6 July 2001
Bunin, Bruce	12 March 2002; 22 March 2002
Coen, Peter	9 October 2001; 25 September 2001
Cook, William	18 June 2001
Czysz, Paul	17 July 2001
Dollyhigh, Samuel	21 July 2001
Dorsey, John	17 January 2002
Draper, Robert	23 January 2002
Driver, Cornelius	28 September 1999; 25 May 2001
Eggers, Alfred	24 February 2000
Evans, Brock	27 March 2001
Fishbach, Lawrence	12 April 2001; 16 November 2001
Fitzsimmons, Richard	7 February 2000
Gilbert, William	26 June 2001; 13 September 2001
Gilkey, Samuel	6 December 2001
Golub, Robert	12 December 2001
Graber, Ed	1 October 2001
Grose, William	3 April 2001
Harris, Roy	7 June 2000; 7 June 2001
Hasel, Lowell	5 September 2000
Hefner, Jerry	5 July 2001; 24 September 2001
Henderson, Michael	25 March 2002
Hergenrother, Paul	20 November 2001
Hertz, Terrence	6 May 2002
Hipskind, Steve	18 December 2001

Hook, Ray	19 July 2002
Hubbard, Harvey	28 July 2000
Johnson, Andrew	23 January 2002
Johnston, Harold	4 December 2000
Jones, Robert	25 June 2001
Kawa, Randy	18 December 2001
Klees, Garry	2 May 2001
Koops, Leigh	12 March 2002
Lavell, Jeff	30 October 2001
Lilley, Geoffrey	7 June 2000
Loftin, Laurence K.	23 November 1999; 15 June 2000
Long-Davis, Mary-Jo	20 December 2001
McCloskey, Michael	5 December 2000
McKinnon, Malcolm	25 February 2002; 4 November 2002
McLean, Edward	1 December 1999
Misra, Ajay	1 April 2002
Niedzwiecki, Richard	22 May 2001; 24 October 2001
Oppenheimer, Michael	13 December 2002
Powers, Albert	12 April 2001
Ricketts, Rodney	27 September 2001
Robins, A. Warner	29 September 2000
Sawyer, Wallace	19 December 2001
Shaw, Robert J.	6 June 2002
Shepherd, Kevin	28 September 2001
Shevell, Richard	1 February 2000
Sobieski, Jaroslaw	25 June 2001
Steiner, John E.	4 December 2000
Stibich, Marv	2 June 2000
Stolarski, Richard	22 December 2001; 26 April 2001
Strack, William	19 July 2002
Swihart, John	5 June 2000; 11 July 2000; 25 July 2000
Tang, Ming	3 December 1999
Tupling, Lloyd	8 December 2000
Welge, Harry R.	9 February 2000
Wesoky, Howard	11 September 2001; 21 November 2002
Whitcomb, Richard T.	12 January 2000; 22 September 2000; 21 January 2001
Whitehead, Robert	17 July 2002
Wilhite, Alan	19 September 2001
Williams, Louis	27 June 2001; 22 October 2001
Withington, Holden R.	6 December 2000; 7 July 2001; 5 June 2001

INDEX

Page numbers in *italics* indicate figures and tables.

A-12 program, 83, 84, 95–96, 108
accidents, space-related, 13
ACEE (Aircraft Energy Efficiency) project, 265, 268
Aeronautics and Space Engineering Board (ASEB), 196–99, 206, 291–92
aeronomers, 161
aerospace industry, 154–55
Aerospatiale, 8–9, 216
AESA (Atmospheric Effects of Stratospheric Aviation), 227, 229–38, 299
Aiken, William S., Jr., 158, 169–70, 202, 224
Airborne Antarctic Ozone Experiment, 229, 230–31
Airbus Industrie: A330/A340 series, 278; Aerospatiale and, 8–9; creation of, 3; international competition and, 193–94; jumbo twin aircraft and, 194–95; ozone issue and, 236; VLA (Very Large Aircraft) and, 290
aircraft business: fiscal issues for, 288–89; free market and, 193, 197, 222–23; as high risk, 2–3, 46–47, 96–97; HSR program and, 255–56; NASA and, 304–5. *See also specific companies*
Aircraft Energy Efficiency (ACEE) project, 265, 268
airfares, 336–37n. 73
airframe for HSR program, 264–71
airframe integrated scramjet, 208–9
Air France, 78, 159, 180, 185
airline industry, 157, 198–99. *See also specific airlines*
air pollution, activism against, 118
Airport Operators Council International, 143

Air Transport Association, 137
Alderson, George, 140, 145
Alford, William, 53
Allen, William: economic issues and, 42, 134; Jackson and, 146; SCAT and, 63; SST program and, 82, 112–13
Ambrose, Stephen, 44
Ames Research Center, 63, 166, 261
Anderson, Clinton, 165
antigovernment ideology, 190–91
Armstrong, Sam, 293–94, 296
arrow-wing planforms, 175–79
ASEB (Aeronautics and Space Engineering Board), 196–99, 206, 291–92
ASHOE/MAESA sampling campaign, 237
AST-100 configuration, 176–79, *177*
Atmospheric Effects of Stratospheric Aviation (AESA), 227, 229–38, 299
Atomic Energy Commission, 161–62
Atwood, Lee, 63, 82
axisymmetric nozzle, 243–44

B-58 "Hustler" bomber, 16, 25–27, 32
B-70 program. *See* XB-70 program
Baals, Donald, 33, 54, 98
Bain, Gordon: evaluation and, 92, 95, 97; sonic boom and, 122; SST program and, 87, 104
Baize, Daniel, 249
balance of payments, 320n. 97
Ballhous, William, 221
Barnes, E. F., 85
Battelle Memorial Institute, conference of, 214–16, 220–21
Baughcum, Steven, 232
Bayh, Birch, 166

Beggs, James: Keyworth and, 210; NASA budget and, 200; noise regulations and, 142; road maps and, 207; SST program and, 148; Steiner committee and, 204
Bell Aircraft Company, 17, 18, 52
Bell Labs, 234
Benzakein, Mike, 293–94
Beranek, Leo, 147–48
Berner, T. Roland, 91
Berry, Philip, 128
Bisplinghoff, Raymond, 107, 113, 148
Black, Eugene, 82
Blackwelder, Brent, 140
blade cooling technologies, 31–32
blade material systems, 275
Blaha, Bernie, 244–45
Boeing: Boeing 747, 134–35, 141–42, 290; Boeing's 2707-100 design, 110–13; 2707-200 design, 111–13; 2707-300 design, 114, 137–38; Dash 100 and 200 and, 111, 112, 116; financial status of, 8, 41–42, 146; floating wing tips and, 29; GE and, 138–39; HiSCAT, 297, 298; HSCT and, 218; HSR and, 224, 261–62, 291, 293–94, 295; Nixon administration and, 153; noise standards and, 285–86; reimbursement of, 153; SCAT and, 63, 64–65; SST design and, 104–5, 106–7; SST design competition and, 87–89, 88, 97–98, 100; SST program and, 145–46, 150; supersonic bomber and, 35–36, 37; Turbine Bypass Engine, 245–46; as winner of design competition, 108–9
Boland, Edward, 200
bombers: B-58 "Hustler," 16, 25–27, 32; intercontinental supersonic, 33–37, 34; nuclear-powered, 28–29, 42–43. See also XB-70 program
Bond, Alan, 209
bond coat, 274–76
Borger, John, 185
Borman, Frank, 195
boundary layer separation, 348n. 14
Bowles, Ann, 252
Boyer, Paul, 11
Breguet factor, 31
Bristol Aircraft, 60, 69, 71
British Aircraft Corporation, 69, 126
British Airways, 159, 180, 185
British Overseas Airways Corporation, 78, 151–52
Brizendine, John C., 182–83
Broderick, Tony, 231
Brooks, Overton, 44, 58
Brower, David, 120, 127–29, 156
Brown, Clarence J., 214

Brown, Clinton, 54
Bryan, Richard, 221
Bullock, Sarah, 99
Burns, Conrad, 257
Busemann, Adolph, 24
Bush, George H. W., 222, 254, 277
Bushnell, Dennis, 270
Butrica, Andrew, 288

C-141 "Starlifter," 43, 106, 132
Caddy, Michael, 172
Callis, Linwood, 229
cambered wing, 54–55
campaign against the SST, 118, 119, 120–21, 131, 153–56
Campbell, John, 52
canard, 111, 283
canard bomber, 33–37, 34
carbon-fiber composites, 264–65
Carlson, Harry, 59–60, 100
Carter, Jimmy, 180, 186, 191–92
Case, Clifford, 131, 151
Cavett, Dick, 152
Cayley, George, 67
ceramic liner, 273–74
Chambers, Joseph, 270
Chapman, Sydney, 161
Chase Smith, Margaret, 152
Chernomyrdin, Viktor, 270
Cherry, George, 160
chlorine chemistry, 167
chlorofluorocarbons (CFCs), 167–68, 188, 228
Church, Frank, 166
Cicerone, Ralph, 167
Citizen's League against the Sonic Boom, 126–27
climate and water vapor, 162–63
Climatic Impact Assessment Program (CIAP), 166, 169, 181, 230, 335n. 56
Clinton administration, 256–57, 286–87
Clipped Wings (Horwitch), 7–8
Coalition against the SST, 120–21, 131, 155–56
Coleman, William T., 181
Colladay, Raymond, 202, 207
commerce and speed, 4–5
Commerce Department, 77
competition: for design, 82, 87–92, 97–98; international, 7, 45–46, 50, 193–94
composites, high-temperature, 264–67
computer, as design tool, 99–100, 179
Concorde: de Gaulle and, 79–80; financial status of, 184–85; international competition and, 7; introduction of, 180–81; Kennedy and, 78–79; Pan Am and,

78–79, 159; politics and, 66–67; popularity of, 181; retirement of, 301; sampling of wake of, 237; Samuelson on, 151; scheduling issues and, 115–16; Supersonic Transport Advisory Group and, 76–77; treaty for development of, 66, 72–73

Concorde 2 studies, 289

Concorde and the Americans (Owen), 7, 8

Condit, Phil, 293

controlled separation, 38

Convair: B-58 "Hustler" bomber, 16, 25–27, 32; commercial SST and, 40; composites study by, 265–66; delta wing and, 21–23; YF-102, 23, 24–25

convection cooling, 32

Cook, William, 87

Cooper, Bob, 210

cooperation, international, 69, 70–71

Copper Canyon, 209, 210

Corn, Joseph, *The Winged Gospel,* 10–11

corporatization of government, x

Cortright, Edgar, 170

cost-sharing plans, 82–83, 186, 225–26

Creighton, Charlotte, 99–100

Crossfield, A. Scott, 212–13

Crutzen, Paul, 162, 165, 167

Cummings, William, 252

Curtiss-Wright, 91–92

Cuthbertson, Bob, 280

Czysz, Paul, 211, 212

Dassault, 39, 70, 196

Davis, Kingsley, 124

deep stall, 106–7

Defense Advanced Research Projects Agency, 209

defense budget, 288

de Gaulle, Charles, 50–51, 66, 72, 79–80

DeHavilland Comet, 323n. 44

DeLauer, Richard, 204

delta wing, 21–23, 63

Department of Transportation (DOT), 137–40, 144, 146–49, 158. *See also* Federal Aviation Administration (FAA)

design: competition for, 82, 87–92, 97–98; computer and, 99–100, 179; delta wing, 21–23, 63; double delta wing, 89; incompatible beliefs about, 302–3; noise regulations and, 9; off design performance, 56; outboard pivot, 53; slender delta, 38–39, 67; Stack and, 17, 18, 53, 114–15; supersonic aircraft evolution, *88;* swept wing, 14; swing-wing, 113–15; variable geometry, 52–54; variable sweep, 53, 113–15; wing-body, simultaneous, 30–31

Dewey, Scott Hamilton, 118

direct operating costs, 335n. 61

Dobson, Gordon, 161

Dollyhigh, Sam, 217, 218

Doniger, David, 221

Donlan, Charles, 52

DOT. *See* Department of Transportation (DOT)

double delta wing configuration, 89

Douglas, Deborah G., ix

Douglas, Donald, Jr., 84–85

Douglas Aircraft Company: D-558, 17, 18, 19; DC-8 stretch, 134; financial status of, 41; McDonnell Corporation and, 136; SST program and, 84–86; SST team at, 47; widebody "jumbo" jet and, 135. *See also* McDonnell-Douglas Corporation

Dow, Marvin, 265

drag: delta wing and, 21–23; engine nozzles and, 26; estimates of, 14; prediction of, 23–24; supersonic area rule and, 29–30; supersonic bomber and, 36–37; transonic area rule and, 24–25; trim drag, 101; wave drag, 98–99

Driver, Neil, 91, 109, 170, 224

Dryden, Hugh, 24, 33, 55–56

Dryden Flight Research Center, 260, 269. *See also* Flight Research Station

DuBridge, Lee, 141

duct-burning turbofan, 246

Dulles International Airport, 180–81

Dunning, John R., 122, 138

DuPont, Tony, 209

Durfee, James, 58

Earth Day, 140, 157

Eastern Airlines, 195

Eccles, Marriner, 191

economic issues: aerospace industry and, 150; aircraft business and, 288–89; airfares, 336–37n. 73; balance of payments, 144, 320n. 97; boom mitigation and, 249; campaign against SST and, 131–32; commercial SST and, 74–75; Concorde and, 184–85; congressional hearings and, 151–53; cost-sharing plan, 82–83, 186, 225–26; HSCT and, 219; HSR and, 286–88; inflation, 186–87, 191–92; market fundamentalism, 190–91, 305; mercantilism, 222–23; NASA, 186–87, 201–2, 254, 287, 293; overland flight and, 125; SCAT program and, 65–66; Senate hearings and, 143–44; SST and, 85–86, 185–86

economic nationalism, 222

economic redevelopment, 3, 50–51

Edsall, John T., 121, 126–27
Eggers, Alfred, 30–31, 36, 46
Ehrlichman, John, 157, 158, 160
Eisenhower, Dwight, 42–44, 58
ejector suppressor, 173, *174*
Elliot, David, 164
Energy Efficient Engine project, 183–84
engines: altitude affecting performance of, 318n. 56; bypass ratios of, 133–34; combustion products of, 162; Curtiss-Wright, 91–92; Energy Efficient Engine project, 183–84; GE, 28, 90–91, 97–98, 133–34, 172–73; high bypass turbofan, 132–34; hypersonic jet, 208–9; J79, 32; mixed-flow turbofan, 246; noise standards for, 132; novel engine concept, 245–46; Pratt & Whitney turbofan, 89, 91, 94, 101; reliability of, 102–3; for SST, 138; SST design competition and, 90–92; takeoff condition and, 349n. 38; test-bed program, 172–73; turbojet, introduction of, 14; Variable Cycle Engine program, 171–73; X279 afterburning turbojet, 35. *See also* nozzles, engine
Environmental Impact (EI) team for HSR program, 278, 281–82, 283–84
environmentalism, 302–3, 305, 326n. 1
environmental politics: air pollution, activism against, 118; Earth Day, 140; Nixon and, 118–19, 149–50, 154–55; overview of, 2; SST and, 9–10, 141–43; symbolic campaigns and, 9–10, 120; technology as out of control and, 119–20
EPNdB (effective perceived noise level in decibels), 137–38
European Community, 277–78, 282
European Joint Aviation Authority, 236
Evans, Brock, 129
Experimental Clean Combustor program, 171, 174–75

F-16XL experiments, 268–70
F-86 "Saber," 22, 23
F-106, 24–25
Fabry, Charles, 161
Fairchild Engine and Aircraft Corporation, 28
Fan-on-blade (FLADE) concept, 246
Federal Aviation Administration (FAA): competition for design and, 82, 87–92, 97–98; creation of, 48–49; developmental work and, 50; noise regulations and, 136–38; SCAT and, 56–58; sonic boom tests and, 123; SST development decision and, 73–79

Federal Express, 215
finite element analysis, 350n. 54
Fishbach, Larry, 172
Fletcher, James, 160, 165
Flight Research Station, 22. *See also* Dryden Flight Research Center
Flippen, J. Brooks, 149
flutter, *284*, 284–85
focused technology program, 169–70
France: Airbus Industrie and, 193–95; Caravelle program, 39, 70; economic redevelopment in, 50–51; nuclear power and, 12–13. *See also* Air France; Concorde
Friedman, Milton, 144
Friends of the Earth, 128–29, 131
Fritzsche, Peter, 11
Frosch, Robert, 180, 184
fuel crisis, 160
fuel tank sealant, 267
Fulbright, William, 143

Galbraith, John Kenneth, 144
Garn, Jake, 200
Garwin, Richard, 141–42
Gates, Thomas, 42–43, 56, 267
GE (General Electric): advertisement by, 4–6, *5;* Boeing and, 138–39; CJ805-23, 133; double bypass engine, 172–73; Earth Day and, 157; HSR and, 227; integrated product development team and, 262; J79 engine, 32; nuclear-powered engine and, 28; SST engine designs by, 90–91, 97–98; TF39, 134; 2-D mixer-ejector nozzle, 244, 245; X279 afterburning turbojet, 35; YJ101 engine, 173
Gelzer, Christian, 4
Germany, 11
Gibson, John, 127
Gilbert, Bill, 281, 297–98
Gilkey, Sam, 225, 235–36, 262
Gilpin, Robert, 66
Gilruth, Robert, 18
Gingrich, Newt, 287
Glennan, T. Keith, 50, 56, 58
Glenn Research Center, 259, 262. *See also* Lewis Research Center
Glickman, Dan, 210
Goddard Space Flight Center, 166, 168, 228
Goldin, Daniel: agenda of, 253–55, 287; HSR program and, 256, 296; NASA budget and, 257; noise regulations and, 285
Goldwater, Barry, 190
Goodmanson, Lloyd, 64
Gordon, Kermit, 86, 94

Gore, Albert, Jr., 221, 223, 270
government: corporatization of, x; development project support and, 189–90; public good issues and, 294–95; SST project support and, 3–6, 47. *See also* politics of aircraft business
Graedel, Thomas, 234, 235
Graf, Donald, 218–19, 225
Great Britain: France and, 51; SST development in, 48, 67–69, 71–73. *See also* British Airways; Concorde
Green Party, 236
Greif, Ken, 130
Gross, Courtland, 82
Guthrie, Woody, 127

Halaby, Najeeb: as activist, 80; airline committee evaluation and, 92; Concorde and, 75–76, 94; FAA and, 50, 73–74; L. Johnson and, 101–2; McNamara and, 83, 86–87; Oklahoma City sonic boom tests and, 123; resignation of, 103; schedule and, 95
Hallion, Richard, *On the Frontier,* 6
Hamilton, William T., 64
Hansen, James, 8, 16–17, 19
Harper, Charles W., 188
Harris, Roy, 98, 99–100
Harris, Wesley, 270
Harrison, Halstead, 162–63
Hartley, W. N., 161
Hatfield, Mark, 145, 152
Hawker Siddeley, 69, 71
Heath, Donald, 166
Hecht, Gabrielle, 12–13, 79–80
Hecht, Ralph, 272
Hefner, Jerry, 225, 268–69
Heldenfels, Richard, 170
Heller, Walter, 77, 144
Henderson, Michael: HSCT and, 285; HSR program and, 10, 225, 235–36, 286, 293; overland flight and, 251
Heppe, Richard, 63–64, 89
Hergenrother, Paul, 266, 267
heterogeneous chemistry, 231
Hibbard, Hall, 40–41, 69
High Energy Fuels program, 43, 46
high-lift devices, 178
High Speed Civil Transport. *See* HSCT (High Speed Civil Transport)
High Speed Research (HSR) program: AESA and, 229–38; airframe and, 264–71; Boeing and, 261–62, 291, 293–94, 295; cancellation of, 295–300; evaluation of, 291–92; low-emission combustor and,

238–42; mission of, 263–64, 276; noise and propulsion system development and, 242–47; noise regulations and, 276–86; overview of, 224–27; Phase II of, 253–57, *254,* 259–60, 286; Phase III/IIA of, 292–94; propulsion technologies, 271–76; sonic booms and, 247–53
Hines, Richard, 225
Hirschfelder, Joseph, 163–64
Hofman, David, 232–33
Hohenemser, Kurt, 127
holding patterns, flying, 108
Hollings, Ernest, 221
Hong, Jim, 63–64, 89
Hook, Ray, 261
Horn, John, 215
Hornig, Donald, 123, 136
Horwitch, Mel, 7–8, 50, 83, 140, 148
HOTOL (Horizontal Take-Off and Landing), 209–10, 216
Hotz, Robert B., 40, 105–6, 145, 184
HSCT (High Speed Civil Transport): Battelle Institute and, 220–21; contract studies of, 216–20; MacKinnon and, 259, 260–61; ozone layer and, 229, 237–38, *238;* politics and, 221–22, 225–26; promotion of idea of, 214–16; scheduling issues and, 263–71, 285, 291–92, 295; sonic boom and, 247–53; speed of, 217–19; thermal profile comparison, *260*
HSR program. *See* High Speed Research (HSR) program
Hubbard, Harvey, 57, 59, 60, 61, 328n. 60
Hudson, Bob, 167, 168
hypersonic jet, 208–10

indirect operating costs, 335–36n. 61
inflation, 186–87, 191–92
Integrated Planning Team, 262, *263,* 292–93
intercontinental airliner, 331n. 107
International Air Transport Association, 74–75
International Civil Aviation Organization, 48, 56, 282

Jackson, Henry M. "Scoop": Allen and, 146; SST program and, 130, 143, 145, 152; stratospheric research and, 166
Jackson, Roy P., 158, 170
Japan, 216
Japan Air Lines, 216
John F. Kennedy International Airport, 180, 181
Johnson, Andrew, 272

Johnson, Clarence "Kelly," 47
Johnson, Lyndon: B-70 program and, 45; candidacy of, 116; environmental issues and, 118; Halaby and, 75, 101–2; noise abatement and, 136; presidential advisory committee to, 86–87, 93–97, 102–3, 109–10, 115, 123; SST program and, 77–78, 97
Johnston, Harold, 163–65, 230
Jones, Bob, 209, 210
Jones, Robert T., 18, 29–30, 101, 209
jumbo twin airliner, 194–95
Junkers aircraft company, 31

Kahn, Alfred E., 191
Kaufman, Sanford, 47
Keck, Frank, 93–94
Kellogg, William, 162
Kennedy, John F.: Concorde and, 76, 78–79; Halaby and, 74, 86; "missile gap" rhetoric of, 45; SST program and, 82
Kent, Richard, 137
Kerrebrock, Jack, 138
Keynes, John Maynard, 191
Keyworth, George A., II, 192–93, 200, 207, 210–11
Kilpatrick, James J., 129
King, James, 168
Kistiakowsky, George, 43
Klees, Garry, 171, 243–44
Kolk, Frank, 135, 193, 194, 328n. 60
Koons, Lee, 293–94
Kotcher, Ezra, 17
Kotz, Nick, 154
Küchemann, Dietrich, 38, 46, 67

Lambright, Henry, 83, 303
laminar flow flight research, 260–61, 268–71
Langley Aeronautical Laboratory, 16–17, 18, 115
Langley Research Center: accelerated testing at, 267; aerodynamic studies at, 62–65; aeronautics and, 49–50; airframe technology and, 175–79; Boeing and, 98; Environmental Quality Office, 168; HSR and, 227, 261; laminar flow control research and, 268–71; manufacturer interest and, 48; ozone issue and, 166; SST program and, 80–81. *See also* Supersonic Cruise Aircraft Research (SCAR) program
Lawrence, James D., 168
LdN curves, 279–80
lean direct injection (LDI) combustor, 239, 240, *240*

lean premixed, prevaporized (LLP) combustor concept, 175, 220, 239
LeMay, Curtis, 311n. 42
Lewis Research Center: Experimental Clean Combustor program, 174–75; HSR and, 227; nozzle branch at, 244–45; rocketry and, 49; SCAR program and, 170–71. *See also* Glenn Research Center
Liebmann, George, 130–31
Life magazine, 20
lift effect and sonic boom, 59–60
lift-to-drag ratio, 26
Lilley, Geoffrey, 59
"Limited Exclusive Rights" doctrine, ix–x
Lindbergh, Charles, 93, 125, 134
Lindsay, Lina, 59
Lippisch, Alexander, 22
Littlewood, William, 74
Lockheed: C-141 "Starlifter," 43, 106, 132; Hawker Siddeley and, 69; L-1011 project, 136; SCAT and, 63–64; SST design and, 105, 107–8, 109; SST design competition and, 89, *90*, 100–101; SST studies by, 40–41
Loftin, Laurence, Jr., 6, 38, 63, 101
Loomis, James P., 214
Lovelace, Alan, 183
Low, George, 170, 338n. 97
low-emission combustor development, 238–42, 273–74
Lumb, Michael, 126
Lundberg, Bo, 121, 124–26

MacDonald, Gordon, 143
MacKinnon, Malcolm: HSCT and, 259, 260–61; HSR and, 294, 295; on laminar flow, 269; on show stopper, 285, 286; sonic boom and, 250, 251; speed and, 218
Maglieri, Dominic, 59, 224
Magnuson, Warren, 130, 143, 145, 152
Magruder, William: Allen and, 146; L-2000 and, 107; SST program and, 141–44, 149–52, 157–58, 329n. 70; Withington and, 138–39
marine mammal research, 252–53
market fundamentalism, 190–91, 305
Marks, Virginia "Gin," ix
Maskell, Eric, 38, 46, 67
Mathauser, Eldon, 57
Maxam, Fred, 111–12
Maxwell, Jewell C., 104, 107–9, 112–13, 146
McCarthy, John, 207
McCloskey, Michael, 128, 129
McCone, John, 93, 95–96, 110

McCormick, M. Patrick, 230
McDonald, Ian, 118
McDonald, James E., 163
McDonnell Corporation, 136
McDonnell-Douglas Corporation: advocacy
  campaign of, 179–80, 181–83; AST-100
  configuration, 176–79, *177;* HSCT and,
  218–19; HSR and, 224; market assessment by, 251–52; SCAR program and,
  160; technology demonstrator concept,
  211, 211–12
McDougall, Walter, 8, 47, 66
McFarlane, Robert, 210
McKay, Rebecca, 275
McKee, William F., 103–4, 109, 136–37
McNamara, Robert: A-12 program and, 96;
  cost-sharing plan and, 83; Halaby and,
  75; Hotz on, 105; ideological position of,
  106; overland flight and, 125; presidential advisory committee of, 86–87, 93–97,
  102–3, 109–10, 115, 123; scheduling
  issues and, 84, 95; SST program and,
  92–93, 118; TFX program and, 54
Melville, Frank, 184
mercantilism, 222–23
Messerschmitt P.1101, 52
military aircraft, 11–12, 14–16
Miller, Bob, 275
Mineta, Norman, 294
Ministry of Supply, 39, 67–68
Mishra, Ajay, 274
missiles, intercontinental ballistic, 12
mixed-flow turbofan, 246
mixer-ejector nozzle, 243–45, 272–73
Molina, Mario, 167
Mondale, Walter, 140
Monroney, A. S. "Mike," 121, 122
Moon, Howard, 7, 185
Morgan, Morien, 67–68, 69, 71, 80
Morris, Owen, 33
Mortlock, Alan, 278
Moss, Laurence, 131
Muroc Dry Lake, 18

Nahum, Andrew, 68
National Academy of Sciences, 122, 123–24
National Advisory Committee on Aeronautics (NACA): demise of, 47; developmental work and, 49; financing for, 19–20;
  research by, 6; supersonic bomber program, 33–37, *34;* technologies and, 14.
  See also Langley Aeronautical Laboratory
National Aeronautics and Space Administration (NASA): Advanced Supersonic
  Transport program and, 158, 159–60;
  aeronautics program and, 192, 196–99;

aircraft industry and, 199–201, 304–5;
  applied technology programs of, 195–96,
  197; budget of, 186–87, 201–2, 254, 287,
  293; competition for design and, 97–98;
  Congress and, 287–88; creation of, 47;
  developmental work and, 49, 50, 304–5;
  earth sciences and, 303; Energy Efficient
  Engine project, 183–84; free market and,
  192; ozone issues and, 165–66; problems
  with, 189–90; stratospheric research and,
  166–69; supersonics road map of, 207–8;
  Upper Atmosphere Research Program,
  161, 168–69, 227, 229. *See also* High
  Speed Research (HSR) program; Langley
  Research Center; Space Shuttle
National AeroSpace Plane program, 212.
  See also Orient Express
nationalism, 10–13
National Oceanic and Atmospheric Admin-
  istration (NOAA), 228–29
National Research Council: Aeronautics
  and Space Engineering Board, 196–99,
  209, 291–92; Committee on Atmospheric
  Chemistry, 233–35; NASA budget and,
  201–2
National Security Act (1948), 42
national security and research, 203–4,
  210–13
Navy-NASA Engine Program software, 188
Nellis Air Force Base, 250
Newhouse, John, 2, 135–36, 195
Nichols, Mark, 56–57
Niedzwiecki, Richard, 174, 175, 238–39
Nihart, Gene, 278
Nimbus G satellite, 168–69
nitrogen oxides, 164, 220
Nittinger, Klaus, 216
Nixon, Richard: antigovernment rhetoric
  and, 190; environmental issues and,
  118–19, 154–55; SST development and,
  157; SST program and, 116, 129–30, 140,
  149–50, 152–53
noise regulations: aircraft design and, 9;
  engines and, 132; EPNdB (effective perceived noise level in decibels), 137–38;
  European Community and, 277–78, 282;
  FAA and, 136–38; fleet size and, *280;*
  HSCT and, 219–20; HSR program and,
  276–86, 297–98; LdN curves and,
  279–80; measurement points for aircraft
  certification, *279;* Part 36 Stage 2, 173,
  242; Part 36 Stage 3, 277; research related
  to, 242–47; SST and, 137–39, 147–49;
  Variable Cycle Engine program and,
  171–73. *See also* sonic boom
Nord Aviation, 39, 70

North American Aviation: commercial SST and, 40; SCAT and, 63; SST design competition and, 89–90; strategic bomber program and, 29; supersonic bomber program and, 33–35, 36–37
Northwest Airlines, 215
novel engine concept, 245–46
nozzles, engine: drag and, 26; mixer-ejector, 243–45, 272–73; noise suppression and, 173, *174*
nuclear fear, culture of, 11–12
nuclear power, 28–29, 42–43
"N" wave shape of sonic boom, 248, *250*

oblique wing, 343n. 2
off design performance, 56
Office of Science and Technology Policy, 200, 202–7, 210
Office of Technology Assessment, 154–55, 305
On the Frontier (Hallion), 6
Oppenheimer, Michael, 230, 252
opponents of SST, 118, 143–44. *See also* campaign against the SST
Orient Express: competition and, 253; origins of, 212–13; Reagan and, 193, 201; rejection of, 219; SST program and, 222, 223
Osborne, Stanley, 82
outboard pivot design, 53
overland flight and sonic boom, 125–26, 251
Owen, Kenneth: on British SST effort, 67, 69; on Concorde, 66; *Concorde and the Americans*, 7, 8; on McNamara, 83
ozone layer: AESA and, 227, 229–38; depletion of, 163–66; existence of, 161; Experimental Clean Combustor program and, 174–75; experiments regarding, 229–31; hole in, 220, 226, 227–29; HSCT and, 237–38, *238;* HSR Phase I program and, 241–42; low-emission combustor development and, 238–42; volcanic eruptions and, 232–33

P-51 Mustang, 19
Pan American World Airways: Boeing 747 and, 134–35; Concorde and, 78–79, 159; financial status of, 157; Office of Technology Assessment study and, 185; on SST, 47
Pennell, Maynard, 64, 87, 104
PETI-5, 266, 267
Phelps, Howard, 107

podded scramjet, 208
polar stratospheric clouds, 230–31
Polhamus, Edward, 53
politics of aircraft business: activist government ideology and, 198–99; B-70 program and, 44–45; Concorde and, 66–67, 75–76; conflict within, 299–300; congressional hearings, 141–44, 151–53; dreams of speed and power in, 302; Eisenhower and, 42–44; HSCT program and, 221–22, 225–26; HSR program and, 253–54; international, 298–99; overview of, 1–2, 3; SST and, 7–10, 145–47. *See also* McNamara, Robert
polymer-matrix composites, 265–66
Prather, Michael, 229–30
Pratt & Whitney: HSR and, 227; J58 engine, 94; turbofan engine, 89, 91, 101; variable steam control engine, 172–73
Preisser, Jack, 243
Preliminary Technology Concept, 282–83, *283, 284*
productivity, aircraft, 4
progress in aviation, 6, 79–80, 305–6
Project Horizon, 74
propulsion technologies, 133, 271–76
prototype construction stage, funding for, 130, 144–46
Proxmire, William, 125, 131, 147, 151, 152

Quesada, Elwood: FAA and, 44, 48, 50; SST program and, 38, 58; Stack and, 56

Radkey, R. L., 176
ramjets, 208
ramp wave shape of sonic boom, 249, *250*
Raymond, Arthur E., 109
Reagan, Ronald: antigovernment rhetoric and, 190; deconstruction of federal civil government by, 192–93; National Aero-Space Plane program and, 212–13; Nixon and, 149; Orient Express and, 201, 219, 222, 223, 253; Star Wars and, 210, 212. *See also* Stockman, David
research institutions, state-funded, 189–90
Reuss, Henry, 131, 151, 163
Rhode, Richard, 58–59
rich-burn, quick quench, lean burn (RQL) combustor, 239–41, *240*
Ricketts, Rodney, 284
Right Stuff, The (Wolfe), 6
risks of aircraft business, 2–3, 46–47, 96–97, 114–15
Robbins, Warner, 218

Robins, A. Warner, 98
rocket plane, 17
Roensch, Robert, 176
Rohrbacher, Dana, 213, 296
Rolls-Royce, 132
Rosen, Cecil, 225, 229, 255, 268–69
Rosen, Robert, 221
route fragmentation, 290
Rowland, F. Sherwood, 167
Royal Aircraft Establishment (RAE), 38–39,
    67, 80–81
Rudd, Ralph, 63
Ruppenthal, Karl, 151

Samuelson, Paul, 144, 151
Sanders, Newell, 138
Sawyer, Wally, 263, 268, 295
SCAR (Supersonic Cruise Aircraft Research)
    program, 160, 169–71, 176–79, 180,
    187–88
SCAT (Supersonic Commercial Air Trans-
    port) program, 54–58, 62–65, 98–100
SCEP study, 162, 164
Schairer, George, 36, 106, 111
scheduling issues: Concorde, 115–16;
    HSCT, 263–71, 285, 291–92, 295; SST
    program, 83–84, 93, 95, 102–3
Schenk, Donald, 216
Schiff, Harold, 164
science state, 189–90
scramjet, 208
SEEDS database, 232
Seitz, Frederick, 122
shaped tube electrolytic machining
    (STEM), 35
Shaw, Robert "Joe," 225, 227, 259, 261, 296
Shepherd, Kevin, 248–49, 252
Sherry, Michael, 11
Shevell, Richard, 85
Shrontz, Frank, 255
Shurcliff, Charles, 127
Shurcliff, William, 121, 125, 126–27, 129
Sierra Club, 119, 127–28, 131, 155
silicon carbide, 273–74
slender delta configuration, 38–39, 67
Smith, C. R., 74
Smith, Floyd, 150
Smith, Fred, 211, 215
Solomon, Susan, 229, 232–33
sonic boom: acceptability of, 123, 248–50,
    251; HSCT and, 247–53; human response
    to, 58–62; Las Vegas and, 60; lift effect
    and, 59–60; Lundberg and, 124–26;
    marine mammal research and, 252–53;

at night, 124; Oklahoma City and,
    121–24; overland flight and, 125–26,
    251; prediction of intensity of, 105;
    SCAT and, 57; St. Louis and, 61, 122
Soros, George, 190
Soviet Union, 16, 23, 39–40. *See also*
    Tupolev TU-144
Space Shuttle, 150, 167–68, 180, 186–87,
    210
space station, international, 257, 293
speed: dreams of, 302; of HSCT, 217–19;
    indirect operating costs and, 335–36n.
    61; progress and, 79–80; SCAR program
    and, 176; of SST aircraft, 65; as virtue in
    American society, 4–6
Spencer, Nelson, 168
Spitzer, Bob, 293–94
Sputnik, 16, 40
SST and Sonic Boom Handbook, The (Shur-
    cliff), 129
Stack, John: activism of, 55–56, 80; aircraft
    design and, 17, 18, 53, 114–15; D-558
    and, 19; departure of, 73; FAA report
    and, 56, 57–58; SCAT and, 54; SST
    research program and, 50; SST sketch
    and, 20, *20;* study group of, 38; Whit-
    comb and, 24
Stanford Research Institute (SRI), 77, 85–86
Star Wars (Strategic Defense Initiative),
    210, 212
Steiner, John, 196, 204–7, 214
Stever, H. Guyford, 196
Stockman, David, 190, 192, 199–200
stoichiometric conditions, 335n. 58
Stolarski, Richard: ASEA and, 235, 236;
    chlorine and, 167; stratospheric research
    and, 168, 232, 233
Stonecipher, Harry, 285, 293, 294
Strack, Bill, 243, 245, 246
Straight, Donald J., 150
Strategic Air Command, 27
Strategic Defense Initiative (Star Wars),
    210, 212
stratospheric research program, 161–62,
    166–69
subsonic jet aircraft, 41–42, 184, 349n. 38
Sud Aviation, 39, 70, 126
Suddreth, Jack, 224
Sullivan, Walter, 165
superjumbo aircraft, 289–90
supersonic aircraft: American imagination
    held by, 301–2; classic shape for, 24;
    design evolution, *88;* early sketch of, *20;*
    SCAT 4, *55;* SCAT 15F, *98*

supersonic area rule, 29–30, 98–99, *99*
Supersonic Commercial Air Transport
    (SCAT) program, 54–58, 62–65, 98–100
Supersonic Cruise Aircraft Research (SCAR)
    program, 160, 169–71, 176–79, 180,
    187–88
supersonic laminar flow control (SLFC),
    268–69, 270–71
Supersonic Transport Advisory Group,
    76–77
supersonic wedge principle, 30–31, 36, 46
swept wing design, 14
swept-wing theory, 18
Swihart, John: on afterburning turbofan,
    57; on Boeing, 135; on nozzle design, 26;
    on risk, 114–15; on Stack, 38; on super-
    sonic transport, 215–16; Supersonic
    Transport Research Committee and, 54
swing-wing design, 113–15
systems analysis, 83

tail problem, 104, 106–7
takeoff condition, 349n. 38
Tang, Ming, 253
Taylor, John, 139
technology: campaign against SST and,
    154–55; environmentalism and, 305;
    focused technology program, 169–70;
    government and, 3–6, 47; interests inter-
    acting in, 1–2; Kennedy and, 79;
    national identity and, 13; progress and,
    305–6; public good and, 294–95; road-
    blocks to, 205; viewed as out of control,
    119–20, 128; World War II and, 14–16
Technology Concept Aircraft (TCA),
    281–82, *283*, 297
technology demonstration, 198, 202–3
technology validation, 198, 202–3, 207
Teitz, Joyce, 131, 140, 143–44
Tenzer, Herbert, 136
TFX (Tactical Fighter, Experimental), 49,
    53–54
thermal acoustic shield nozzle, 173
thermal barrier coat (TBC), 274–76
Thompson, Floyd, 49, 63
Thorneycroft, Peter, 50, 71, 72
Tillinghast, Charles, 94
titanium sandwich, 178–79, 264
Toll, Thomas, 33, 57
Train, Russell, 142–43
transonic area rule, 24–25
transpiration cooling, 32
trim drag, 101
Trippe, Juan, 78–79, 93, 94, 134
Tupolev TU-144, 7, 116, 117, 185, 270
turbine materials, long-life, 274–76

turbojet engines, 14
Twining, Nathan, 44

Udall, Stewart, 125, 127
United Nations Environment Program,
    228, 231
"unstart" problem, 96
U.S. Air Force: cargo plane, 43; commercial
    SST and, 40, 42; Muroc Dry Lake, 18;
    North American Aviation and, 37;
    Nuclear Energy Propulsion for Aircraft
    program, 28; publicity stunt of, 61–62;
    quest for speed and, 6, 12; strategic
    bomber and, 27–28; Wright Air Develop-
    ment Center, 14, 25; WS-110 program,
    33–37, 38; XS-1 and, 19
U.S. Army Air Forces: financing for, 19–20;
    Materiel Command, 17; quest for speed
    and, 14; strategic bomber and, 11; super-
    sonic aircraft and, 21
U.S. Navy Combat Air Patrol, 53–54

Variable Cycle Engine program, 160,
    171–73, 188, 242, 246
variable geometry configurations, 52–54
variable sweep design, 53, 113–15
Vincenti, Walter, 17, 23
VLA (Very Large Aircraft) study group, 289,
    290
volcanic eruptions and ozone, 232–33
Volpe, John, 147, 150
von Kann, Clifton, 199
vortex lift theory, 38, 46

Wallach, Henry, 144
Wallis, Barnes, 53
water vapor, 162–63
Watkinson, Harold, 68–69
Watson, Robert, 225, 229
wave drag, 98–99
Webb, Jim, 95
weight issues: Boeing 2707-100, 110–12;
    SST prototype, 148; takeoff gross weight,
    245–46
Weiss, Gus, 212–13
Welge, Bob, 176, 219, 259, 299
Welliver, Bert, 171, 255, 256
Wellock, Thomas, 128, 155
Wells, Ed, 112, 135
Wesoky, Howard, 229, 231, 233, 235, 236
West Germany, 216
Whitcomb, Richard T.: on linear theory,
    54; Lockheed evaluation and, 101; SCAT
    4 and, 55, *55*; transonic area rule and,
    24–25, 29–30
Whitehead, Allen, 227, 250, 278

Whitehead, Robert, 256
widebody "jumbo" jet, 134–35
Wilderness Society, 127
Wilhite, Alan, 259, 261, 271
Williams, Bob, 209
Williams, Louis J., 225, 227, 229, 252, 270
Wilson, Thornton, 106, 145–46, 153
wind tunnels, 23
wing-body design, simultaneous, 30–31
Winged Gospel, The (Corn), 10–11
wing translation, 52
Winner, Langdon, 119
Withington, Holden R. "Bob": Magruder and, 138–39; Parallel Group and, 113; on SST production, 148–49, 155; swing-wing and, 107; tail problem and, 106; WS-110A program and, 37
Wolfe, Tom, *The Right Stuff,* 6
Wright, Ray, 19
Wright brothers, 10–11

Wright Field, 17, 173
WS-110 program, 33–37, 38. *See also* XB-70 program

XB-70 program: commercial SST and, 40; cost of, 46; Halaby and, 84; L. Johnson and, 44–45; McNamara and, 83; North American and, 36–37; reorientation of, 43
XF-92A, 22–23
X-plane research, 15
XS-1 rocket plane, 17, 18, 19, 23

Yates, Ronald, 291
Yates, Sidney, 130, 131, 151
Yeager, Charles "Chuck," 19
YF-102, 23, 24–25
YJ-93 engine, 90, 91
YJ101 engine, 173